T0214819

Lecture Notes in Computer Science 11600

Commenced Publication in 1973
Founding and Former Series Editors:
Gerhard Goos, Juris Hartmanis, and Jan van Leeuwen

FoLLI Publications on Logic, Language and Information

Subline of Lectures Notes in Computer Science

More information about this series at http://www.springer.com/series/7407

Md. Aquil Khan · Amaldev Manuel (Eds.)

Logic and
Its Applications

8th Indian Conference, ICLA 2019
Delhi, India, March 1–5, 2019
Proceedings

 Springer

Editors
Md. Aquil Khan
Indian Institute of Technology Indore
Madhya Pradesh, India

Amaldev Manuel
Indian Institute of Technology Goa
Goa, India

ISSN 0302-9743   ISSN 1611-3349 (electronic)
Lecture Notes in Computer Science
ISBN 978-3-662-58770-6   ISBN 978-3-662-58771-3   (eBook)
https://doi.org/10.1007/978-3-662-58771-3

Library of Congress Control Number: 2019930849

LNCS Sublibrary: SL1 – Theoretical Computer Science and General Issues

This Springer imprint is published by the registered company Springer-Verlag GmbH, DE
part of Springer Nature
The registered company address is: Heidelberger Platz 3, 14197 Berlin, Germany

Preface

The eighth edition of the Indian Conference on Logic and Its Applications (ICLA 2019) was held during March 3–5, 2019, at IIT Delhi. The conference was preceded by two pre-conference workshops (March 1–2) — the Workshop on Paraconsistent Logic and the Workshop on Logic and Cognition. This volume contains the papers that were accepted to be published in the proceedings of ICLA 2019.

The ICLA is the biennial conference of ALI, the Association for Logic in India. The aim of this conference series is to bring together researchers working in areas of mathematics, computer science, and philosophy in which formal logic plays a significant role. Areas of interest include mathematical and philosophical logic, computer science logic, foundations and philosophy of mathematics, use of formal logic in areas of theoretical computer science and artificial intelligence, logic and linguistics, and the relationship between logic and other branches of knowledge. Systems of logic in the Indian tradition and their history are of special interest to the conference.

The conference received 34 submissions this year. These submissions underwent a single-blind peer-review by at least three Program Committee members, and by external experts in some cases. We thank all those who submitted papers to ICLA 2019. Following a discussion on the reviews, the Program Committee decided to accept 13 papers for publication and presentation. The accepted papers span a wide variety of areas including model theory, modal and fixpoint logics, sequent calculus, rough set theory, and computer-aided theorem proving. In addition to these selected papers, authors of some submissions were offered to present their work at the conference. We would like to extend our gratitude to the Program Committee members for their hard work, patience, and knowledge in putting together an excellent technical program. We also extend our thanks to the external reviewers for their efforts in providing expert opinions and valuable feedback to the authors.

The program also included six invited talks. We are grateful to Johann A. Makowsky, Carolin Antos, Philippe Balbiani, Martin Lange, Ian Pratt-Hartmann, and Mike Prest for accepting our invitation to speak at ICLA 2019 and for contributing to this proceedings volume.

On behalf of ALI, we thank the Department of Computer Science and Engineering, IIT Delhi, for hosting the conference. The local Organizing Committee comprised S. Arun-Kumar, Sanjiva Prasad, and Subodh V Sharma of IIT Delhi. We extend our sincere gratitude to them for their commitment and efforts in ensuring the smooth running of the conference. We are indebted to Divyanjali Sharma (IIT Delhi) for setting up and maintaining the conference website. We thank Anil Seth for his great help in the organization of ICLA 2019 and R. Ramanujam for his help in the discussions with Springer. To Sujata Ghosh and Sanjiva Prasad, PC chairs of ICLA 2017, we extend our sincere gratitude for their continuous help throughout the organization of this conference in every phase. We also express our appreciation of the tireless efforts of all the volunteers who contributed to making the conference a success.

We thank the EasyChair conference management system, which we used from managing the submissions to producing these proceedings. We are grateful to the Editorial Board at Springer for publishing this volume in the LNCS series.

January 2019

Md. Aquil Khan
Amaldev Manuel

Organization

Program Committee

Sreejith Ajithkumar	IIT Goa, India
S. Akshay	IIT Bombay, India
S. Arun-Kumar	IIT Delhi, India
Sankha Basu	IIIT Delhi, India
Benedikt Bollig	CNRS, LSV, ENS Paris-Saclay, France
Torben Braüner	Roskilde University, Denmark
Mihir Chakraborty	Jadavpur University, India
Kaustuv Chaudhuri	École Polytechnique, France
Anuj Dawar	University of Cambridge, UK
Ivo Duntsch	Fujian Normal University, China
Soma Dutta	University of Warmia and Mazury, Poland
Sujata Ghosh	ISI Chennai Centre, India
Davide Grossi	University of Groningen, The Netherlands
Stefan Göller	University of Kassel, Germany
Md. Aquil Khan	IIT Indore, India
Astrid Kiehn	IIT Mandi, India
Amit Kuber	IIT Kanpur, India
Denis Kuperberg	CNRS, LIP, ENS Lyon, France
Benedikt Loewe	University of Amsterdam, The Netherlands
Minghui Ma	ILC, Sun Yat-Sen University, China
Amaldev Manuel	IIT Goa, India
Benjamin Monmege	CNRS, LIF, Aix-Marseille Université, France
Ramchandra Phawade	IIT Dharwad, India
M. Praveen	Chennai Mathematical Institute, India
Gabriele Puppis	CNRS, LaBRI, Bordeaux, France
Arnaud Sangnier	IRIF, Université Paris Diderot, France
Abhisekh Sankaran	University of Cambridge, UK
Katsuhiko Sano	Hokkaido University, Japan
Sunil Easaw Simon	IIT Kanpur, India
Smita Sirker	Jawaharlal Nehru University, India
Hans van Ditmarsch	CNRS, LORIA, University of Lorraine, France
Richard Zach	University of Calgary, Canada

Additional Reviewers

Baskar, A.

Bisht, Harshit

Chapman, Peter

D'Souza, Deepak

Emmeche, Claus

Lahiri, Utpal

Mathew, Anup Basil

S., Sheerazuddin

Srivastava, Shashi

Suresh, S. P.

Tarafder, Sourav

Thinniyam, Ramanathan

Verbrugge, Rineke

Short Papers

On the Average Complexity of SAT

J. A. Makowsky

Department of Computer Science, Technion - Israel Institute of Technology,
Haifa, Israel
janos@cs.technion.ac.il

In 1995 I co-authored with A. Sharell the paper [MS95] which I will review in this talk from today's point of view. The results were discussed and interpreted by. S. Cook and D. Mitchel in [CM97], but otherwise were little noticed.

In the paper we investigated *natural* distributions of clauses for the satisfiability problem (SAT) of prepositional logic with input in conjunctive normal form, using concepts previously introduced by to study the average-case complexity of NP-complete problems.

Average case complexity was introduced in 1986 by L. Levin, [Lev86] who also defined complete problems for the complexity class DistNP. In 1991 Y. Gurevich, [Gur91], showed that a problem with a flat distribution is not DistNP complete (for deterministic reductions), unless DEXPTime is different from NEXPTime. We expressed the known results concerning fixed size and fixed density distributions for CNF in the framework of average-case complexity and show that all these distributions are flat. We introduced the family of symmetric distributions, which generalizes those mentioned before, and showed that bounded symmetric distributions on ordered tuples of clauses (CNFTupIes) and on k-CNF (sets of k-literal-clauses), are flat.

This eliminated all these distributions as candidates for *provably hard* (i.e. DistNP complete) distributions for SAT, if one considered only deterministic reductions. Given the (presumed) naturalness and generality of symmetric distributions, this result supported evidence that (at least polynomial-time, no-error) randomized reductions are appropriate in average-case complexity.

We also observed, that there are non-flat distributions for which SAT is polynomial on the average, but that this is due to the particular choice of the size functions. In [CS88] V. Chvátal and E. Szemerédi (1988) have shown that for certain fixed size distributions (which are also flat) resolution is exponential for almost all instances. We used this to show that every resolution algorithm will need at least subexponential time on the average. In other words, resolution-based algorithms will not establish that SAT, with these distributions, is in AverP.

In later developments U. Feige established in 2002 [Fei02] a relationship between average-case complexity and approximation complexity which also applies to SAT. More recent papers [AV01, CDA+03, VSTER16] analyse random SAT more closely and discuss phase transitions by letting the distribution of the instances vary along various parameters. Experimental studies on the meaning of these parameters were studied in [ZMW+17]. Phase transitions are also analyzed in the literature [MIDV07] for HornSAT and 2SAT, which are polynomial time solvable in the worst case.

There is a big discrepancy between the complexity theory applied to SAT and engineering practice. SAT-engineering is capable of solving very large problems in practice. Theoreticians tried to explain this with structural properties of a typical SAT-instance such as the width (tree-width or clique-width), cf. [FMR08]. However, it seems that the density of the distribution, i.e. the ratio between the number of clauses and the number of boolean variable, of the SAT-instances to be analyzed bears more relevant information than other structural properties. For the better understanding of average-case complexity still much has to be done, see also [Var14].

References

[AV01]   Aguirre, A.S.M., Vardi, M.: Random 3-SAT and BDDs: the plot thickens further. In: Walsh, T. (ed.) Principles and Practice of Constraint Programming—CP 2001. CP 2001. LNCS, vol 2239, pp. 121–136. Springer, Berlin, Heidelberg (2001). https://doi.org/10.1007/3-540-45578-7_9

[CDA+03]   Coarfa, C., Demopoulos, D.D., Aguirre, A.S.M., Subramanian, D., Vardi, M.Y.: Random 3-sat: the plot thickens. Constraints 8(3), 243–261 (2003)

[CM97]   Cook, S.A., Mitchell, D.G.: Finding hard instances of the satisfiability problem. In: Satisfiability Problem: Theory and Applications: DIMACS Workshop, vol. 35, pp 1–17 (1997)

[CS88]   Chvátal, V., Szemerédi, E.: Many hard examples for resolution. J. ACM (JACM) 35 (4), 759–768 (1988)

[Fei02]   Feige, U.: Relations between average case complexity and approximation complexity. In: Proceedings of the Thirty-Fourth Annual ACM Symposium on Theory of Computing, pp. 534–543. ACM (2002)

[FMR08]   Fischer, E., Makowsky, J.A., Ravve, E.V.: Counting truth assignments of formulas of bounded tree-width or clique-width. Discrete Appl. Math. 156(4), 511–529 (2008)

[Gur91]   Gurevich, Y.: Average case completeness. J. Comput. Syst. Sci. 42(3), 346–398 (1991)

[Lev86]   Levin, L.A.: Average case complete problems. SIAM J. Comput. 15(1), 285–286 (1986)

[MIDV07]   Moore, C., Istrate, G., Demopoulos, D., Vardi, M.Y.: A continuous–discontinuous second-order transition in the satisfiability of random horn-sat formulas. Random Struct. Algorithms 31(2), 173–185 (2007)

[MS95]   Makowsky, J.A., Sharell, A.: On average case complexity of sat for symmetric distribution. J. Logic Comput. 5(1), 71–92 (1995)

[Var14]   Vardi, M.Y.: Boolean satisfiability: theory and engineering. Commun. ACM 57(3), 5 (2014)

[VSTER16] Varga, M., Sumi, R., Toroczkai, Z., Ercsey-Ravasz, M.: Order-to-chaos transition in the hardness of random boolean satisfiability problems. Phys. Rev. E 93(5), 052211 (2016)

[ZMW+17]   Zulkoski, E., Martins, R., Wintersteiger, C., Robere, R., Liang, J., Czarnecki, K., Ganesh, V.: Relating complexity-theoretic parameters with sat solver performance. arXiv preprint arXiv:1706.08611 (2017)

What Did Tarski Have in Mind with Elementary Geometry and Its Decidability?

J. A. Makowsky

Department of Computer Science, Technion - Israel Institute of Technology,
Haifa, Israel
janos@cs.technion.ac.il

We survey the status of decidability of the first order consequences in various axiomatizations of Hilbert-style Euclidean geometry. We draw attention to a widely overlooked result by Martin Ziegler from 1980, which proves Tarski's conjecture on the undecidability of finitely axiomatizable theories of fields. We elaborate on how to use Ziegler's theorem to show that the consequence relations for the first order theory of the Hilbert plane and the Euclidean plane are undecidable. As new results we add:

(A) The first order consequence relations for Wu's orthogonal and metric geometries (Wen-Tsün Wu, 1984), and for the axiomatization of Origami geometry (J. Justin 1986, H. Huzita 1991) are undecidable.

It was already known that the universal theory of Hilbert planes and Wu's orthogonal geometry is decidable. We show here using elementary model theoretic tools that

(B) the universal first order consequences of any geometric theory T of Pappian planes which is consistent with the analytic geometry of the reals is decidable.

The techniques used were all known to experts in mathematical logic and geometry in the past but no detailed proofs are easily accessible for practitioners of symbolic computation or automated theorem proving.
We also discuss the status of projective and hyperbolic geometry.
The talk is based on

J. A. Makowsky,
Can one design a geometry engine?
On the (un)decidability of certain affine Euclidean geometries.
Annals of Mathematics and Artificial Intelligence, 2019 (in press).
https://doi.org/10.1007/s10472-018-9610-1

---

[1] Partially supported by a grant of Technion Research Authority. Work done in part while the author was visiting the Simons Institute for the Theory of Computing in Fall 2016.

Contents

Unification in Modal Logic

Philippe Balbiani[⊠]

Institut de recherche en informatique de Toulouse,
CNRS — Toulouse University, Toulouse, France
balbiani@irit.fr

Keywords: Modal logics · Unification problem ·
Elementary unification · Unification with constants ·
Computability of unification · Unification type

Let $\varphi_1, \ldots, \varphi_n$ and ψ be some formulas. The figure $\frac{\varphi_1, \ldots, \varphi_n}{\psi}$ is the inference rule which for all substitutions σ, derives $\sigma(\psi)$ from $\sigma(\varphi_1), \ldots, \sigma(\varphi_n)$. It is admissible in a propositional logic L whenever for all substitutions σ, $\sigma(\psi) \in L$ if $\sigma(\varphi_1), \ldots, \sigma(\varphi_n) \in L$. It is derivable in L whenever there is a derivation of ψ in L from the hypothesis $\varphi_1, \ldots, \varphi_n$. It is evident that every derivable rule is also admissible. L is called structurally complete when the converse holds. Some propositional logics — such as Classical Propositional Logic — are structurally complete. Others — like Intuitionistic Propositional Logic — are not. See [14, Chap. 2]. When L is not structurally complete, owing to the importance of the admissibility problem in the mechanization of propositional logics, it is crucial to be able to recognize whether a given inference rule is admissible. The question of the existence of a decidable modal logic with an undecidable admissibility problem has been negatively answered by Wolter and Zakharyaschev [41] within the context of normal modal logics between K and $K4$ enriched with the universal modality — see also the pioneering article of Chagrov [13] for an earlier example of a decidable modal logic with an undecidable admissibility problem. In some other cases, for instance Intuitionistic Propositional Logic and transitive normal modal logics like $K4$, the question of the decidability of the admissibility problem has been positively answered by Rybakov [33–35]. See also [15,25,27,28,32]. The truth is that Rybakov's decidability results are related to the fact that the propositional logics he has considered are finitary [22–24].

The finitariness character of a propositional logic L originates in its unification problem. The unification problem in L is to determine, given a formula φ, whether there exists a substitution σ such that $\sigma(\varphi)$ is in L. In that case, σ is a unifier of φ. A formula φ is filtering if for all unifiers σ and τ of φ, there exists a unifier of φ which is more general than σ and τ. We shall say that a set of unifiers of a formula φ is complete if for all unifiers σ of φ, there exists a unifier τ of φ in that set such that τ is more general than σ. An important question is the following: when a formula is unifiable, has it a minimal complete set of unifiers? When the answer is "no", the formula is nullary. When the answer is "yes", the formula is either infinitary, or finitary, or unitary depending on the cardinalities

Md. A. Khan and A. Manuel (Eds.): ICLA 2019, LNCS 11600, pp. 1–5, 2019.
https://doi.org/10.1007/978-3-662-58771-3_1

of its minimal complete sets of unifiers. Filtering formulas are always unitary, or nullary. A propositional logic is called nullary if it possesses a nullary formula. Otherwise, it is called either infinitary, or finitary, or unitary depending on the types of its unifiable formulas. When every formula is filtering, the propositional logic is called filtering. See [3,17,26] for details. It is evident that if L is consistent then its unification problem can be reduced to its admissibility problem: a given formula φ possesses a unifier in L if and only if the inference rule $\frac{\varphi}{\bot}$ is not admissible. Reciprocally, if L is unitary, or finitary and the elements of the minimal complete sets of unifiers given rise by its unifiable formulas can be computed then its admissibility problem can be reduced to its unification problem: a given inference rule $\frac{\varphi_1,\ldots,\varphi_n}{\psi}$ is admissible in L if and only if either $\varphi_1 \wedge \ldots \wedge \varphi_n$ is not unifiable, or the substitutions that belong to a minimal complete set of unifiers of $\varphi_1 \wedge \ldots \wedge \varphi_n$ are unifiers of ψ.

In accordance with their capacity to talk about relational structures, normal modal logics like epistemic logics, or temporal logics are playing a fundamental role in many applications. By virtue of its close relationships with the admissibility problem, the unification problem lies at the heart of their mechanization. For this reason, as advocated by Babenyshev *et al.* [7], investigations in the unification problem in normal modal logics can greatly contribute to the development of their applications. Regarding the unification problem in normal modal logics, we usually distinguish between elementary unification and unification with constants. In elementary unification, all variables are likely to be replaced by formulas when one applies a substitution. In unification with constants, some variables — called constants — remain unchanged. In normal modal logics extending KD, the elementary unification problem can be decided in nondeterministic polynomial time, seeing that one can easily decide whether a given variable-free formula is equivalent to $\bot$, or is equivalent to $\top$. In many transitive normal modal logics like $K4$, solving the elementary unification problem is more difficult. In some other normal modal logics like K, it is even unknown whether the elementary unification problem is decidable. As for the unification problem with constants, it is not known to be decidable even for a simple normal modal logic such as Alt_1. The truth is that the unification problem with constants is only known to be decidable for a limited number of transitive normal modal logics like $K4$. In many cases, the decidability of the unification problem is a consequence of the decidability of the corresponding admissibility problem. See [23,24,29,36–40] for details.

Concerning the unification types of normal modal logics, it is known that $S5$ is unitary [3], KT is nullary [8], KD is nullary [9], Alt_1 is nullary [11], $S4.3$ is unitary [19], transitive normal modal logics like $K4$ are finitary [24] and K is nullary [30], though the nullariness character of KT and KD has only been obtained within the context of unification with constants. For some other normal modal logics such as the normal extensions of $K4$, they are filtering — therefore they are unitary, or nullary — if and only if they contain the modal translation of the weak law $\neg p \vee \neg\neg p$ of the excluded middle [26]. Taking a look at the literature about unification types in normal modal logics [3,17,26], one will quickly notice that much remains to be done. For instance, the types of symmetric normal modal logics like KB, KTB and KDB and the types of Church-Rosser normal modal logics like KG, KTG and KDG are unknown. Even the types of simple normal modal logics such as $K + \Box^k \bot$ are unknown when $k \in \mathbb{N}$ is such that $k \geq 2$. In his proof that K is nullary, Jeřábek [30] has taken the formula $(x \to \Box x)$ and has shown that it has no minimal complete set of unifiers. In this respect, he has mainly used the following properties of the modality $\Box$: for all $k, l \in \mathbb{N}$, if $(\Box^l \bot \to \Box^k \bot) \in K$ then $l \leq k$; for all formulas ψ and for all $k \in \mathbb{N}$, if $(\psi \to \Box \psi) \in K$ and $\deg(\psi) \leq k$ then $\psi \in K$, or $(\psi \to \Box^k \bot) \in K$. Since for all $k \in \mathbb{N}$, $\neg \Box^k \bot \in KD$, therefore the adaptation of Jeřábek's argument to KT and KD is not straightforward. It has been done in [8,9] by taking the formula $(x \to p) \wedge (y \to q) \wedge (x \to \Box(q \to y)) \wedge (y \to \Box(p \to x))$ in the case of KT and by taking the formula $(x \to p) \wedge (x \to \Box(p \to x))$ in the case of KD.

In this talk, we will give a survey of the results on unification in modal logic and we will present some of the open problems whose solution will have a great impact on the future of the area. After an introductory part about unification in equational theories, we will consider the case of Boolean unification [1,31], we will study the unification problem in Intuitionistic Propositional Logic and transitive normal modal logics like $K4$ [21–23,36,37], we will introduce the notions of projective and transparent unifiers [16–19,24] and we will define filtering unification [26]. Then, we will present the latest results obtained within the context of unification in description logics [2,4,5,20] and in multimodal, tense and epistemic logics [6,10,12,16].

Special acknowledgement is heartily granted to Çiğdem Gencer (Istanbul Aydın University, Turkey), Mojtaba Mojtahedi (Tehran University, Iran), Maryam Rostamigiv (Toulouse University, France) and Tinko Tinchev (Sofia University, Bulgaria) for their feedback: their useful suggestions have been essential for improving a preliminary versions of this talk.

References

1. Baader, F.: On the complexity of Boolean unification. Inf. Process. Lett. **67**, 215–220 (1998)
2. Baader, F., Borgwardt, S., Morawska, B.: Extending unification in $\mathcal{EL}$ towards general TBoxes. In: Brewka, G. et al. (eds.) Principles of Knowledge Representation and Reasoning, pp. 568–572. AAAI Press (2012)
3. Baader, F., Ghilardi, S.: Unification in modal and description logics. Log. J. IGPL **19**, 705–730 (2011)
4. Baader, F., Morawska, B.: Unification in the description logic $\mathcal{EL}$. In: Treinen, R. (ed.) RTA 2009. LNCS, vol. 5595, pp. 350–364. Springer, Heidelberg (2009). https://doi.org/10.1007/978-3-642-02348-4_25
5. Baader, F., Narendran, P.: Unification of concept terms in description logics. J. Symb. Comput. **31**, 277–305 (2001)
6. Babenyshev, S., Rybakov, V.: Unification in linear temporal logic LTL. Ann. Pure Appl. Log. **162**, 991–1000 (2011)

7. Babenyshev, S., Rybakov, V., Schmidt, R., Tishkovsky, D.: A tableau method for checking rule admissibility in $S4$. Electron. Notes Theor. Comput. Sci. **262**, 17–32 (2010)
8. Balbiani, P.: Remarks about the unification type of some non-symmetric non-transitive modal logics. Log. J. IGPL (2018, to appear)
9. Balbiani, P., Gencer, Ç.: KD is nullary. J. Appl. Non Class. Log. **27**, 196–205 (2017)
10. Balbiani, P., Gencer, Ç.: Unification in epistemic logics. J. Appl. Non Class. Log. **27**, 91–105 (2017)
11. Balbiani, P., Tinchev, T.: Unification in modal logic Alt_1. In: Advances in Modal Logic, pp. 117–134. College Publications (2016)
12. Balbiani, P., Tinchev, T.: Elementary unification in modal logic $KD45$. J. Appl. Log. IFCoLog J. Log. Appl. **5**, 301–317 (2018)
13. Chagrov, A.: Decidable modal logic with undecidable admissibility problem. Algebra i Logika **31**, 83–93 (1992)
14. Chagrov, A., Zakharyaschev, M.: Modal Logic. Oxford University Press, Oxford (1997)
15. Cintula, P., Metcalfe, G.: Admissible rules in the implication-negation fragment of intuitionistic logic. Ann. Pure Appl. Log. **162**, 162–171 (2010)
16. Dzik, W.: Unitary unification of $S5$ modal logics and its extensions. Bull. Sect. Log. **32**, 19–26 (2003)
17. Dzik, W.: Unification Types in Logic. Wydawnicto Uniwersytetu Slaskiego, Katowice (2007)
18. Dzik, W.: Remarks on projective unifiers. Bull. Sect. Log. **40**, 37–46 (2011)
19. Dzik, W., Wojtylak, P.: Projective unification in modal logic. Log. J. IGPL **20**, 121–153 (2012)
20. Fernández Gil, O.: Hybrid Unification in the Description Logic $\mathcal{EL}$. Master thesis of Technische Universität Dresden (2012)
21. Gencer, Ç.: Description of modal logics inheriting admissible rules for $K4$. Log. J. IGPL **10**, 401–411 (2002)
22. Gencer, Ç., de Jongh, D.: Unifiability in extensions of $K4$. Log. J. IGPL **17**, 159–172 (2009)
23. Ghilardi, S.: Unification in intuitionistic logic. J. Symb. Log. **64**, 859–880 (1999)
24. Ghilardi, S.: Best solving modal equations. Ann. Pure Appl. Log. **102**, 183–198 (2000)
25. Ghilardi, S.: A resolution/tableaux algorithm for projective approximations in IPC. Log. J. IGPL **10**, 229–243 (2002)
26. Ghilardi, S., Sacchetti, L.: Filtering unification and most general unifiers in modal logic. J. Symb. Log. **69**, 879–906 (2004)
27. Iemhoff, R.: On the admissible rules of intuitionistic propositional logic. J. Symb. Comput. **66**, 281–294 (2001)
28. Iemhoff, R., Metcalfe, G.: Proof theory for admissible rules. Ann. Pure Appl. Log. **159**, 171–186 (2009)
29. Jeřábek, E.: Complexity of admissible rules. Arch. Math. Log. **46**, 73–92 (2007)
30. Jeřábek, E.: Blending margins: the modal logic K has nullary unification type. J. Log. Comput. **25**, 1231–1240 (2015)
31. Martin, U., Nipkow, T.: Boolean unification – the story so far. J. Symb. Comput. **7**, 275–293 (1989)
32. Rozière, P.: Règles admissibles en calcul propositionnel intuitionniste. Thesis of the University Paris VII (1993)

33. Rybakov, V.: A criterion for admissibility of rules in the model system $S4$ and the intuitionistic logic. Algebra Log. **23**, 369–384 (1984)
34. Rybakov, V.: Bases of admissible rules of the logics $S4$ and *Int*. Algebra Log. **24**, 55–68 (1985)
35. Rybakov, V.: Admissibility of Logical Inference Rules. Elsevier, Amsterdam (1997)
36. Rybakov, V.: Construction of an explicit basis for rules admissible in modal system $S4$. Math. Log. Q. **47**, 441–446 (2001)
37. Rybakov, V., Gencer, Ç., Oner, T.: Description of modal logics inheriting admissible rules for $S4$. Log. J. IGPL **7**, 655–664 (1999)
38. Rybakov, V., Terziler, M., Gencer, Ç.: An essay on unification and inference rules for modal logics. Bull. Sect. Log. **28**, 145–157 (1999)
39. Rybakov, V., Terziler, M., Gencer, Ç.: Unification and passive inference rules for modal logics. J. Appl. Non Class. Log. **10**, 369–377 (2000)
40. Rybakov, V., Terziler, M., Gencer, Ç.: On self-admissible quasi-characterizing inference rules. Stud. Logica. **65**, 417–428 (2000)
41. Wolter, F., Zakharyaschev, M.: Undecidability of the unification and admissibility problems for modal and description logics. ACM Trans. Comput. Log. **9**, 25:1–25:20 (2008)

Propositional Modal Logic with Implicit Modal Quantification

Anantha Padmanabha[(✉)] and R. Ramanujam

Institute of Mathematical Sciences, HBNI, Chennai, India
{ananthap,jam}@imsc.res.in

Abstract. Propositional term modal logic is interpreted over Kripke structures with unboundedly many accessibility relations and hence the syntax admits variables indexing modalities and quantification over them. This logic is undecidable, and we consider a variable-free propositional bi-modal logic with implicit quantification. Thus $[\forall]\alpha$ asserts necessity over all accessibility relations and $[\exists]\alpha$ is classical necessity over some accessibility relation. The logic is associated with a natural bisimulation relation over models and we show that the logic is exactly the bisimulation invariant fragment of a two sorted first order logic. The logic is easily seen to be decidable and admits a complete axiomatization of valid formulas. Moreover the decision procedure extends naturally to the 'bundled fragment' of full term modal logic.

Keywords: Term modal logic · Implicitly quantified modal logic · Bisimulation invariance · Bundled fragment

1 Introduction

Propositional multi-modal logics [4,13] are used extensively in the context of multi-agent systems, or to reason about labelled transition systems. In the former case, $\Box_i\alpha$ might refer to knowledge or belief of agent i that α holds. In the latter case, $\Diamond_a\alpha$ may assert the existence of an a-labelled transition from the current state to one in which α holds. Such applications include epistemic reasoning [6,7], games [12], system verification [1,5] and more.

In either of the settings, the indices of modalities come from a fixed finite set. However, the applications themselves admit systems of unboundedly many agents, or infinite alphabets of actions. The former is the case in dynamic networks of processes, and the latter in the case of systems handling unbounded data. In fact, the set of agents relevant for consideration may itself be dynamic, changing with state.

Such motivations naturally lead to modal logics with unboundedly many modalities, and indeed quantification over modal indices. Grove and Halpern [10,11] discuss epistemic logics where the agent set is not fixed and the agent names are not common knowledge. Khan et al. [14] use unboundedly many modalities and allow quantification over them to model information systems

© Springer-Verlag GmbH Germany, part of Springer Nature 2019
Md. A. Khan and A. Manuel (Eds.): ICLA 2019, LNCS 11600, pp. 6–17, 2019.
https://doi.org/10.1007/978-3-662-58771-3_2

in approximation spaces. Other works on indexed modalities include Passy and Tinchev [20], Gargov and Goranko [9], Blackburn [3].

Term Modal logic (TML), introduced by Fitting, Voronkov and Thalmann [8] offers a natural solution to these requirements. It extends first order logic with modalities of the form $\Box_x \alpha$ where x is a variable (and hence can be quantified over). Thus we can write a formula of the form: $\forall x \Box_x(p(x) \supset \exists y \Diamond_y q(x,y))$. Kooi [15] considers the expressivity of TML in epistemic setting. Wang and Seligman [23] introduce a restricted version of TML where we have assignments in place of quantifiers (formulas of the form $[x := b]K_x(\alpha)$ where b is a constant, whose interpretation as an agent will be assigned to x).

Note that TML extends first order logic, and hence its satisfiability problem is undecidable. In [17] we prove that the problem is undecidable even when the atoms are restricted to boolean propositions (PTML). Hence the question of finding decidable fragments of PTML is well motivated. In [17] we prove that the *monodic fragment* of PTML is decidable. The monodic fragment is a restriction allowing at most one free variable within the scope of a modality. i.e, every subformula of the form $\Box_x \alpha$ has $FV(\alpha) \subseteq \{x\}$.

Orlandelli and Corsi [16] consider two decidable fragments: (1) When quantifier occurrence is restricted to the form: $\exists x \Box_x \alpha$ (denoted by $[\exists]\alpha$); (2) Quantifiers appear in a restricted guarded form: $\forall x(P(x) \Rightarrow \Box_x \alpha)$ and $\exists x(P(x) \wedge \Box_x \alpha)$ (and their duals). The corresponding first order modal logic counterparts of the first of these fragments is studied by Wang [22]. Shtakser [21] considers a monadic second order version of the guards (with propositional atoms) of the form $\forall X(P(X) \Rightarrow \Box_X \alpha)$ and $\exists X(P(X) \wedge \Box_X \alpha)$ where X is quantified over subsets of indices and P is interpreted appropriately. These fragments are semantically motivated from their interest in epistemic logic to model notions like 'everyone knows', 'someone knows' and community knowledge (e.g. All eye-witnesses know who killed Mary).

Note that when modalities and quantifiers are 'bundled' together and atomic formulas are propositional, $\exists x \Box_x \alpha$ can be replaced by a variable free modality $[\exists]\alpha$, and similarly $\forall x \Box_x \alpha$ by $[\forall]$. In some sense this is the most natural variable free fragment of PTML with modalities being *implicitly quantified*. This is the logic IQML studied in this paper.

Just as propositional modal logic is the bisimulation-invariant fragment of first order logic, we show that IQML is the bisimulation-invariant fragment of an appropriate two-sorted first order logic. The notion of bisimulation needs to be carefully re-defined to account for quantification over edge labels. Other natural questions on IQML such as decidability of satisfiability and complete axiomatization of valid formulas are answered easily. Interestingly, the natural tableau procedure for the logic can be extended to the 'bundled fragment' of TML with predicates of arbitrary arity, by an argument similar to the one developed in [19] (for a 'bundled fragment' of first order modal logic).

2   Implicitly Quantified Modal Logic (IQML)

The *implicitly quantified modal logic* (IQML) is the variable free fragment of PTML, as discussed above. For more details on PTML, refer [17].

Definition 1 (IQML syntax). *Let $\mathcal{P}$ be a countable set of propositions. The syntax of* IQML *is given by:*

$$\varphi := p \in \mathcal{P} \mid \neg\varphi \mid \varphi \wedge \varphi \mid [\exists]\varphi \mid [\forall]\varphi$$

Note that, $[\exists]\varphi$ translates to $\exists x \Box_x \varphi$ in PTML. Similarly $[\forall]\varphi$ translates to $\forall x \Box_x \varphi$. Since there are no variables in IQML, it is closer to classical propositional modal logics where the set of modal indices is not fixed a priori.

The boolean operators $\vee$ and $\supset$ are defined in the standard way. Also we define $\langle\forall\rangle\varphi = \neg[\exists]\neg\varphi$ and $\langle\exists\rangle\varphi = \neg[\forall]\neg\varphi$ to be the respective duals of the modal operators.

In classical modal logics, the Kripke structure for n modalities is given by $\mathcal{M} = (\mathcal{W}, \mathcal{R}_1, \cdots, \mathcal{R}_n, \rho)$ where each $R_i \subseteq (\mathcal{W} \times \mathcal{W})$ is the accessibility relation for the corresponding index and ρ is the valuation of propositions at every world. But in case of IQML, the modal index set is specified along with the model.

Definition 2 (IQML structure). *An* IQML *structure is given by* $\mathcal{M} = (\mathcal{W}, \mathbb{R}_{\mathbb{I}}, \rho)$ *where* $\mathcal{W}$ *is a non-empty set of worlds,* $\mathbb{I}$ *is a non-empty countable index set and* $\mathbb{R} = \{\mathcal{R}_i \mid i \in \mathbb{I}\}$ *where each* $\mathcal{R}_i \subseteq (\mathcal{W} \times \mathcal{W})$ *and* $\rho : \mathcal{W} \mapsto 2^{\mathcal{P}}$ *is the valuation function.*

Note that $\mathbb{I}$ is the agent set and could be finite or countably infinite. Hence we assume $\mathbb{I}$ to be some initial segment of $\mathbb{N}$ or $\mathbb{N}$ itself. Thus we often denote the model as $\mathcal{M} = (\mathcal{W}, [\mathcal{R}_1, \mathcal{R}_2, \cdots], \rho)$ when $\mathbb{I}$ is clear from the context. Given a model $\mathcal{M}$, we refer to $\mathcal{W}^{\mathcal{M}}$ etc. to denote its corresponding components. The semantics is defined naturally as follows:

Definition 3 (IQML semantics). *Given a model $\mathcal{M}$, a formula φ, $w \in \mathcal{W}^{\mathcal{M}}$, define $\mathcal{M}, w \models \varphi$ inductively as follows:*

$\mathcal{M}, w \models p$	$\Leftrightarrow p \in \rho(w)$
$\mathcal{M}, w \models \neg\varphi$	$\Leftrightarrow \mathcal{M}, w, \not\models \varphi$
$\mathcal{M}, w \models (\varphi \wedge \psi)$	$\Leftrightarrow \mathcal{M}, w \models \varphi$ and $\mathcal{M}, w \models \psi$
$\mathcal{M}, w \models [\exists]\varphi$	$\Leftrightarrow$ *there is some $i \in \mathbb{I}$ such that for all $u \in \mathcal{W}$ if $(w, u) \in R_i$ then $\mathcal{M}, u \models \varphi$*
$\mathcal{M}, w \models [\forall]\varphi$	$\Leftrightarrow$ *for all $i \in \mathbb{I}$ and for all $u \in \mathcal{W}$ if $(w, u) \in R_i$ then $\mathcal{M}, u \models \varphi$*

The formula $\varphi \in$ IQML is *satisfiable* if there is some model $\mathcal{M}$ and $w \in \mathcal{W}$ such that $\mathcal{M}, w \models \varphi$. A formula φ is said to be *valid* if $\neg\varphi$ is not satisfiable. Table 1 gives a complete axiom system for the valid formulas of IQML.

In the sequel we adopt the following convention. Given any model $\mathcal{M}, w \in \mathcal{W}$ and a formula of the form $[\exists]\varphi$, if $\mathcal{M}, w \models [\exists]\varphi$ and $i \in \mathbb{I}$ is the corresponding witness then we write $\mathcal{M}, w \models \Box_i\varphi$ (similarly we have $\mathcal{M}, w \models \Diamond_i\varphi$ for $\langle\exists\rangle\varphi$).

Table 1. IQML axiom system ($\mathcal{AX_A}$)

$\vdash_{\mathcal{AX_A}}$
A0.   All instances of propositional validities.
A1. $[\forall](\varphi \supset \psi) \supset ([\forall]\varphi \supset [\forall]\psi)$
A2. $[\forall](\varphi \supset \psi) \supset (\langle\forall\rangle\varphi \supset \langle\forall\rangle\psi)$
(MP) $\dfrac{\varphi \supset \psi, \varphi}{\psi}$
($[\forall]$Nec) $\dfrac{\varphi}{[\forall]\varphi}$
($[\exists]$Nec) $\dfrac{\varphi}{[\exists]\varphi}$

The axioms and inference rules are standard. Axiom $A2$ describes the interaction between $[\forall]$ and $\langle\forall\rangle$ operators. The ($[\exists]$Nec) rule is sound since $\mathbb{I}$ is non-empty. Note that the axiom system is similar to the one in [11], except for ($[\forall]$Nec) and ($[\exists]$Nec). This is because IQML has no names, as opposed to the logic considered in [11].

Theorem 1. $\vdash_{\mathcal{AX_A}}$ *is sound and complete for* IQML.

The proof of completeness is by construction of canonical model for any consistent formula. The details are presented in [18].

3   IQML Bisimulation and Elementary Equivalence

Modal logics are naturally associated with bisimulations. If two pointed models are bisimilar, the related worlds agree on propositions and satisfy the so-called "back and forth" property [4]. However, when we come to PTML, since the agent set is not fixed, we need to have the notion of 'world bisimilarity' as well as 'agent bisimilarity'. Towards this, in [17], we introduce a notion of bisimulation for propositional term modal logic and show that it preserves PTML formulas. Similar definitions of bisimulations for first order modal logics can be found in [2,22].

Now we introduce the notion of bisimulation for IQML. Here the idea is that two worlds are bisimilar if they agree on all propositions and every index in one structure has a *corresponding index* in the other. The following definition of bisimulation formalizes the notion of 'corresponding index'.

Definition 4. *Given two* IQML *models* $\mathcal{M}_1$ *and* $\mathcal{M}_2$, *an* IQML-*bisimulation on them is a non-empty relation* $G \subseteq (\mathcal{W}_1 \times \mathcal{W}_2)$ *such that for all* $(w_1, w_2) \in G$ *the following conditions hold:*
Val. $\rho_1(w_1) = \rho_2(w_2)$.

$[\exists]$*forth. For all* $i \in \mathbb{I}_1$ *there is some* $j \in \mathbb{I}_2$ *such that, for all* u_2 *such that* $w_2 \xrightarrow{j} u_2$, *there is some* u_1 *such that* $w_1 \xrightarrow{i} u_1$ *and* $(u_1, u_2) \in G$.

$[\exists]back.$ *For all* $j \in \mathbb{I}_2$ *there is some* $i \in \mathbb{I}_1$ *such that, for all* u_1 *such that* $w_1 \xrightarrow{i} u_1$, *there is some* u_2 *such that* $w_2 \xrightarrow{j} u_2$ *and* $(u_1, u_2) \in G$.

$\langle\exists\rangle forth.$ *For all* $i \in \mathbb{I}_1$ *and for all* u_1 *such that* $w_1 \xrightarrow{i} u_1$, *there is some* $j \in \mathbb{I}_2$ *and some* u_2 *such that* $w_2 \xrightarrow{j} u_2$ *and* $(u_1, u_2) \in G$.

$\langle\exists\rangle back.$ *For all* $j \in \mathbb{I}_2$ *and for all* u_2 *such that* $w_2 \xrightarrow{j} u_2$, *there is some* $i \in \mathbb{I}_1$ *and some* u_1 *such that* $w_1 \xrightarrow{i} u_1$ *and* $(u_1, u_2) \in G$.

Given two models $\mathcal{M}_1$ and $\mathcal{M}_2$ we say that w_1, w_2 are IQML bisimilar if there is some IQML bisimulation G on the models such that $(w_1, w_2) \in G$ and denote it $(\mathcal{M}_1, w_1) \leftrightarrow (\mathcal{M}_2, w_2)$. Also, we say $(\mathcal{M}_1, w_1) \equiv_{\mathsf{IQML}} (\mathcal{M}_2, w_2)$ if they agree on all IQML formulas i.e, for all $\varphi \in$ IQML, $\mathcal{M}_1, w_1 \models \varphi$ iff $\mathcal{M}_2, w_2 \models \varphi$.

Theorem 2. *For any two models* $\mathcal{M}_1$ *and* $\mathcal{M}_2$ *and any* $w_1 \in \mathcal{W}_1$ *and* $w_2 \in \mathcal{W}_2$, *if* $\mathcal{M}_1, w_1 \leftrightarrow \mathcal{M}_2, w_2$ *then* $\mathcal{M}_1, w_1 \equiv_{\mathsf{IQML}} \mathcal{M}_2, w_2$.

The proof follows along the standard lines and is provided in [18]. Now we prove that the converse holds over image finite models with finite index set ($\mathbb{I}$). $\mathcal{M}$ is said to be (index, image) finite if $\mathbb{I}$ is finite and $N^i(w) = \{u \mid (w, u) \in R_i\}$ is finite for all $w \in \mathcal{W}$ and $i \in \mathbb{I}$.

Theorem 3. *Suppose* $\mathcal{M}_1$ *and* $\mathcal{M}_2$ *are (index, image) finite models then* $\mathcal{M}_1, w_1 \leftrightarrow \mathcal{M}_2, w_2$ *iff* $\mathcal{M}_1, w_1 \equiv_{\mathsf{IQML}} \mathcal{M}_2, w_2$.

Proof. ($\Rightarrow$) follows from Theorem 2.

For ($\Leftarrow$) suppose $\mathcal{M}_1, w_1 \equiv_{\mathsf{IQML}} \mathcal{M}_2, w_2$, then define $G = \{(v_1, v_2) \mid \mathcal{M}_1, v_1 \equiv_{\mathsf{IQML}} \mathcal{M}_2, v_2\}$. Note that $(w_1, w_2) \in G$. Hence it suffices to show that G is indeed an IQML bisimulation. For this, choose any $(v_1, v_2) \in G$. Clearly $[Val]$ holds since v_1, v_2 agree on all IQML propositions. Now we verify the other conditions:

Now suppose that the ($[\exists]$forth) condition does not hold. Then there is some $\mathbf{i} \in \mathbb{I}_1$ such that for all $j \in \mathbb{I}_2$ there is some $u_j(*)$ such that $v_2 \xrightarrow{j} u_j$ and for all $v_1 \xrightarrow{i} u'$ we have $u' \not\equiv_{\mathsf{IQML}} u_j$. Let $\mathbb{I}_2 = \{j_1 \cdots j_n\}$ and let u_l be the corresponding (*) for every j_l. Also let $\mathbf{i}$-successors of v_1 be $N_{\mathbf{i}}(v_1) = \{s_1 \cdots s_m\}$. By above argument, we have $u_l \not\equiv_{\mathsf{IQML}} s_d$ for all $l \leq n$ and $d \leq m$. Hence for every u_l and every $s_d \in N_{\mathbf{i}}(v_1)$ there is a formula φ_d^l such that $\mathcal{M}_1, s_d \models \varphi_d^l$ but $\mathcal{M}_2, u_l \models \neg\varphi_d^l$. Now consider the formula $\alpha = [\exists](\bigwedge_l \bigvee_d \varphi_d^l)$. Note that for all l and for all $\mathbf{i}$-successors $s_d \in N_{\mathbf{i}}(v_1)$ we have $\mathcal{M}_1, s_d \models \varphi_d^l$ and hence $\mathcal{M}_1, v_1 \models \Box_{\mathbf{i}}(\bigwedge_l \bigvee_d \varphi_d^l)$ which implies $\mathcal{M}_1, w_d \models \alpha$. On the other hand for every $j_l \in \mathbb{I}_2$ at u_l we have $\mathcal{M}_2, u_l \models \bigwedge_d \neg\varphi_d^l$ and hence $\mathcal{M}_2, v_2 \models \langle\forall\rangle(\bigvee_l \bigwedge_d \neg\varphi_d^l)$ which contradicts $v_1 \equiv_{\mathsf{IQML}} v_2$.

The ($[\exists]$back) condition is argued symmetrically.

Suppose that the ($\langle\exists\rangle$back) condition does not hold. Then there is some $\mathbf{j} \in \mathbb{I}_2$ and some $w_2 \xrightarrow{j} \mathbf{u_2}$ such that for all $i \in \mathbb{I}_1$ and for all $w_1 \xrightarrow{i} u'$ we have

$u' \not\equiv_{\mathsf{IQML}} \mathbf{u_2}$. Let $\mathcal{R} = \bigcup_{i \in \mathbb{I}_1} \}R_i$ and let $N(w_1) = \{u' \mid (v_1, u') \in \mathcal{R}\}$ be the set of all successors of w_1. Since $\mathcal{M}_1$ is (index, image) finite, let $N(w_1) = \{t_1 \cdots t_r\}$. By above argument, for every $t_d \in N(w_1)$ there is a formula ψ_d such that $\mathcal{M}_1, t_d \models \psi_d$ and $\mathcal{M}_2, \mathbf{u_2} \models \neg\psi_d$. Hence $\mathcal{M}_2, w_2 \models \Diamond_{\mathbf{j}}(\bigwedge_d \neg\psi_d)$. Now consider $\beta = \langle\exists\rangle(\bigwedge_d \neg\psi_d)$. Clearly $\mathcal{M}_2, w_2 \models \beta$ (with $\mathbf{j}$ and $\mathbf{u_2}$ as witnesses). On the other hand, for any successor t_d of w_1 since $\mathcal{M}_1, t_d \models \psi_d$ we have $\mathcal{M}_1, w_1 \models [\forall](\bigvee_d \psi_d)$ which contradicts our assumption that w_1 and w_2 satisfy the same formulas.

The ($\langle\exists\rangle$forth) is argued symmetrically.

An important consequence of the theorem above is that we can confine ourselves to tree models for IQML formulas, since it is easily seen that an IQML model is bisimilar to its tree unravelling.

Given a tree model $\mathcal{M}$ we define its restriction to level n in the obvious manner: $\mathcal{M}|n$ is simply the same as $\mathcal{M}$ up to level n and the remaining nodes in $\mathcal{M}$ are 'thrown away'.

We can now sharpen the result above: we can define a notion of n-bisimilarity and show that it preserves IQML formulas with modal depth at most n.

Definition 5. *Given two tree models $\mathcal{M}_1$ and $\mathcal{M}_2$, and w_1 in M_1, w_2 in M_2, we say w_1 and w_2 are 0-bisimilar if $\rho_1(w_1) = \rho_2(w_2)$.*

For $n > 0$, we say w_1 and w_2 are n-bisimilar if the following conditions hold:

n-$[\exists]forth$. *For all $i \in \mathbb{I}_1$ there is some $j \in \mathbb{I}_2$ such that for all $w_2 \xrightarrow{j} u_2$ there is some $w_1 \xrightarrow{i} u_1$ such that u_1 and u_2 are $(n-1)$-bisimilar.*

n-$[\exists]back$. *For all $j \in \mathbb{I}_2$ there is some $i \in \mathbb{I}_1$ such that for all $w_1 \xrightarrow{i} u_1$ there is some $w_2 \xrightarrow{j} u_2$ such that u_1 and u_2 are $(n-1)$-bisimilar.*

n-$\langle\exists\rangle forth$. *For all $i \in \mathbb{I}_1$ and for all $w_1 \xrightarrow{i} u_1$ there is some $j \in \mathbb{I}_2$ and some $w_2 \xrightarrow{j} u_2$ such that u_1 and u_2 are $(n-1)$-bisimilar.*

n-$\langle\exists\rangle back$. *For all $j \in \mathbb{I}_2$ and for all $w_2 \xrightarrow{j} u_2$ there is some $i \in \mathbb{I}_1$ and some $w_1 \xrightarrow{i} u_1$ such that u_1 and u_2 are $(n-1)$-bisimilar.*

We can now speak of an n-bisimulation relation between models and speak of models being n-bisimilar, and employ the notation $(\mathcal{M}_1, w_1) \leftrightarroweq_n (\mathcal{M}_2, w_2)$. Clearly, for tree models $(\mathcal{M}_1, w_1) \leftrightarroweq_n (\mathcal{M}_2, w_2)$ iff $(M_1|n, w_1) \leftrightarroweq (M_2|n, w_2)$.

A routine re-working of the proof of Theorem 2 shows that when two tree models are n-bisimilar, they satisfy the same formulas of modal depth at most n. That is, $(\mathcal{M}_1, w_1) \leftrightarroweq_n (\mathcal{M}_2, w_2)$ we have $(\mathcal{M}_1, w_1) \equiv_{\mathsf{IQML}}^n (\mathcal{M}_2, w_2)$. We can go further and show that every n-bisimulation class is represented by a single formula of modal depth at most n. For this, we assume (as is customary in modal logic), that we have only finitely many atomic propositions.

Lemma 1. *Suppose that $\mathcal{P}$ is a finite set. Then for any n and for any $\mathcal{M}, w$ there is a formula $\chi_{[\mathcal{M},w]}^n \in \mathsf{IQML}$ of modal depth n such that for any $(\mathcal{M}', w') \models \chi_{[\mathcal{M},w]}^n$ iff $(\mathcal{M}', w') \leftrightarroweq_n (\mathcal{M}, w)$.*

Proof. Note that ($\Leftarrow$) follows from Theorem 2 specialized to n-bisimulation. For the other direction, the proof is by induction on n. For $n = 0$, since $\mathcal{P}$ is finite, $\chi^0_{[\mathcal{M},w]} = \bigwedge_{p \in \rho(w)} p \wedge \bigwedge_{q \notin \rho(w)} \neg q$ is the required formula.

Let $\mathcal{R} = \bigcup \mathcal{R}_i$ and let $\Gamma^n_{\mathcal{M}} = \{\chi^n_{[\mathcal{M},w]} \mid w \in \mathcal{W}\}$. Inductively $\Gamma^n_{\mathcal{M}}$ is finite. For any $S \subseteq \Gamma^n_{\mathcal{M}}$ let $\overset{\vee}{S}$ denote the disjunction $\bigvee_{\varphi \in S} S$. For the induction step, the characteristic formula is given by:

$$\chi^{n+1}_{[\mathcal{M},w]} = \chi^0_{[\mathcal{M},w]} \wedge \overbrace{\bigwedge_{i \in \mathbb{I}} [\exists] (\bigvee_{(w,u) \in \mathcal{R}_i} \chi^n_{[\mathcal{M},u]})}^{[\exists] \text{forth}} \wedge \overbrace{\bigwedge_{S \subseteq \Gamma^n_{\mathcal{M}}} ([\exists](\overset{\vee}{S}) \supset \bigvee_{i \in \mathbb{I}} \bigwedge_{(w,u) \in \mathcal{R}_i} [\forall](\chi^n_{[\mathcal{M},u]} \supset \overset{\vee}{S}))}^{[\exists] \text{back}}$$

$$\underbrace{\bigwedge_{(w,u) \in \mathcal{R}} \langle \exists \rangle \chi^n_{[M,u]}}_{\langle \exists \rangle \text{forth}} \wedge \underbrace{[\forall] (\bigvee_{(w,u) \in \mathcal{R}} \chi^n_{[M,u]})}_{\langle \exists \rangle \text{back}}$$

Note that the formula remains finite even if $\mathbb{I}$ is infinite or the number of successors of w is infinite since inductively there are only finitely many characteristic formulas of depth n. Showing that $\chi^n_{[\mathcal{M},w]}$ indeed captures n-bisimulation classes is proved in [18].

4   Bisimulation Games and Invariance Theorem

Like every propositional modal logic, IQML is also a fragment of first order logic. However, implicit quantification over domain elements in IQML needs to be made explicit as well as quantification over worlds. Since these serve different purposes in the semantics, we use a two sorted first order logic.

Definition 6 (2Sor.FO syntax). *Let $\mathcal{V}_X$ and $\mathcal{V}_\tau$ be two countable and disjoint sorts of variables and R a ternary predicate. The two sorted FO (2Sor.FO), corresponding to IQML is given by:*

$$\alpha ::= Q_p(x) \mid R(x, \tau, y) \mid \neg \alpha \mid \alpha \wedge \alpha \mid \exists \tau \, \alpha \mid \exists x \, \alpha$$

where Q_p is the corresponding monadic predicate for every $p \in \mathcal{P}$ and $x, y \in \mathcal{V}_X$ and $\tau \in \mathcal{V}_\tau$.

A 2Sor.FO structure is given by $\mathfrak{M} = [(\mathcal{W}, \mathbb{I}), (\hat{R}, \hat{\rho})]$ where $(\mathcal{W}, \mathbb{I})$ is the two sorted domain and $(\hat{R}, \hat{\rho})$ are interpretations with $\hat{R} \subseteq (\mathcal{W} \times \mathbb{I} \times \mathcal{W})$ and $\hat{\rho} : \mathcal{W} \mapsto 2^{Q_\mathcal{P}}$ where $Q_\mathcal{P} = \{Q_p \mid p \in \mathcal{P}\}$. The semantics $\Vdash$ is defined for 2Sor.FO in the standard way where the variables in $\mathcal{V}_X$ range over the first sort $(\mathcal{W})$ and variables of $\mathcal{V}_\tau$ range over second $(\mathbb{I})$.

Given an IQML structure $\mathcal{M} = (\mathcal{W}, \mathcal{R}_\mathbb{I}, \rho)$ the corresponding 2Sor.FO structure is given by $\mathfrak{M} = [(\mathcal{W}, \mathbb{I}), (\hat{R}, \hat{\rho})]$ where $(w, i, v) \in \hat{R}$ iff $(w, v) \in \mathcal{R}_i$ and $Q_p \in \hat{\rho}(w)$ iff $p \in \rho(w)$. Similarly given any 2Sor.FO structure, it can be interpreted as an IQML structure. Thus there is a natural correspondence between IQML structures and 2Sor.FO structures. For any IQML structure $\mathcal{M}$ let the corresponding 2Sor.FO structure be denoted by $\mathfrak{M}$.

Definition 7 (IQML to 2Sor.FO translation). *The translation of $\varphi \in$ IQML into a 2Sor.FO parametrized by $x \in \mathcal{V}_X$ is given by:*
$\mathsf{Tr}(p : x) = Q_p(x)$
$\mathsf{Tr}(\neg\varphi : x) = \neg\mathsf{Tr}(\varphi : x)$
$\mathsf{Tr}(\varphi \wedge \psi : x) = \mathsf{Tr}(\varphi : x) \wedge \mathsf{Tr}(\psi : x)$
$\mathsf{Tr}([\exists]\varphi : x) = \exists\tau\forall y \, (R(x,\tau,y) \supset \mathsf{Tr}(\varphi : y))$
$\mathsf{Tr}([\forall]\varphi : x) = \forall\tau\forall y \, (R(x,\tau,y) \supset \mathsf{Tr}(\varphi : y))$

Proposition 1. *For any formula $\varphi \in$ IQML and any IQML structure $\mathcal{M}$*
$\quad \mathcal{M}, w \models \varphi$ *iff* $\mathfrak{M}, [x \mapsto w] \Vdash \mathsf{Tr}(\varphi : x)$.

Hence IQML can be translated into 2Sor.FO with 2 variables of $\mathcal{V}_X$ sort and one variable of $\mathcal{V}_\tau$ sort. Given two IQML models $\mathcal{M}_1$ and $\mathcal{M}_2$, the notion of IQML bisimulation naturally translates to bisimulation over the corresponding 2Sor.FO models $\mathfrak{M}_1$ and $\mathfrak{M}_2$.

Now we state the van Benthem type characterization theorem: bisimulation invariant 2Sor.FO formulas can be translated back into IQML. We say that $\alpha(x) \in$ 2Sor.FO is *bisimulation invariant* if for all $\mathcal{M}_1, w_1 \leftrightarrow \mathcal{M}_2, w_2$ we have $\mathfrak{M}_1, [x \mapsto w_1] \Vdash \alpha(x)$ iff $\mathfrak{M}_2, [x \mapsto w_2] \Vdash \alpha(x)$. We can similarly speak of $\alpha(x)$ being n-bisimulation invariant as well. Also, $\alpha(x)$ is equivalent to some IQML formula if there is some formula $\varphi \in$ IQML such that for all $\mathcal{M}$ we have $\mathfrak{M}, [x \mapsto w] \Vdash \alpha(x)$ iff $\mathcal{M}, w \models \varphi$.

Theorem 4. *Let $\alpha(x) \in$ 2Sor.FO with one free variable $x \in \mathcal{V}_X$. Then $\alpha(x)$ is bisimulation invariant iff $\alpha(x)$ is equivalent to some IQML formula.*

Note that $\Leftarrow$ follows from Theorem 2. To prove ($\Rightarrow$) it suffices to show that if $\alpha(x)$ is bisimulation invariant then, for some n it is n-bisimulation invariant, since we have already shown in the last section that n-bisimulation classes are defined by IQML formulas.

Towards proving this, we introduce a notion of locality for 2Sor.FO formulas. For any tree model $\mathcal{M}$ and let $\mathfrak{M}|n$ be the corresponding 2Sor.FO model of $\mathcal{M}$ restricted to n depth.

Definition 8. *We say that a formula $\alpha(x)$ is n-local if for any tree model $(\mathcal{M}, w)$, $\mathfrak{M} \Vdash \alpha(w)$ iff $\mathfrak{M}|n \Vdash \alpha(w)$.*

Lemma 2. *For any $\alpha(x) \in$ 2Sor.FO formula which is bisimulation invariant with $x \in \mathcal{V}_X$ then $\alpha(x)$ is n-local for $n = 2^q$ where $q = q_x + q_\tau$ where q_x is the quantifier rank of $\mathcal{V}_X$ sort in $\alpha(x)$ and q_τ is the quantifier rank of $\mathcal{V}_\tau$ in $\alpha(x)$.*

Assuming this lemma, consider a 2Sor.FO formula $\alpha(x)$ which is bisimulation invariant. It is n-local for a syntactically determined n. We now claim that $\alpha(x)$ is n-bisimulation invariant. To prove this, consider $\mathcal{M}_1, w_1 \leftrightarrow_n \mathcal{M}_2, w_2$. We need to show that $\mathfrak{M}_1, [x \mapsto w_1] \Vdash \alpha(x)$ iff $\mathfrak{M}_2, [x \mapsto w_2] \Vdash \alpha(x)$.

Suppose that $\mathfrak{M}_1, [x \mapsto w_1] \Vdash \alpha(x)$. By locality, $\mathfrak{M}_1|n, [x \mapsto w_1] \Vdash \alpha(x)$. Now observe that $\mathcal{M}_1|n, w_1 \leftrightarrow \mathcal{M}_2|n, w_2$. By bisimulation invariance of $\alpha(x)$, $\mathfrak{M}_2|n, [x \mapsto w_2] \Vdash \alpha(x)$. But then again by locality, $\mathfrak{M}_2, [x \mapsto w_2] \Vdash \alpha(x)$, and we are done.

Thus it only remains to prove the locality lemma. For this, it is convenient to consider the *Ehrenfeucht-Fraisse* (EF) game for 2Sor.FO. In this game we have two types of pebbles, one for $\mathcal{W}$ and the other for $\mathbb{I}$.

The game is played between two players Spoiler (**Sp**) and Duplicator (**Dup**) on two 2Sor.FO structures. A configuration of the game is given by $[(\mathfrak{M}, \bar{s}); (\mathfrak{M}', \bar{t})]$ where $\bar{s} \in (\mathcal{W} \cup \mathbb{I})^*$ is a finite string $(\mathcal{W} \cup \mathbb{I})$ and similarly $\bar{t} \in (\mathcal{W}' \cup \mathbb{I}')^*$.

Suppose the current configuration is $[(\mathfrak{M}, \bar{s}); (\mathfrak{M}', \bar{t})]$. In a $\mathcal{W}$ round, **Sp** places a $\mathcal{W}$ pebble on some $\mathcal{W}$ sort in one of the structures and **Dup** responds by placing a $\mathcal{W}$ pebble on a $\mathcal{W}$ sort in the other structure. In a $\mathbb{I}$ round, similarly **Sp** picks one structure and places an $\mathbb{I}$ pebble on some $\mathbb{I}$ sort and **Dup** responds by placing an $\mathbb{I}$ pebble on some $\mathbb{I}$ sort in the other structure. In both cases, the new configuration is updated to $[(\mathfrak{M}, \bar{s}s); (\mathfrak{M}', \bar{t}t)]$ where s and t are the new elements (either $\mathcal{W}$ or $\mathbb{I}$ sort) picked in the corresponding structures.

A (q_x, q_τ) round game is one where q_x many pebbles of type $\mathcal{W}$ are used and q_τ many pebbles of type $\mathbb{I}$ is used. Player **Dup** wins after (q_x, q_τ) if after (q_x, q_τ) rounds, if in $[(\mathfrak{M}, \bar{s}); (\mathcal{M}', \bar{t})]$ the mapping $f(s_i) = t_i$ forms a partial isomorphism over $\mathfrak{M}$ and $\mathfrak{M}'$. Otherwise **Sp** wins.

It can be easily shown that **Dup** has a winning strategy in the (q_x, q_τ) round game over two structures iff they agree on all formulas with quantifier rank of $\mathcal{V}_X$ sort $\le q_x$ and quantifier rank of $\mathcal{V}_\tau$ sort $\le q_\tau$.

Let $\mathcal{M}, w$ be any tree structure. To prove Lemma 2, we need to prove that $\mathfrak{M}, w \models \alpha(x)$ iff $\mathfrak{M}|n \models \alpha(x)$.

Let $q = q_x + q_\tau$ and $\mathfrak{N}$ be q disjoint copies of $\mathfrak{M}$ and $\mathfrak{M}|n$. Note that *inclusion* relation G over $\mathfrak{M}$ and $\mathfrak{M}|n$ forms a bisimulation. Also note that G continues to be a bisimulation over the disjoint union of $\mathfrak{N} \uplus \mathfrak{M}, w$ and $\mathfrak{N} \uplus \mathfrak{M}|n, w$. Moreover, notice that $(\mathfrak{M}, w)$ is bisimilar to $(\mathfrak{N} \uplus \mathfrak{M}, w)$ and further $(\mathfrak{M}|n, w)$ is bisimilar to $(\mathfrak{N} \uplus \mathfrak{M}|n, w)$.

Now since $\alpha(x)$ is bisimulation invariant, it is enough to show that **Dup** has a winning strategy in the game starting from $[(\mathfrak{N} \uplus \mathfrak{M}, w), (\mathfrak{N} \uplus \mathfrak{M}|n, w)]$.

The winning strategy for **Dup** is to ensure that at every round $m < (q_x + q_\tau)$ the critical distance $d_m = 2^{q-m}$ is respected:

If **Sp** places $\mathcal{W}$ pebble on a $\mathcal{W}$ sort which is within d_m of an already pebbled $\mathcal{W}$ pebble, **Dup** plays according to a local isomorphism in the d_m- neighbourhoods of previously pebbled elements (exists since $n = 2^q$ and $m < q$); if **Sp** places a $\mathcal{W}$ pebble somewhere beyond 2^{q-m} distance from all $\mathcal{W}$ pebbles previously used, then, **Dup** responds in a fresh isomorphic copy of type $\mathfrak{M}$ or $\mathfrak{M}|n$ correspondingly (again, it is guaranteed to exist since previously at most $m - 1(< q)$ would have been used).

If **Sp** decides to use an $\mathbb{I}$ pebble and places it on some $\mathbb{I}$ sort i in one structure, then **Dup** responds by placing an $\mathbb{I}$ pebble on i in the mirror copy in the other structure, where by mirror copy we mean: for $\mathfrak{M}$ or $\mathfrak{M}|n$ in $\mathfrak{N}$ then the mirror copy in the other structure is itself and the original $\mathfrak{M}$ and $\mathfrak{M}|n$ are mirror copies of each other.

5   Satisfiability Problem

The satisfiability problem for IQML can be solved by sharpening the completeness proof of the axiom system by showing that every consistent formula is satisfied in a model of bounded size. Indeed, a PSPACE decision procedure can be given along the lines of Grove and Halpern [11]. However, we give a tableau procedure for IQML which is instructive, and as we will observe later, neatly generalizes to more expressive logics.

Given a formula φ, we set $I = \{c_\alpha \mid \langle \exists \rangle \alpha \in SF(\varphi)\} \cup \{d_\beta \mid [\exists]\beta \in SF(\varphi)\}$ where $SF(\varphi)$, the set of subformulas of φ is defined in the standard way. This forms the index set where c_α and d_β act as witnesses for the corresponding formulas.

We construct a tableau tree structure $T = (W, V, E, \lambda)$ where W is a finite set, (V, E) is a rooted tree and $\lambda : V \mapsto L$ is a labelling map. Each element in L is of the form $(w : \Gamma, i_\chi)$, where $w \in W$, Γ is a finite set of formulas and $i_\chi \in I$. The intended meaning of the label is that the node constitutes a world w that satisfies the formulas in Γ and i_χ is the incoming label edge of w.

The tableau rules for IQML are inspired from the tableau procedure for the bundled fragment of first order modal logic introduced in [19]. The $(\wedge)$ and $(\vee)$ tableau rules are standard. For the modalities, the intuition for the corresponding tableau rule is the following: Suppose that we are in an intermediate step of tableau construction when we have formulas $\{\langle \exists \rangle \alpha, [\exists]\beta, \langle \forall \rangle \varphi, [\forall]\psi\}$ to be satisfied at a node w. For this, first we need to add a new c_α successor node wv_α where α holds; this new node inherits not only α but also ψ. Also, we need a d_β successor which inherits β, φ and ψ. Finally for each $e_\gamma \in I$ we need a φ-successor which also inherits ψ.

The (BR) tableau rule extends this idea when there are multiple occurrences of each kind of formulas above. In general if the set of formulas considered at node $w : (A, B, C, D)$ where $A = \{\langle \exists \rangle \alpha_1 .. \langle \exists \rangle \alpha_{n_1}\}$; $B = \{[\exists]\beta_1 .. [\exists]\beta_{n_2}\}$; $C = \{\langle \forall \rangle \varphi_1 .. \langle \forall \rangle \varphi_{m_1}\}$ and $D = \{[\forall]\psi_1 .. [\forall]\psi_{m_2}\}$. Let $D' = \{\psi \mid [\forall]\psi \in D\}$. The BR rule is given as follows:

$$\frac{w : (A, B, C, D)}{\begin{array}{c} \{\langle wv_{\alpha_i} : (\alpha_i, D'), c_{\alpha_i}\rangle \mid i \le n_1\} \cup \\ \{\langle wv_{\beta_j}^k : (\beta_j, \varphi_k, D'), d_{\beta_j}\rangle \mid k \le m_1, j \le n_2\} \cup \\ \{\langle wv_{e_\chi}^k : (\varphi_k, D), e_\chi\rangle \mid l \le m_1, \chi \notin (A \cup B)\} \end{array}} (BR)$$

From an 'open tableau' we can construct a model for φ, along the lines of [19]. Conversely it can be proved that every satisfiable formula has an open tableau.

This tableau construction can be extended to the 'bundled fragment' of full TML where we have predicates of arbitrary arity and the quantifiers and modalities occur (only) in the form $\forall x \Box_x \varphi$ and $\exists x \Box \varphi$ (and their duals). The proof follows the lines of [19].

6 Discussion

We have studied the variable-free fragment of PTML, with implicit modal quantification. We could also consider more forms of implicit quantification such as $\Box\forall$ and $\Diamond\forall$ modalities, though there is no obvious semantics to them. These logics are the obvious variable free versions of monadic 'bundled' fragments of TML. One could consider a similar exercise for 'bundled' fragments of first order modal logic (FOML). As [19] shows, this is a decidable logic for increasing domain semantics.

Our study suggests that there are other forms of implicitly quantified modal logics. For instance, is there an implicit *hybrid* version of the logic studied by Wang and Seligman [23]?

A natural question is the delimitation of expressiveness of these logics: which are the properties of models expressed only by $\exists\Box$ or only by $\forall\Box$ modalities? How does nesting of these modalities increase expressive power? We believe that the model theory of implicit modal quantification may offer interesting possibilities for abstract specifications of some infinite-state systems. However, for such study, we will need to consider transitive closures of accessibility relations, and this seems to be quite challenging.

Recent developments in tools for model checking and other decision procedures for fragments of FOML offer a promising direction to develop similar practical frameworks for IQML and other decidable fragments of term-modal logics. Such tools can be of help in the synthesis and verification of some classes of systems with unboundedly many agents.

Acknowledgement. We thank Yanjing Wang for his insightful and extensive discussions on the theme of this paper. Also, we thank the anonymous reviewers for their comments that helped us improve the presentation and quality of the paper.

References

1. Alur, R., Henzinger, T.A., Kupferman, O.: Alternating-time temporal logic. J. ACM **49**(5), 672–713 (2002). https://doi.org/10.1145/585265.585270
2. van Benthem, J., et al.: Frame correspondences in modal predicate logic. Proofs, categories and computations: essays in honor of Grigori Mints, pp. 1–14 (2010)
3. Blackburn, P.: Nominal tense logic. Notre Dame J. Form. Log. **34**(1), 56–83 (1993). https://doi.org/10.1305/ndjfl/1093634564
4. Blackburn, P., de Rijke, M., Venema, Y.: Modal logic. Cambridge Tracts in Theoretical Computer Science. Cambridge University Press, Cambridge (2001)
5. Clarke, E.M., Emerson, E.A., Sistla, A.P.: Automatic verification of finite-state concurrent systems using temporal logic specifications. ACM Trans. Programm. Lang. Syst. (TOPLAS) **8**(2), 244–263 (1986)
6. van Ditmarsch, H., van der Hoek, W., Kooi, B.: Dynamic Epistemic Logic. Synthese Library, vol. 337. Springer, Dordrecht (2007). https://doi.org/10.1007/978-1-4020-5839-4
7. Fagin, R., Halpern, J.Y., Moses, Y., Vardi, M.: Reasoning About Knowledge. A Bradford Book, Cambridge (2004)

8. Fitting, M., Thalmann, L., Voronkov, A.: Term-modal logics. Stud. Logica **69**(1), 133–169 (2001). https://doi.org/10.1023/A:1013842612702
9. Gargov, G., Goranko, V.: Modal logic with names. J. Philos. Log. **22**(6), 607–636 (1993). https://doi.org/10.1007/BF01054038
10. Grove, A.J.: Naming and identity in epistemic logic Part II: a first-order logic for naming. Artif. Intell. **74**(2), 311–350 (1995)
11. Grove, A.J., Halpern, J.Y.: Naming and identity in epistemic logics Part I: the propositional case. J. Log. Comput. **3**(4), 345–378 (1993)
12. van der Hoek, W., Pauly, M.: 20 modal logic for games and information. In: Studies in Logic and Practical Reasoning, vol. 3, pp. 1077–1148. Elsevier (2007)
13. Hughes, M., Cresswell, G.: A New Introduction to Modal Logic. Routledge, London and New York (1996)
14. Khan, M.A., Banerjee, M., Rieke, R.: An update logic for information systems. Int. J. Approximate Reasoning **55**(1), 436–456 (2014). https://doi.org/10.1016/j.ijar.2013.07.007
15. Kooi, B.: Dynamic term-modal logic. In: A Meeting of the Minds, pp. 173–186 (2007)
16. Orlandelli, E., Corsi, G.: Decidable term-modal logics. In: Belardinelli, F., Argente, E. (eds.) EUMAS/AT -2017. LNCS (LNAI), vol. 10767, pp. 147–162. Springer, Cham (2018). https://doi.org/10.1007/978-3-030-01713-2_11
17. Padmanabha, A., Ramanujam, R.: The monodic fragment of propositional term modal logic. Stud. Logica 1–25 (2018). https://doi.org/10.1007/s11225-018-9784-x
18. Padmanabha, A., Ramanujam, R.: Propositional modal logic with implicit modal quantification. arXiv preprint arXiv:1811.09454 (2018)
19. Padmanabha, A., Ramanujam, R., Wang, Y.: Bundled fragments of first-order modal logic: (un)decidability. In: 38th IARCS Annual Conference on Foundations of Software Technology and Theoretical Computer Science, FSTTCS 2018, 11–13 December 2018, Ahmedabad, pp. 43:1–43:20 (2018). https://doi.org/10.4230/LIPIcs.FSTTCS.2018.43
20. Passy, S., Tinchev, T.: Quantifiers in combinatory PDL: completeness, definability, incompleteness. In: Budach, L. (ed.) FCT 1985. LNCS, vol. 199, pp. 512–519. Springer, Heidelberg (1985). https://doi.org/10.1007/BFb0028835
21. Shtakser, G.: Propositional epistemic logics with quantification over agents of knowledge. Stud. Logica **106**(2), 311–344 (2018)
22. Wang, Y.: A new modal framework for epistemic logic. In: Proceedings Sixteenth Conference on Theoretical Aspects of Rationality and Knowledge, TARK 2017, Liverpool, 24–26 July 2017, pp. 515–534 (2017). https://doi.org/10.4204/EPTCS.251.38
23. Wang, Y., Seligman, J.: When names are not commonly known: epistemic logic with assignments. In: Advances in Modal Logic, vol. 12, pp. 611–628. College Publications (2018)

Infinite Liar in a (Modal) Finitistic Setting

Michał Tomasz Godziszewski[1] and Rafal Urbaniak[2,3]($\boxtimes$)

[1] University of Warsaw, Warsaw, Poland
mtgodziszewski@gmail.com
[2] University of Gdańsk, Gdańsk, Poland
rfl.urbaniak@gmail.com
[3] Ghent University, Ghent, Belgium
https://uw.academia.edu/MichalGodziszewski
https://ugent.academia.edu/RafalUrbaniak

Abstract. Yablo's paradox results in a set of formulas which (with local disquotation in the background) turns out consistent, but ω-inconsistent. Adding either uniform disquotation or the ω-rule results in inconsistency. One might think that it doesn't arise in finitary contexts. We study whether it does. It turns out that the issue turns on how the finitistic approach is formalized.

Keywords: Axiomatic theories of truth · Paradoxes · Yablo's paradox · Finitism · Potential infinity

1 Introduction

[10] provided a by now famous example of a semantic paradox which, according to the author, does not involve self-reference. Recall the paradox arises when one considers the following sequence of sentences:

Y_0 For any $k > 0$, Y_k is false.
Y_1 For any $k > 1$, Y_k is false.
Y_2 For any $k > 2$, Y_k is false.

$\vdots$

Y_n For any $k > n$, Y_k is false.

$\vdots$

Take any Y_n and suppose it is true. Then, for any $j > n$ Y_j is false. In particular Y_{n+1} is false and also for any $j > n + 1$ Y_j is false. But the second

This research has been supported by the FWO postdoctoral research grant and the National Science Centre SONATA BIS research grant number 2016/22/E/HS1/00304. The second author has been supported by the National Science Centre OPUS research grant number 2014/13/B/HS1/02892.

Md. A. Khan and A. Manuel (Eds.): ICLA 2019, LNCS 11600, pp. 18–29, 2019.
https://doi.org/10.1007/978-3-662-58771-3_3

conjunct is exactly what Y_{n+1} states, so Y_{n+1} is true after all. So Y_n is false. So for some $k > n$, Y_k is true. But then, we obtain a contradiction by repeating for Y_k the same reasoning.

A fruitful study of the paradox formalized over arithmetic performed e.g. in [1,3] has revealed that in order to derive the contradiction one needs a strong assumption: "for all n, Y_n if and only if $\ulcorner Y_n \urcorner$ is true." $\forall n \ (Y_n \equiv Tr(Y_n))$. If we wanted to replace this *uniform disquotation* with an infinity of *local disquotation* instances, contradiction could be obtained only if we used some infinitary inference rule (requiring an infinite number of premises) such as the ω-rule.

So far, the story is rather well-known. What is somewhat less known, is that there is a way of handling the paradox which relies on finitistic assumptions. After all, if the world is finite, there aren't enough things in the world to interpret all sentences from the Yablo sequence, and the last interpreted one is vacuously true without any threat of paradox.

The finitist owes us a story about how they make sense of arithmetic, and how the whole thing should be studied by formal methods. Formal tools for this task have already been developed [4–6]. In what follows we'll explain what it is, and we'll use it to study the Yablo paradox in the finitistic setting. On this approach, it will turn out that things are as we expected: Yablo sentences are all false in potentially infinite domains, despite the fact that the framework is rich enough to incorporate sufficiently strong arithmetic.

There is, however, a glitch. We'll argue that the way quantifiers are handled in this finitistic setting results in a somewhat scary arithmetical theory.

If your goal, as a finitist, is not to revise current mathematics, but to make sense of it in terms of potential infinity, this approach isn't for you.

There is another formal approach to potential infinity developed in [9], which has already been used to obtain standard arithmetic, and to make sense of abstraction principles (in the neologicist sense). In the third part of this paper we study how this framework handles Yablo's paradox. It turns out that the price of making potential infinity digestible to classical mathematicians is that the Yablo paradox strikes back, even with more power than in the standard arithmetical setting.

2   Arithmetization of Yablo Sentences

Let's start with going over the results pertaining to Yablo sentences obtained in the standard arithmetical setting.

One might ask how we actually know that Yablo sequences exist in formal theories. This is a legitimate question since we're moving from the paradox as formulated hand-wavily in natural language to its properly defined formalized counterpart. It is possible to construct a Yablo sequence within a given theory (but in order to do so, we need to use a general version of the diagonal lemma for formulae with two free variables in the language containing the truth predicate).

Definition 1 (Yablo Formula). $Y(x)$ *is a Yablo formula in a theory* T *iff it satisfies the* Yablo condition, *i.e.:*

$$\text{T} \vdash \forall x(Y(x) \equiv \forall w > x \neg Tr(\overline{\ulcorner Y(\dot{w})\urcorner})).$$ ⊣

This also gives rise to a natural way of defining sentences belonging to a Yablo sequence.

Definition 2 (Yablo Sentence). φ *is a Yablo sentence in a theory* T *iff it is obtained by substituting a numeral for x in Yablo formula $Y(x)$.* ⊣

Theorem 1 (Existence of Yablo Formula [8]). *Let* T *be a theory in a language $\mathcal{L}_{Tr}$ containing Robinson arithmetic Q. Then there exists a Yablo formula in* T.

An interesting question one might ask is exactly what principles about Yablo sentences lead to the inconsistency of a formal theory. Despite the fact that on the level of natural language it is not difficult to derive contradiction from the definition of Yablo sequence, the formal counterpart (to be specified below) is only ω-inconsistent and consistent. Most of the results from this section were originally obtained by [3].

Definition 3 (ω-consistency). *Let* T *be a first-order theory in the arithmetical language $\mathcal{L}$.* T *is ω-consistent if there is no $\varphi(x) \in Frm_{\mathcal{L}}$ such that simultaneously:*

$$\forall n \in \omega \ \text{T} \vdash \neg\varphi(\overline{n})$$
$$\text{T} \vdash \exists x \varphi(x)$$ ⊣

If there is such a formula φ, then we say that T is an ω-inconsistent theory.

Definition 4 (Local Arithmetical Disquotation).

$$AD = \{Tr(\overline{\ulcorner \varphi \urcorner}) \equiv \varphi : \varphi \in Sent_{\mathcal{L}}\}$$ ⊣

Definition 5 (Local Yablo Disquotation).

$$YD = \{Tr(\overline{\ulcorner Y(\overline{n})\urcorner}) \equiv Y(\overline{n}) : Y(\overline{n}) \text{ belongs to the Yablo sequence}\}. \quad ⊣$$

Definition 6 (PA$_D$). *Let* PAT *be a theory obtained from* PA *by extending the language of arithmetic with a truth predicate Tr (this means that induction scheme applies also to formulae containing Tr). Let* PA$_D$ = PAT $\cup$ AD $\cup$ YD *and let* PA$_D^-$ *be* PA$_D$ *with the induction axiom scheme restricted to formulae without the truth predicate.*

Theorem 2. PA$_D$ *is a consistent, yet ω-inconsistent conservative extension of* PA.

It is however still possible to derive a contradiction from the Yablo sequence, but in order to achieve this we would have to use a generalized version of the truth principle governing the Yablo sequence—it would be necessary to add to our arithmetical theory principles that would be strong enough to prove a version of disquotation schema *uniform* for all Yablo sentences.

Definition 7 (Uniform Yablo Disquotation).

$$\forall x (Tr(\ulcorner Y(\dot{x}) \urcorner) \equiv Y(x)) \qquad \text{(UYD)}$$

Theorem 3. *Let* S = PAT + UYD. *S is inconsistent.*

Definition 8 (ω-rule). *The ω-rule is the following infinitary inference rule defined for arithmetical formulae φ:*

$$\frac{\varphi(\overline{0}),\ \varphi(\overline{1}),\ \varphi(\overline{2}),\ ...,\ \varphi(\overline{n}),\ ...}{\forall x\, \varphi(x)}$$

Definition 9 (T^ω). *Let* T *be an axiomatizable first-order theory in the language $\mathcal{L}_{Tr}$. If α is an ordinal, a sequence $(\varphi_0, ..., \varphi_\alpha)$ of formulae is a derivation in ω-logic (ω-derivation) from* T *if and only if, for each ordinal $\beta \leq \alpha$:*

1. *φ_β is an axiom of* T *or*
2. *there are $\gamma < \beta$ and $\delta < \beta$, such that $\varphi_\delta = (\varphi_\gamma \rightarrow \varphi_\beta)$, or*
3. *$\varphi_\beta = \forall x \psi(x)$ for some $\psi(x) \in Frm_{\mathcal{L}}$ with exactly one free variable and there is an injective function $f : \omega \rightarrow \beta$ such that $\forall n \in \omega\ \varphi_{f(n)} = \psi(n)$ (this condition means that φ_β has been introduced by means of the ω-rule).*

We say that φ is a theorem of T *in ω-logic if there is an ω-derivation of φ from* T*. We also call such φ an ω-consequence of* T *and denote it as follows:* $T^\omega \vdash \varphi$. *T^ω is a set of sentences that are theorems of* T *in ω-logic.* $\dashv$

Theorem 4. *Let* $\text{PA}_D^{\omega^-} = (\text{PAT}^- \cup \text{AD} \cup \text{YD})^\omega$. $\text{PA}_D^{\omega^-}$ *is inconsistent.*

Let's sum up the situation so far. We have a paradox in natural language. Formalizing it, as long as we work in a disquotational theory of truth without any other assumptions, even those theories which prove the existence of Yablo sentences are still consistent (albeit ω-inconsistent). To obtain a contradiction, we need either inferential rules that go beyond the standard means of first-order logic (namely: the ω-rule), or a stronger uniform principle of disquotation for Yablo sentences.

Since the paradox somehow involves the concept of infinity (as the essential role of the ω-rule suggests), we'll turn our attention to a formalization of Yablo's paradox in a setting that takes a somewhat different approach to infinity. It is the framework developed by the late Marcin Mostowski and others in [4–7], meant as a formalization of the concept of potential infinity. We'll describe how one could go about avoiding the paradox by taking the distinction between potential and actual infinity seriously.

3   Potentially Infinite Domains and sl-Semantics

In this section we consider the so-called sl-semantics of FM-domains. This framework was motivated by considerations in computational foundations of mathematics and the search for the semantics under which first-order sentences would be interpreted in potentially infnite domains.[1]

Potentially infinite domains are in this framework understood as sequences of finite models increasing without any finite bound. Once we turn to actually finite domains, we need to overcome a small technical obstacle concerning the arithmetical language: it seems that the reference of some singular terms should come out undefined. One way to go would be to employ the apparatus of partial functions. A simpler method, however, uses relational symbols instead of function symbols. This is the one that we'll employ.

Let $R \subseteq \omega^r$ be an r-ary relation on natural numbers. By $R^{(n)}$ we denote $R \cap \{0, 1, ..., n-1\}^r$, the restriction of this relation to first n natural numbers. For any model $\mathcal{A}$ over some fixed signature $\sigma = (R_1, ..., R_k)$ (in particular, one can take the signature to comprise the relational counterpart of standard arithmetical functions such as addition or multiplication) we define the FM-domain of $\mathcal{A}$ by saying $FM(\mathcal{A}) = \{\mathcal{A}_n \colon n = 1, 2, ...\}$ where $\mathcal{A}_n = (\{0, 1, ..., n-1\}, R_1^{(n)}, ..., R_k^{(n)})$.

By $\mathbb{N}$ we denote the standard model of arithmetic $(\omega, +, \times, 0, s, <)$, where the arithmetical functions are interpreted as relations, so we have that: $FM(\mathbb{N}) = \{\mathbb{N}_n \colon n = 1, 2, ...\}$, where $\mathbb{N}_n = (\{0, 1, ..., n-1\}, +^{(n)}, \times^{(n)}, 0^{(n)}, s^{(n)}, <^{(n)})$.

Definition 10 $(sl(FM(\mathbb{N})))$. *For any $\varphi \in Sent_{\mathcal{L}}$ we say that φ is sl-true in $FM(\mathbb{N})$ (true in sufficiently large models, hence the shortcut sl):*

$$FM(\mathbb{N}) \models_{sl} \varphi \text{ if and only if } \exists m \, \forall k \ (k \geq m \ \Rightarrow \ \mathbb{N}_k \models \varphi).$$

Let us then denote:

$$sl(FM(\mathbb{N})) = \{\varphi \in Sent_{\mathcal{L}} \colon FM(\mathbb{N}) \models_{sl} \varphi\}.$$

$sl(FM(\mathbb{N}))$ is called the sl-theory of $FM(\mathbb{N})$. Obviously, for a given vocabulary, for any class $\mathcal{K}$ of finite models and for any set of sentences Δ we say that $\mathcal{K} \models_{sl} \Delta$ if and only if $\mathcal{K} \models_{sl} \varphi$ for any $\varphi \in \Delta$. ⊣

We consider the language obtained by adjoining the truth predicate Tr to the arithmetical language $\mathcal{L}$ and we modify $(FM(\mathbb{N}))$ by adding to its every element $\mathbb{N}_k$ an interpretation T_k of the truth predicate Tr.

Definition 11 $(FM(\mathbb{N})^T)$. *Let $\mathcal{K} = \{(\mathbb{N}_k, T_k) \colon k \in \omega \text{ and } T_k \subseteq \{0, ..., k-1\}\}$. An $FM(\mathbb{N})^T$-domain is any subset of $\mathcal{K}$ such that for any natural m it contains exactly one model of the cardinality m.* ⊣

---

[1] The methods we develop here – especially the semantics of quantifiers in FM-domains – is similar to the framework of modal potentialism developed independently in [2].

We obviously have to ensure that Yablo sentences exist in $sl((FM(\mathbb{N})^T))$ - as in the case of axiomatic truth theories over arithemtic, this is not problematic thanks to diagonal lemma.

Corollary 1 (Existence of Yablo sentences in $FM(\mathbb{N})^T$-domains). *There exists a formula $Y(x)$ such that for any $FM(\mathbb{N})^T$-domain we have:*

$$\forall n \in \omega \; FM(\mathbb{N})^T \models_{sl} Y(n) \equiv \forall x \, (x > n \Rightarrow \neg Tr(\ulcorner Y(\dot{x})\urcorner)),$$

i.e. Yablo sentences exist in $sl(FM(\mathbb{N})^T)$.

Proof. Obvious by the fact that the diagonal lemma is sl-true in $FM(\mathbb{N})^T$.

Theorem 5 (Yablo sentences are false in the limit). *For any class $\mathcal{K}$ of finite models, if $\mathcal{K} \models_{sl} AD + YD$, then for all $n \in \omega$ $\mathcal{K} \models_{sl} \neg Y(n)$. In fact, AD isn't essential (it's only added to ensure Tr behaves like a truth predicate): for any $n \in \omega$ we have $YD \models_{sl} \neg Y(n)$.*

Proof. Let us fix a class $\mathcal{K}$. For the sake of contradiction, suppose that there is a Yablo sentence that is not false at sufficiently large models, that is which for any size of a model is true at some model in $\mathcal{K}$ of at least that size, that is:

$$\exists n \forall k \exists \mathcal{M} \in \mathcal{K} \; (card(\mathcal{M}) \geq k \wedge \mathcal{M} \models Y(n)).$$

Let us take take such n. Let us fix k and take $\mathcal{M} \in \mathcal{K}$ (without loss of generality we could assume that $\mathcal{M} = \mathbb{N}_k^T$ for some natural k) with $card(\mathcal{M}) \geq k$ such that $\ulcorner Y(n+1) \urcorner \in |\mathcal{M}|$ and:

(i) $\mathcal{M} \models Y(n)$.
(ii) $\mathcal{M} \models Y(n+1) \equiv \forall x(x > n+1 \Rightarrow \neg Tr(\ulcorner Y(\dot{x})\urcorner))$.
(iii) $\mathcal{M} \models Y(n+1) \equiv Tr(\ulcorner Y(n+1)\urcorner)$.

(i) will be satisfied by the assumption of the proof, (ii) follows from Theorem 1 and (iii) results from the assumption that $\mathcal{K} \models_{sl} YD$ once we have enough numbers to code all the formulas needed for the claims to hold.

From (i) and the definition of Yablo sentences we obtain:

$$\mathcal{M} \models \neg Tr(\ulcorner Y(n+1)\urcorner), \tag{1}$$

as well as:

$$\mathcal{M} \models \forall x(x > n+1 \rightarrow \neg Tr(\ulcorner Y(\dot{x})\urcorner)). \tag{2}$$

Now, from (1), by (iii) we have that:

$$\mathcal{M} \models \neg Y(n+1),$$

and from (2), by (ii) we have that:

$$\mathcal{M} \models Y(n+1),$$

which gives a contradiction that ends the proof.

We will now show a construction of a class $\mathcal{K}$ such that $\mathcal{K} \models_{sl} YD + AD$, which means that Theorem 5 holds non-vacuously.

Definition 12 ($FM(\mathbb{N})^Y$ and $sl(FM(\mathbb{N})^Y)$). *We fix a formula $Y(x)$ satisfying the condition specified in Corollary 1. A family of models $FM(\mathbb{N})^Y$ is an $FM(\mathbb{N})^T$-domain $\{\mathbb{N}_k^Y : k \in \omega\}$ such that $\mathbb{N}_k^Y = (\mathbb{N}_k, T_k)$, where $T_k = TA_k \cup TY_k$, and:*

$$TA_k = \{\ulcorner\varphi\urcorner : \varphi \in Sent_{\mathcal{L}} \text{ and } \mathbb{N}_k \models \varphi\} \cap |\mathbb{N}_k|$$
$$TY_k = \{\ulcorner Y(m)\urcorner : \ulcorner Y(m)\urcorner \in |\mathbb{N}_k| \wedge \ulcorner Y(m+1)\urcorner \notin |\mathbb{N}_k|\}.$$

Naturally, $sl(FM(\mathbb{N})^Y) = \{\varphi : FM(\mathbb{N})^Y \models_{sl} \varphi\}$. ⊣

Corollary 2. *For any class $\mathcal{K}$, if $\mathcal{K} \models_{sl} YD$, then for sufficiently large $\mathcal{M} \in \mathcal{K}$, there is exactly one $n \in \omega$ s.t. $\ulcorner Y(n)\urcorner \in T_{\mathcal{M}}$.*

Theorem 6. $\forall n \in \omega \ \ FM(\mathbb{N})^Y \models_{sl} \neg Y(n)$.

Proof. We claim that for any n there exists m such that for any $k > m$:

$$\mathbb{N}_k^Y \models \neg Y(n).$$

Indeed, let us fix n and take m such that $\ulcorner Y(n+1)\urcorner \in |\mathbb{N}_m^Y|$ and for any $x < n+1$ and any $k > m$ we have:

$$\mathbb{N}_k^Y \models Y(x) \equiv \forall w > x \ \neg Tr(\ulcorner Y(\dot{w})\urcorner).$$

Let $k > m$. Then obviously $\ulcorner Y(n+1)\urcorner \in |\mathbb{N}_k^Y|$. Let j be the greatest number such that $\ulcorner Y(j)\urcorner \in |\mathbb{N}_k^Y|$. Such a number exists since our FM-domain is infinite and every model in it is finite. Obviously, $n < j$. From the definition of the class $FM(\mathbb{N})^Y$ and by the choice of j we obtain:

$$\mathbb{N}_k^Y \models Tr(\ulcorner Y(j)\urcorner),$$

and for any $x < j$ we get:

$$\mathbb{N}_k^Y \models \exists w(w > x \ Tr(\ulcorner Y(\dot{w})\urcorner)).$$

hence for any $x < j$, by the definition of Yablo sentence $Y(x)$ the following holds:

$$\mathbb{N}_k^Y \models \neg Y(x).$$

Thus, since $n < j$, we obtain:

$$\mathbb{N}_k^Y \models \neg Y(n),$$

which ends the proof.

Corollary 3. $sl(FM(\mathbb{N})^Y)$ *is ω-inconsistent.*

Proof. We have just shown that

$$\forall n \in \omega \ (\neg Y(n) \in sl(FM(\mathbb{N})^{Y})).$$

We obviously also have that there is m such that for all $k > m$ there is j such that $\mathbb{N}_k^Y \models Y(j)$, so by existential generalization we obtain that there is m such that for all $k > m$ $\mathbb{N}_k^Y \models \exists x Y(x)$ and thus:

$$\exists x Y(x) \in sl(FM(\mathbb{N})^{Y}).$$

So, it seems, there is a finitistic approach to arithmetic, according to which all Yablo sentences are false. The cost of this move, however, isn't negligible: the set of arithmetical formulae true in the intended model is ω-inconsistent. Therefore, we pursue the topic further, looking at another way to think finitistically about the issue.

4   Modal Finitistic Semantics

Definition 13 (Accessibility relation in FM-domains). *Let $\mathcal{K}$ be an FM-domain. For any $M, N \in \mathcal{K}$ N is accessible from M $(R(M,N))$ if $M \subseteq N$. For $m, n \in \omega$ and elements $\mathbb{N}_m$, $\mathbb{N}_n$ of the FM-domain $FM(\mathbb{N})$ this boils down to the condition $m \leq n$.*

Definition 14 (Modal semantics for FM-domains (*m*-semantics)). *Let $\mathcal{K}$ be an FM-domain over some structure $\mathbb{A}$ (i.e. $\mathcal{K} = FM(\mathbb{A})$) and $M \in \mathcal{K}$:*

- *If φ is atomic, then $(\mathcal{K}, M) \models_m \varphi$, if $M \models \varphi$.*
- *Satisfaction clauses for boolean connectives and negation are standard.*
- *$(\mathcal{K}, M) \models_m \exists x \varphi(x)$ iff there are $N \in \mathcal{K}$ and $a \in N$ s.t. $R(M,N)$ and $(\mathcal{K}, N) \models_m \varphi[a]$.*

Thus we also have that $(\mathcal{K}, M) \models_m \forall x \varphi(x)$ iff for all $N \in \mathcal{K}$ s.t. $R(M,N)$ and for all $a \in N$ $(\mathcal{K}, N) \models_m \varphi[a]$.

The intuition behind this semantics is as follows. '$\exists x \, \varphi(x)$' reads 'there could be enough objects so that for some a, $\varphi(a)$' and '$\forall x \, \varphi(x)$' reads 'however many more objects there could be, it still would be the case that for any a, $\varphi(a)$'.

Now, let $msl(FM(\mathbb{N})) = \{\varphi : \exists n \, \forall k \, k \geq n \Rightarrow (FM(\mathbb{N}), \mathbb{N}_k) \models_m \varphi\}$. That is, intuitively, $msl(FM(\mathbb{N}))$ is the set of those formulas, which are true in all sufficiently large models, where the notion of truth involves the modal reading of quantifiers.

As an example, let $\varphi = \exists x \, \forall y \, x \geq y$—that is, φ says: *there exists a maximal element*. While, as we remember, $\varphi \in sl(FM(\mathbb{N}))$, things are different under *m*-semantics—φ is false in every possible world of the FM-domain.

Before we move on, let us emphasize that just as with *sl*-semantics, we work with a relational arithmetical language. While in the case of *sl*-semantics this wasn't too important, it becomes crucial when we turn to *msl*-semantics.

For otherwise, we need to treat, say, addition and successor as total functions. This being the case, for each finite initial segment we'd need to identify the candidates for the values of functions which intuitively should surpass the capabilities of that segment. The least unnatural way to do this would be to plug in loops at ends of segments, so that $s(max(\mathbb{N}_k)) = max(\mathbb{N}_k)$ etc. But then we would run into problems. For instance, take $\varphi := \exists x\, x + x = x \wedge x \neq 0$. If we evaluate atomic sentences in the elements of our FM-domain, then for any k we have:

$$\mathbb{N}_k \models_m \varphi \text{ iff } \exists j \geq k\, \exists a < j\, \mathbb{N}_j \models a + a = a \wedge a \neq 0.$$

However, the above would come out true. after all, let $a = max(\mathbb{N}_j)$. Then, $a + a = \underbrace{s \ldots s}_{a} a = a$ and we have $\mathbb{N}_j \models a \neq 0 \wedge a + a = a$ and so $\mathbb{N}_k \models_m \varphi$ for any k. Thus, we would have $\varphi \in msl(FM(\mathbb{N}))$, while $\mathsf{PA} \vdash \neg\varphi$.

The underlying cause of the issue is that when we work with a functional language we cannot think about the initial segments as submodels of larger initial segments, because the functions are not preserved when we move to superstructures. The problem disappears when we abandon function symbols and use a relational language instead. So this is what we'll do in what follows.

In particular, we'll be working towards a theorem according to which the resulting arithmetic is the classical arithmetic, unlike in the previous case. We'll start with two lemmata.

Lemma 1. *For any $k > 0$, $\mathbb{N}_k$ is a submodel of $\mathbb{N}$.*

Proof. This holds because our language is relational, and for any r-ary relation symbol R^r we have $(R^r)^{\mathbb{N}} \cap \mathbb{N}_k^r = (R^r)^{\mathbb{N}_k}$.

Lemma 2. *For any quantifier-free $\varphi(x_1, \ldots, x_n)$ and for any choice of parameters $a_1, \ldots a_n \in \mathbb{N}$, if we have $a_1, \ldots, a_n \in \mathbb{N}_k$, then it holds that:*

$$\mathbb{N}_k \models \varphi[a_1, \ldots, a_n] \text{ iff } \mathbb{N} \models \varphi[a_1, \ldots, a_n]$$

Proof. Immediate by Lemma 1.

Theorem 7. *Let $msl(FM(\mathbb{N}))$ denote the msl theory (i.e. the sl-theory of the FM-domain of natural numbers with the modal interpretation of quantifiers). Then we have: $msl(FM(\mathbb{N})) = Th(\mathbb{N})$.*

Proof. The proof is by induction on formula complexity.

For the basic case of quantifier-free formulae, the claim holds by Lemma 2. For boolean connectives, the equivalence is trivial. The only interesting case is for $\varphi := \exists x \psi(x)$.

$\supseteq$: Suppose $\varphi \in Th(\mathbb{N})$, that is, $\mathbb{N} \models \exists x \psi(x)$. Then there is a witness $a \in \mathbb{N}$, such that $\psi[a] \in Th(\mathbb{N})$. By IH, $\psi[a] \in msl(FM(\mathbb{N}))$. This means:

$$\exists k\, \forall l \geq k\, \mathbb{N}_l \models_m \psi[a] \tag{3}$$

and from this it follows that:

$$\exists k \forall l \geq k \, \exists j \geq l \, \exists a < j \, \mathbb{N}_j \models_m \exists x \, \psi(x). \tag{4}$$

which means that $\varphi \in msl(FM(\mathbb{N}))$.

$\subseteq$: Say $\varphi := \exists x \, \psi(x) \in msl(FM(\mathbb{N}))$. So (4) holds as well and there is an a such that (3) also holds (this step essentially depends on the language being relational). This means $\psi[a] \in msl(FM(\mathbb{N}))$, and so by the IH, $\psi[a] \in Th(\mathbb{N})$, and therefore $\varphi \in Th(\mathbb{N})$.

5   Yablo Sequences and Modal Interpretation of Quantifiers

The semantics in sufficiently large models in potentially infinite domains under modal interpretation of quantifiers presented above entails that Yablo sentences stay paradoxical even for the finitist, if she interprets the quantifiers in the modal manner. That is, we'll be arguing that not even local Yablo Disquotation can be included in an msl-theory. Let's start with a lemma.

Lemma 3. *For any $FM(\mathbb{N})^Y$-domain with $YD \subseteq msl(FM(\mathbb{N})^Y)$ it holds that:* $\forall n \in \omega \, Y(n) \notin msl(FM(\mathbb{N})^Y)$.

Proof. Suppose some Yablo sentence is in the msl-theory, that is

$$\exists n \, Y(n) \in msl(FM(\mathbb{N}^Y)).$$

This means:

$$\exists l \forall k \geq l \, \mathbb{N}_k \models_m Y(n).$$

Pick an l witnessing this. By the definition of Yablo sentences this entails:

$$\forall k \geq l \, \mathbb{N}_k \models_m \forall x \, (x > n \rightarrow \neg Tr(Y(x))).$$

By the semantics, this means:

$$\forall k \geq l \forall p \geq k \forall a < p \, \mathbb{N}_p \models_m a > n \rightarrow \neg Tr(Y(a)).$$

But then:

$$\forall p \geq l \, \forall a \in (n, p) \, \mathbb{N}_p \models \neg Tr(Y(a)). \tag{5}$$

So, by Yablo Disquotation, for sufficiently large p, we have that models of size p fail to satisfy Yablo sentences for numbers between n and p:

$$\forall p \geq l \, \forall a \in (n, p) \, \mathbb{N}_p \models_m \neg Y(a),$$

Now, fix p and a. By the definition of $Y(x)$ the above means:

$$\mathbb{N}_p \models_m \exists x > a \, Tr(Y(x)).$$

By our definition of $\models_m$ this is equivalent to:

$$\exists q \geq p \, \exists b < q \, \mathbb{N}_q \models_m b > a \wedge Tr(Y(b))$$

Hence:

$$\exists q \geq p \, \exists b \in (a,q) \, \mathbb{N}_q \models_m Tr(Y(b)).$$

This, however, contradicts (5), which completes the argument.

With this lemma at hand, we can proceed to the theorem which tells us that not only Yablo sentences are not in any msl-theory, but also that no msl-domain can (modally) satisfy the local Yablo Disquotation principles either.

Theorem 8. *There is no $FM(\mathbb{N})^Y$-domain such that $YD \subseteq msl(FM(\mathbb{N})^Y)$.*

Proof. Suppose there is an $msl(FM(\mathbb{N})^Y)$ which contains all Local Yablo Disquotation sentences.

By Lemma 3, we know that $\forall n \, Y(n) \notin msl(FM(\mathbb{N})^Y)$. We therefore have:

$$\forall n \, \forall l \, \exists k \geq l \, \mathbb{N}_k \not\models_m Y(n).$$

By the definition of the Yablo sentences we infer:

$$\forall n \forall l \exists p \geq l \exists a > n \, \mathbb{N}_p \models_m Tr(Y(a)).$$

Which, by Yablo Disquotation, yields:

$$\forall n \forall l \exists p \geq l \exists a > n \, \mathbb{N}_p \models_m Y(a).$$

The claim holds for any n and l. For us, it is enough to look at $n = l = 0$. By the definition of $Y(a)$ we obtain:

$$\exists p, a > 0 \forall q \geq p \, \mathbb{N}_q \models_m \forall x > a \, \neg Tr(Y(x)).$$

Pick an $a > 0$ witnessing the above claim. By the definition of msl-theory we now have

$$Y(a) \in msl(FM(\mathbb{N})^Y),$$

which is impossible by Lemma 3.

So, when we consider Yablo sequences with different treatments of infinity in the background, the following observations come to mind:

1. In the standard setting, without potential infinity, Local Arithmetical Disquotation and Local Yablo Disquotation are consistent, yet ω-inconsistent with the background arithmetical theory. Once ω-rule or Uniform Yablo Disquotation are introduced, the theory is inconsistent.
2. Under sl-semantics, Yablo sentences are all false (in the limit), yet the sl-theory of a given FM-domain is consistent, but ω-inconsistent. This is a particular case of a general flaw of sl-semantics, because $sl(FM(\mathbb{N}))$ itself is ω-inconsistent.
3. Under msl-semantics, i.e. semantics in sufficiently large models in potentially infinite domains under the modal interpretation of quantifiers, even adding only the Local Arithmetical Disquotation and Yablo Disquotation results in an inconsistent theory. Uniform Yablo Disquotation or ω-rule are not needed to ensure this.

References

1. Godziszewski, M.T.: Yablo sequences in potentially infinite domains and partial semantics (2018). [submitted]
2. Hamkins, J.D., Linnebo, Ø.: The modal logic of set-theoretic potentialism and the potentialist maximality principles. Rev. Symb. Logic (2018, to appear). http://wp.me/p5M0LV-1zC, arXiv:1708.01644
3. Ketland, J.: Yablo's paradox and ω-inconsistency. Synthese **145**, 295–302 (2005)
4. Mostowski, M.: On representing concepts in finite models. Math. Logic Q. **47**, 513–523 (2001)
5. Mostowski, M.: On representing semantics in finite models. In: Rojszczak, A., Cachro, J., Kurczewsk, G. (eds.) Philosophical Dimensions of Logic and Science, pp. 15–28. Kluwer Academic Publishers, Dordrecht (2001)
6. Mostowski, M., Zdanowski, K.: FM-representability and beyond. In: Cooper, S.B., Löwe, B., Torenvliet, L. (eds.) CiE 2005. LNCS, vol. 3526, pp. 358–367. Springer, Heidelberg (2005). https://doi.org/10.1007/11494645_45
7. Mostowski, M.: Truth in the limit. Rep. Mathe. Logic **51**, 75–89 (2016)
8. Priest, G.: Yablo's paradox. Analysis **57**, 236–242 (1997)
9. Urbaniak, R.: Potential infinity, abstraction principles and arithmetic (Leśniewski style). Axioms **5**(2), 18 (2016)
10. Yablo, S.: Paradox without self-reference. Analysis **53**, 251–252 (1993)

The Finite Embeddability Property
for Topological Quasi-Boolean Algebra 5

Zhe Lin[1(✉)] and Mihir Kumar Chakraborty[2]

[1] Institute of Logic and Cognition, Sun Yat-sen University, Guangzhou, China
pennyshaq@gmail.com
[2] School of Cognitive Science, Jadavpur University, Kolkata, India
mihir4@gmail.com

Abstract. In this paper we study some basic algebraic structures of rough algebras. We proved that the class of topological quasi-Boolean algebra 5s (tqBa5s) has the finite embeddability property (FEP). Further we also extend this result to some related classes of algebras.

1 Introduction

The algebraic structures called Pre-rough algebras [1,2] arose as a natural abstraction from the calculus of rough sets proposed by Pawlak in 1983 [3]. A Pre-rough algebra is a topological quasi-Boolean algebra (tqBa) which is quasi-Boolean algebra (see Definition 1) endowed with a topological (interior) operator. Although quasi Boolean algebra and topological Boolean algebra are presented in the similar work of Rasiowa [4], Topological quasi-Boolean algebra was first introduced in [1]. As a further step towards Pre-rough algebras, an axiom corresponding to modal logic axiom S5 has been added to tqBas resulting in tqBa5s [2]. Further studies with tqBas, tqBa5s and related algebras have been carried out in [1,2,5–7]. However, the finite embeddability property of these algebraic structures have not been investigated before.

On the study of the connection between logical systems and classes of algebras in general, one important and natural question is whether a given class of algebras has a decidable equational or even universal theory. The finite embeddability property (or FEP for short) i.e., every finite partial subalgebra of an algebra in the class is isomorphic to a subalgebra of a finite algebra in the class of algebras, entails the decidability of its universal theory if this class of algebras is finitely axiomatizable.

The study on FEP of classes of algebras dated back to Henkin [1956]. Henkin proves that the class of abelian groups has FEP. It is also well-known that the class HA of Heyting algebras has FEP. Block and Van Alten [8,9] show that various integral residuated lattices (groupoids) have FEP. Farulewski [10]

The work of the first author was supported by Chinese National Funding of Social Sciences (No. 17CZX048).

shows that the integral condition is not necessary and proves FEP for residuated groupoids. Buszkowski [11] also proves that various lattice extensions of residuated groupoids (including Heything and Boolean extensions) have FEP. The first author of the current paper in [12] extends Buszkowski's results to various lattice extensions of residuated groupoids with modalities. In [13] the present authors investigate residuated Pre-rough algebras and show that residuated Pre-rough algebras have decidable quasiequational theories (see definition in Sect. 1), which entails that the pre-rough algebras have decidable quasiequational theories. Indeed residuated Pre-rough algebras have FEP. However FEP for Pre-rough algebras still remain open. In the present paper we study the class of the basic algebraic structures in Pre-rough algebras (the class of tqBa5s) and prove that it has FEP. This results can also be extended to Pre-rough and some other modal extensions of quasi-Boolean algebras.

The method we developed in the present paper is inspired by [11] and [12]. A sequent calculus which admits the interpolant lemma (see Lemma 1) is introduced and it plays a essential role in proof of FEP. Meanwhile a sequent calculus which does not admit the interpolant lemma for tqBa5 was earlier introduced in [14]. Our method can be regarded as an algebraic substitute of the filtration method for Kripke frames [15].

The paper is organized as below. In the next section we recall some basic algebraic definitions. Then in Sect. 3, we develop a sequent system for tqBa5s and prove the interpolant lemma. In Sect. 4, we present the main results and show FEP for the class of tqBa5s. In Sect. 5, we conclude our paper and make some simple extensions to some related classes of algebras. Hereafter, the class of all topological quasi-Boolean algebra 5 will be denoted by tqBa5 also.

2   Some Basic Definitions

Definition 1. A *quasi-Boolean algebra* (qBa) is an algebra $\mathbb{A} = (A, \wedge, \vee, \neg, 0, 1)$ where $(A, \wedge, \vee, 0, 1)$ is a bounded distributive lattice, and $\neg$ is an unary operation on A such that the following conditions hold for all $a, b \in A$:

$$\text{(DN)} \quad \neg\neg a = a, \quad \text{(DM)} \quad \neg(a \vee b) = \neg a \wedge \neg b$$

A *topological quasi-Boolean algebra* (tqBa) is an algebra $\mathbb{A} = (A, \wedge, \vee, \neg, 0, 1, \Box)$ where $(A, \wedge, \vee, \neg, 0, 1)$ is a quasi-Boolean algebra, and $\Box$ is an unary operation on A such that for all $a, b \in A$:

$$\text{(K}_\Box) \quad \Box(a \wedge b) = \Box a \wedge \Box b, \quad \text{(N}_\Box) \quad \Box\top = \top$$

$$\text{(T}_\Box) \quad \Box a \leq a, \quad \text{(4}_\Box) \quad \Box a \leq \Box\Box a$$

A *topological quasi-Boolean algebra 5* (tqBa5) is a topological quasi-Boolean algebra $\mathbb{A} = (A, \wedge, \vee, \neg, \Box, 0, 1)$ such that for all $a \in A$:

$$\text{(5)} \quad \Diamond a \leq \Box\Diamond a,$$

where $\Diamond$ is an unary operation on A defined by $\Diamond a := \neg\Box\neg a$.

Proposition 1. *For any tqBa5* $\mathbb{A} = (A, \wedge, \vee, \neg, \square, 0, 1)$ *and* $a, b \in A$, *the following hold:*

(1) $\neg 0 = 1$ *and* $\neg 1 = 0$.
(2) $\neg(a \wedge b) = \neg a \vee \neg b$.
(3) If $a \leq b$, *then* $\neg b \leq \neg a$.
(4) $\Diamond 0 = 0$ *and* $\Diamond(a \vee b) = \Diamond a \vee \Diamond b$.
(5) $\square a = \square\square a$ *and* $\Diamond a = \Diamond\Diamond a$.
(6) $\Diamond a = \square\Diamond a$ *and* $\square a = \Diamond\square a$.
(7) $\Diamond a \leq b$ *if and only if* $a \leq \square b$.

The proof of Proposition 1 can be found in [6,7].

We now recall some concepts from universal algebra. *Equation* (identity) and *quasi-equation* (quasi-identity) are defined in standard manner (see Chap. 1 [16]). For any set of equations or quasi-equations Σ, let $\mathbb{A}(\Sigma)$ be the class of all algebras which validate all equations or quasi-equations in Σ. A class of algebras $\mathbb{K}$ is a *variety* if there is a set of equations Σ such that $\mathbb{K} = \mathbb{A}(\Sigma)$. A class of algebras $\mathbb{K}$ is a *quasi-variety* if there is a set of quasi-equations Θ such that $\mathbb{K} = \mathbb{A}(\Theta)$.

Theorem 1. *The class of tqBa5s is a variety.*

Due to the definition of tqBa5, the class of tqBa5s can be classified by a set of equations. Thus by Theorem 1.19, [16]. The class of tqBa5s is a variety.

Corollary 1. *The class of tqBa5s is a quasi-variety.*

Let $\mathbb{A} = \langle A, \langle f_i^{\mathbb{A}} \rangle_{i \in I} \rangle$ be an algebra of fixed type and $B \subseteq A$. Then $\mathbb{B} = \langle B, \langle f_i^{\mathbb{B}} \rangle_{i \in I} \rangle$ is a partial subalgebra of $\mathbb{A}$ where for every $n \in N$, every n-ary function symbol $f_i^{\mathbb{A}}$ with $i \in I$, and for every $b_1, \ldots, b_n \in B$, one defines $f_i^{\mathbb{B}}(b_1, \ldots, b_n) = f_i^{\mathbb{A}}(b_1, \ldots, b_n)$ if $f_i^{\mathbb{A}}(b_1, \ldots, b_n) \in B$, otherwise, the value is not defined. If $\mathbb{A}$ is ordered, then $\leq^{\mathbb{B}} = \leq^{\mathbb{A}} |B$, the restriction of $\leq^{\mathbb{A}}$ to $\mathbb{B}$. $f_i^{\mathbb{A}}$ denotes the operation interpreting the symbol f_i in the algebra $\mathbb{A}$. However we write f_i for $f_i^{\mathbb{A}}$, if it does not cause confusion.

By an embedding from a partial algebra $\mathbb{B}$ into an algebra $\mathbb{C}$, we mean an injection $h : B \mapsto C$ such that if $b_1, \ldots, b_n, f^{\mathbb{B}}(b_1, \ldots, b_n) \in B$, then

$$h(f^{\mathbb{B}}(b_1, \ldots, b_n)) = f^{\mathbb{C}}(h(b_1), \ldots, h(b_n)).$$

If $\mathbb{B}$ and $\mathbb{C}$ are ordered, then h is required to be an order embedding i.e. $a \leq^{\mathbb{B}} b \Leftrightarrow h(a) \leq^{\mathbb{C}} h(b)$.

A class $\mathbb{K}$ of algebras has *the finite embeddability property* (FEP), if every finite partial subalgebra of a member of $\mathbb{K}$ can be embedded into a finite member of $\mathbb{K}$. FEP usually has some consequences on finite model property. FEP implies *the strong finite model property* (SFMP) i.e. every quasi-identity which fails to hold in a class $\mathbb{K}$ of algebras can be falsified in a finite member of $\mathbb{K}$. SFMP and FEP are equivalent in quasivarieties of finite type.

Lemma 1 (Lemma 6.40 [16]**).** *For any quasivariety* $\mathbb{K}$ *of finite type the following are equivalent:*

(1) $\mathbb{K}$ *has FEP*
(2) $\mathbb{K}$ *have SFMP*
(3) $\mathbb{K}$ *is generated as a quasivarieties by its finite members*

Remark 1. If a formal system S is strongly complete with respect to a class $\mathbb{K}$ of algebras, then it yields, actually, an axiomatization of the quasiequational theory of $\mathbb{K}$; hence SFMP for S with respect to $\mathbb{K}$ yields SFMP for $\mathbb{K}$. By SFMP for S, we mean that for any finite set of sequents Φ, if $\Phi \not\vdash_S \Gamma \Rightarrow A$, then there exists a finite $\mathbb{A} \in \mathbb{K}$ and a valuation σ such that all sequents from Φ are true in $(\mathbb{A}\ \sigma)$, but $\Gamma \Rightarrow A$ is not.

3   Sequent Calculus of tqBa5

In this section we develop a sequent system G5 for tqBa5 following the tradition [17]. The language of the logic of tqBa5 is defined as follows

$$\alpha ::= p \mid \bot \mid \top \mid \alpha \wedge \beta \mid \alpha \vee \beta \mid \neg\alpha \mid \Diamond\alpha \mid \Box\alpha,$$

where $p \in \mathbf{Prop}$, the set of propositional variables.

   Formula structure are defined as follows with a unary structural operation $\langle\rangle$:

– α is a formula structure if α is a formula
– $\langle\Gamma\rangle^i$ is a formula structure if Γ is a formula structure

   Hereafter we abbreviate $\underbrace{\langle \ldots \langle}_{n} \alpha \underbrace{\rangle \ldots \rangle}_{n}$ by $\langle\alpha\rangle^n$. Clearly if Γ is a formula structure, then it is of the form $\langle\alpha\rangle^i$ for some formula α and number $i \geq 0$. We use $\langle\alpha\rangle^{i_1}, \langle\beta\rangle^{i_2}, \ldots$ where $i_1, i_2 \geq 0$ to denote formula structures. A *sequent* is an expression of the form $\langle\alpha\rangle^i \Rightarrow \beta$ where $i \geq 0$ for some formulae α and β.

Definition 2. *The Gentzen sequent calculus* G5 *consists of the following axioms and inference rules:*

(1) Axioms:

$$\text{(Id)}\ \varphi \Rightarrow \varphi \qquad (\bot)\ \langle\bot\rangle^i \Rightarrow \varphi \qquad (\top)\ \langle\varphi\rangle^i \Rightarrow \top$$

$$\text{(D)}\ \varphi \wedge (\psi \vee \chi) \Rightarrow (\varphi \wedge \psi) \vee (\varphi \wedge \chi) \qquad \text{(DN)}\ \varphi \Leftrightarrow \neg\neg\varphi$$

(2) Connective rules:

$$\frac{\langle\varphi\rangle^i \Rightarrow \chi}{\langle\varphi \wedge \psi\rangle^i \Rightarrow \chi}(\wedge L) \qquad \frac{\langle\chi\rangle^i \Rightarrow \varphi \quad \langle\chi\rangle^i \Rightarrow \psi}{\langle\chi\rangle^i \Rightarrow \varphi \wedge \psi}(\wedge R)$$

$$\frac{\langle\varphi\rangle^i \Rightarrow \chi \quad \langle\psi\rangle^i \Rightarrow \chi}{\langle\varphi \vee \psi\rangle^i \Rightarrow \chi}(\vee L) \qquad \frac{\langle\chi\rangle^i \Rightarrow \psi}{\langle\chi\rangle^i \Rightarrow \psi \vee \varphi}(\vee R)$$

(3) *Modal rules*

$$\frac{\langle\varphi\rangle^{i+1} \Rightarrow \psi}{\langle\Diamond\varphi\rangle^i \Rightarrow \psi}(\Diamond L) \qquad \frac{\langle\varphi\rangle^i \Rightarrow \psi}{\langle\varphi\rangle^{i+1} \Rightarrow \Diamond\psi}(\Diamond R)$$

$$\frac{\langle\varphi\rangle^i \Rightarrow \psi}{\langle\Box\varphi\rangle^{i+1} \Rightarrow \psi}(\Box L) \qquad \frac{\langle\varphi\rangle^{i+1} \Rightarrow \psi}{\langle\varphi\rangle^i \Rightarrow \Box\psi}(\Box R)$$

$$\frac{\langle\varphi\rangle^{i+1} \Rightarrow \psi}{\langle\varphi\rangle^i \Rightarrow \psi}(T) \qquad \frac{\langle\varphi\rangle^{i+1} \Rightarrow \psi}{\langle\varphi\rangle^{i+2} \Rightarrow \psi}(4) \qquad \frac{\langle\varphi\rangle^i \Rightarrow \psi}{\langle\neg\psi\rangle^i \Rightarrow \neg\varphi}(\Diamond\Box)$$

(4) *Cut rule*

$$\frac{\langle\varphi\rangle^i \Rightarrow \chi \quad \langle\chi\rangle^j \Rightarrow \psi}{\langle\varphi\rangle^{i+j} \Rightarrow \psi}(\text{Cut})$$

where $i, j \geq 0$. By $\vdash_{G5} \langle\alpha\rangle^i \Rightarrow \beta$, we mean the sequent $\langle\alpha\rangle^i \Rightarrow \beta$ is provable in G5. A sequent is called *simple sequent* if it is of the form $\alpha \Rightarrow \beta$ for some formulae α and β. Let Φ be a finite set of simple sequents. By $\Phi \vdash_{G5} \langle\alpha\rangle^i \Rightarrow \beta$, we mean that sequent $\langle\alpha\rangle^i \Rightarrow \beta$ is derivable from Φ in G5.

Proposition 2. *In G5, the following holds:*

- $\vdash_{G5} \Diamond(\alpha \vee \beta) \Rightarrow \Diamond\alpha \vee \Diamond\beta$
- $\vdash_{G5} \Box(\alpha \wedge \beta) \Rightarrow \Box\alpha \wedge \Box\beta$
- $\vdash_{G5} \Box\alpha \Rightarrow \Box\Box\alpha$
- $\vdash_{G5} \Box\alpha \Rightarrow \alpha$
- $\vdash_{G5} \Diamond\alpha \Rightarrow \Box\Diamond\alpha$
- $\vdash_{G5} \Diamond\alpha \Rightarrow \neg\Box\neg\alpha$
- $\vdash_{G5} \neg\Box\neg\alpha \Rightarrow \Diamond\alpha$

Let F be a finite set of formulae closed under subformulae. Define F^{qb} be the closure of F under $\wedge, \vee, \neg$. A set of formula T is called qb-closed if $T = F^{qb}$ for some finite set F which is closed under subformulae. A sequent $\langle\alpha\rangle^i \Rightarrow \beta$ is called a T sequent if $\alpha, \beta \in T$. A derivation from Φ in G5 of a T-sequent $\langle\alpha\rangle^i \Rightarrow \beta$ is called a T-derivation if all sequents appearing in the derivation are T-sequents, which is denoted by $\Phi \vdash_{G5} \langle\alpha\rangle^i \Rightarrow_T \beta$. Assume that $\Phi \vdash_{G5} \langle\varphi\rangle^{i+j} \Rightarrow_T \psi$. A formula γ is called a T interpolant of $\langle\varphi\rangle^i$ if $\gamma \in T$, $\Phi \vdash_{G5} \langle\varphi\rangle^i \Rightarrow_T \gamma$ and $\Phi \vdash_{G5} \langle\gamma\rangle^j \Rightarrow_T \psi$ and additionally $\Phi \vdash_{G5} \langle\gamma\rangle \Rightarrow_T \gamma$ if $i \geq 1$.

Lemma 2 (Interpolant). *If* $\Phi \vdash_{G5} \langle\varphi\rangle^{i+j} \Rightarrow_T \psi$, *then* $\langle\varphi\rangle^i$ *has a* T *interpolant.*

Proof. We proceed by induction on the length of derivation. Axiom is trivial. (Cut) is easy. Assume that the end sequent is obtained by a rule (R). If $i = 0$ then obviously φ is a required interpolant. Let $i \geq 1$ Here we consider three cases. Others can be treated similarly.

($\vee$L) Assume the premise are $\langle\delta\rangle^{i+j} \Rightarrow_T \psi$ and $\langle\chi\rangle^{i+j} \Rightarrow \psi$ and $\varphi = \delta \vee \chi$. Then by induction hypothesis, there are $\gamma_1, \gamma_2 \in T$ such that (1) $\Phi \vdash_{G5} \langle\delta\rangle^i \Rightarrow_T \gamma_1$, (2) $\Phi \vdash_{G5} \langle\chi\rangle^i \Rightarrow_T \gamma_2$, (3) $\Phi \vdash_{G5} \langle\gamma_1\rangle^j \Rightarrow_T \psi$, (4) $\Phi \vdash_{G5} \langle\gamma_2\rangle^j \Rightarrow_T \psi$, (5) $\Phi \vdash_{G5} \langle\gamma_1\rangle \Rightarrow \gamma_1$ and (6) $\Phi \vdash_{G5} \langle\gamma_2\rangle \Rightarrow \gamma_2$. By applying ($\vee$R) to (1) and (2), one gets (7) $\Phi \vdash_{G5} \langle\delta\rangle^i \Rightarrow_T \gamma_1 \vee \gamma_2$, (8) $\Phi \vdash_{G5} \langle\chi\rangle^i \Rightarrow_T \gamma_1 \vee \gamma_2$. Then by applying $(\vee L)$ to (7) and (8), one obtains (9) $\Phi \vdash_{G5} \langle \delta \vee \chi \rangle^i \Rightarrow_T \gamma_1 \vee \gamma_2$. Further by applying $(\vee L)$ to (3) and (4) one gets applying $\Phi \vdash_{G5} \langle \gamma_1 \vee \gamma_2 \rangle^i \Rightarrow_T \psi$. If $i \geq 1$, then applying $(\vee R)$ and $(\vee L)$ to (5) and (6), one gets $\Phi \vdash_{G5} \langle \gamma_1 \vee \gamma_2 \rangle \Rightarrow \gamma_1 \vee \gamma_2$. Thus $\gamma_1 \vee \gamma_2$ is a required interpolant.

$(\Diamond\Box)$ Assume that the premise is $\langle \varphi \rangle^{i+j} \Rightarrow_T \psi$. By induction hypothesis there is $\gamma \in T$ such that $\Phi \vdash_{G5} \langle \varphi \rangle^i \Rightarrow_T \gamma$, $\Phi \vdash_{G5} \langle \gamma \rangle^j \Rightarrow_T \psi$ and $\Phi \vdash_{G5} \langle \gamma \rangle \Rightarrow \gamma$. Then by rule $(\Diamond\Box)$, one gets (1) $\Phi \vdash_{G5} \langle \neg\gamma \rangle^i \Rightarrow \neg\varphi$, (2) $\Phi \vdash_{G5} \langle \neg\psi \rangle^i \Rightarrow \neg\gamma$ and (3) $\Phi \vdash_{G5} \langle \neg\gamma \rangle \Rightarrow_T \neg\gamma$. Thus $\Phi \vdash_{G5} \langle \neg\gamma \rangle^k \Rightarrow_T \neg\gamma$ for any $k \geq 0$. Hence one gets (5) $\Phi \vdash_{G5} \langle \neg\gamma \rangle^i \Rightarrow_T \neg\gamma$ and (6) $\Phi \vdash_{G5} \langle \neg\gamma \rangle^j \Rightarrow_T \neg\gamma$. By applying (T) to (1) and (2) one gets (7) $\Phi \vdash_{G5} \neg\psi \Rightarrow_T \neg\gamma$ and (8) $\Phi \vdash_{G5} \neg\gamma \Rightarrow_T \neg\varphi$. Then by (Cut) to (7) and (5), one gets $\Phi \vdash_{G5} \langle \neg\psi \rangle^i \Rightarrow_T \neg\gamma$. By (Cut) to (8) and (6), one gets $\Phi \vdash_{G5} \langle \neg\gamma \rangle^j \Rightarrow_T \neg\varphi$. Obviously γ is a required interpolant. Hence $\neg\gamma \in T$ is a required interpolant.

(4) Assume that the premise is $\langle \varphi \rangle^{k+1} \Rightarrow_T \psi$ and the conclusion is $\langle \varphi \rangle^{k+2} \Rightarrow_T \psi$. Let $i + j = k + 2$. By induction hypothesis, there is $\gamma \in T$ such that $\Phi \vdash_{G5} \langle \varphi \rangle^i \Rightarrow_T \gamma$, $\Phi \vdash_{G5} \langle \gamma \rangle^{j-1} \Rightarrow_T \psi$ and $\Phi \vdash_{G5} \langle \gamma \rangle \Rightarrow_T \gamma$. Hence by (Cut) one gets $\Phi \vdash_{G5} \langle \gamma \rangle^j \Rightarrow_T \psi$. Hence γ is a required interpolant.

Notice that we did not assume that the set T is closed under any modal operations. Hence the above interpolant lemma with respect to this kind of T is based on the fact that in our sequent calculus we introduce modal structural operation and interpolate modal axioms by structural rules. Further the additional condition is required for the proof of case (4). Without the additional condition, one can not prove the case (4) when $k = 0$ and $i = 1$.

An algebraic model of G5 is a pair $(\mathbb{G}, \sigma)$ such that $\mathbb{G}$ is a **tqBa5**, and σ is a mapping from Prop into $\mathbb{G}$, called a *valuation*, which is extended to formulae and formula trees as follows:

$$\sigma(\Box\alpha) = \Box\sigma(\alpha), \sigma(\Diamond\alpha) = \Diamond\sigma(\alpha)$$
$$\sigma(\alpha \wedge \beta) = \sigma(\alpha) \wedge \sigma(\beta), \quad \sigma(\alpha \vee \beta) = \sigma(\alpha) \vee \sigma(\beta),$$
$$\sigma(\neg\alpha) = \neg\sigma(\alpha), \quad \sigma(\langle \alpha \rangle^{i+1}) = \Diamond\sigma(\langle \alpha \rangle^i).$$

A sequent $\langle \alpha \rangle^i \Rightarrow \beta$ is said to be *true* in a model $(\mathbb{G}, \sigma)$ written $\mathbb{G}, \sigma \models \langle \alpha \rangle^i \Rightarrow \beta$, if $\sigma(\langle \alpha \rangle^i) \leq \sigma(\beta)$ (here $\leq$ is the lattice order in $\mathbb{G}$). It is *valid* in $\mathbb{G}$, if it is true in $(\mathbb{G}, \sigma)$, for any valuation σ. It is valid in a class of algebras $\mathbb{K}$, if it is valid in all algebras from $\mathbb{K}$. $\Phi \models \langle \alpha \rangle^i \Rightarrow \beta$ with respect to $\mathbb{K}$, if $\langle \alpha \rangle^i \Rightarrow \beta$ is true in all models $(\mathbb{G}, \sigma)$ such that $\mathbb{G} \in \mathbb{K}$ and all sequents from Φ are true in $(\mathbb{G}, \sigma)$.

Remark 2. G5 is *strongly complete* with respect to class tqBa5: for any set of sequents Φ and any sequent $\langle \alpha \rangle^i \Rightarrow \beta$, $\Phi \vdash_{G5} \langle \alpha \rangle^i \Rightarrow \beta$ if and only if $\Phi \models \langle \alpha \rangle^i \Rightarrow \beta$ with respect to tqBa5. So it follows that G5 is *weakly complete* with respect to tqBa5: the sequents provable in G5 are precisely the sequents valid in tqBa5. The proof of *strongly completeness* of G5 with respect to tqBa5 follows from the same proof of strong finite model property (definition see Sect. 4) of G5 in Sect. 4.

4   FEP for tqBa5

Given a qb-closed set of formula T and a set of simple T-sequents Φ, we define an order $\leq_T$ on formula structures. The set of T formula structures denote by T^s, consist of all formula structures whose formulae appearing in them belong to T. Let $\langle \alpha_1 \rangle^i, \langle \alpha_2 \rangle^j \in T^s$, we say $\langle \alpha_1 \rangle^i \leq_T \langle \alpha_2 \rangle^j$ if $\Phi \vdash_{\mathrm{G5}} \langle\langle \alpha_2 \rangle^j \rangle^t \Rightarrow_T \beta$ implies $\Phi \vdash_{\mathrm{G5}} \langle\langle \alpha_1 \rangle^i \rangle^t \Rightarrow_T \beta$ for any context $\langle\rangle^t$ where $t \geq 0$ and T formula β. Let $\langle \alpha_1 \rangle^i \approx_T \langle \alpha_2 \rangle^j$ if $\langle \alpha_1 \rangle^i \leq_T \langle \alpha_2 \rangle^j$ and $\langle \alpha_2 \rangle^j \leq_T \langle \alpha_1 \rangle^i$. Obviously $\approx_T$ is a equivalence relation on T formula structures.

We define

$$\{\langle \alpha \rangle^i\}_T^{\approx} = \{\langle \beta \rangle^j | \langle \beta \rangle^j \approx_T \langle \alpha \rangle^i\}(i, j \geq 0)$$

Obviously

$$\{\alpha\}_T^{\approx} = \{\langle \beta \rangle^j | \langle \beta \rangle^j \approx_T \alpha\}(j \geq 0)$$

Let $T^s/_{\approx_T}$ denote the set of all $\{\langle \alpha \rangle^i\}_T^{\approx}$ where $\langle \alpha \rangle^i \in T^s$ where $i \geq 0$ and $T/_{\approx_T}$ denote the set of all $\{\alpha\}_X^{\approx}$ where $\alpha \in T$. Define $\{\langle \alpha \rangle^i\}_T^{\approx} \preceq_T \{\langle \beta \rangle^j\}_T^{\approx}$ if $\langle \alpha \rangle^i \leq_T \langle \beta \rangle^j$. This is well defined since if $\{\langle \alpha \rangle^i\}_T^{\approx} \preceq_T \{\langle \beta \rangle^j\}_T^{\approx}$, $\langle \varphi \rangle^p \in \{\langle \alpha \rangle^i\}_T^{\approx}$ and $\langle \psi \rangle^q \in \{\langle \alpha \rangle^i\}_T^{\approx}$, then $\langle \varphi \rangle^p \leq_T \langle \psi \rangle^q$.

We define a closure operation C on $T^{\approx}$ as follows:

$$C(\{\langle \alpha \rangle^i\}_T^{\approx}) = \{ \underbrace{\bigwedge_{1 \leq j \leq n} \beta_j}\}_T^{\approx} \quad \text{for any } \{\beta_j\}_T^{\approx} \in T^{\approx} \text{ s.t. } \{\langle \alpha \rangle^i\}_T^{\approx} \preceq_T \{\beta_j\}_T^{\approx}$$

Remark 3. For any $\{\langle \alpha \rangle^i\}_T^{\approx}$, $\{\langle \alpha \rangle^i\}_T^{\approx} \preceq_T \{\top\}_T^{\approx}$. Thus $C(\{\langle \alpha \rangle^i\}_T^{\approx})$ always exists. Since T is closed under $\wedge$, for any $C(\{\langle \alpha \rangle^i\}_T^{\approx})$ there exists a $\{\beta\}_T^{\approx}$ such that $C(\{\langle \alpha \rangle^i\}_T^{\approx}) = \{\beta\}_T^{\approx}$. Let $C(T^s/_{\approx_T})$ denoted the sets of all $C(\{\langle \alpha \rangle^i\}_T^{\approx})$. Clearly $C(T^s/_{\approx_T}) \subseteq T/_{\approx_T}$. Since T is qb-closed, $T/_{\approx_T}$ is finite. Thus $C(T^s/_{\approx_T})$ is finite.

Lemma 3. *For any $\{\langle \alpha \rangle^i\}_T^{\approx}, \{\langle \beta \rangle^j\}_T^{\approx} \in T^{\approx}$, the following hold:*

(1) $\{\langle \alpha \rangle^i\}_T^{\approx} \preceq_T C(\{\langle \alpha \rangle^i\}_T^{\approx})$.
(2) if $\{\langle \alpha \rangle^i\}_T^{\approx} \preceq_T \{\langle \beta \rangle^j\}_T^{\approx}$, then $C(\{\langle \alpha \rangle^i\}_T^{\approx}) \preceq_T C(\{\langle \beta \rangle^j\}_T^{\approx})$.
(3) $C(C(\{\langle \alpha \rangle^i\}_T^{\approx})) \subseteq C(\{\langle \alpha \rangle^i\}_T^{\approx})$

Proof. (1) Let $C(\{\langle \alpha \rangle^i\}_T^{\approx}) = \{\varphi_1 \wedge \ldots \wedge \varphi_n\}_T^{\approx}$ such that $\{\langle \alpha \rangle^i\}_T^{\approx} \preceq_T \{\varphi_j\}_T^{\approx}$ where $1 \leq j \leq n$. We suffice to show that $\langle \alpha \rangle^i \leq_T \varphi_1 \wedge \ldots \wedge \varphi_n$. Since $\{\langle \alpha \rangle^i\}_T^{\approx} \preceq_T \{\varphi_j\}_T^{\approx}$ where $1 \leq i \leq n$, $\langle \alpha \rangle^i \leq_T \varphi_j$ for all $1 \leq j \leq n$. Assume that $\Phi \vdash_{\mathrm{G5}} \langle \varphi_1 \wedge \ldots \wedge \varphi_n \rangle^t \Rightarrow_T \psi$. Further by $(\wedge R)$, one gets $\Phi \vdash_{\mathrm{G5}} \langle \alpha \rangle^i \Rightarrow_T \varphi_1 \wedge \ldots \wedge \varphi_n$. Then by (Cut) $\Phi \vdash_{\mathrm{G5}} \langle\langle \alpha \rangle^i \rangle^t \Rightarrow_T \psi$. Thus $\langle \alpha \rangle^i \leq_T \varphi_1 \wedge \ldots \wedge \varphi_n$. Consequently $\{\langle \alpha \rangle^i\}_T^{\approx} \preceq_T C(\{\langle \alpha \rangle^i\}_T^{\approx})$.

(2) Let $C(\{\langle \beta \rangle^j\}_T^{\approx}) = \{\varphi\}_T^{\approx}$. By (1) $\{\langle \beta \rangle^j\}_T^{\approx} \preceq_T \{\varphi\}_T^{\approx}$. Since $\{\langle \alpha \rangle^i\}_T^{\approx} \preceq_T \{\langle \beta \rangle^j\}_T^{\approx}$, $\{\langle \alpha \rangle^i\}_T^{\approx} \preceq_T \{\varphi\}_T^{\approx}$. Hence by definition $C(\{\langle \alpha \rangle^i\}^{\approx}) = \{\varphi \wedge \chi\}_T^{\approx}$ for some $\chi \in T$. We suffice to show that $\varphi \wedge \chi \leq_T \varphi$. Assume that $\Phi \vdash_{\mathrm{G5}} \langle \varphi \rangle^t \Rightarrow_T \psi$. Since $\Phi \vdash_{\mathrm{G5}} \varphi \wedge \chi \Rightarrow_T \varphi$, by (Cut), one gets $\Phi \vdash_{\mathrm{G5}} \langle \varphi \wedge \chi \rangle^t \Rightarrow_T \psi$. Thus $\varphi \wedge \chi \leq_T \varphi$. Consequently $C(\{\langle \alpha \rangle^i\}_T^{\approx}) = \{\varphi \wedge \chi\}_T^{\approx} \preceq_T \{\varphi\}_T^{\approx} = C(\{\langle \beta \rangle^j\}_T^{\approx})$.

(3) First we show that $C(\{\varphi\}_T^{\approx}) \preceq_T \{\varphi\}_T^{\approx}$. Since $\{\varphi\}_T^{\approx} \preceq_T \{\varphi\}_T^{\approx}$, by definition $C(\{\varphi\}_T^{\approx}) = \{\varphi \wedge \chi\}_T^{\approx}$ for some $\varphi \in T$. Obviously $\varphi \wedge \chi \leq_T \varphi$. Thus $\{\varphi \wedge \chi\}_T^{\approx} \preceq_T \{\varphi\}_T^{\approx}$. Hence $C(\{\varphi\}_T^{\approx}) =\preceq_T \{\varphi\}_T^{\approx}$. Clearly for any $C(\{\langle\alpha\rangle^i\}_T^{\approx})$, there is a $\{\varphi\}_T^{\approx}$ where $\varphi \in T$ such that $C(\{\varphi\}_T^{\approx}) = \{\varphi\}_T^{\approx}$. Consequently $C(C(\{\varphi\}_T^{\approx})) \subseteq C(\{\varphi\}_T^{\approx})$.

We defined a interitor operation I on $T^{\approx}$ as follows:

$$I(\{\langle\alpha\rangle^i\}_T^{\approx}) = \{ \bigvee_{1 \leq j \leq n} \beta_j\}_T^{\approx} \quad \text{for any } \{\beta_j\}_T^{\approx} \text{ s.t. } \{\beta_j\}_T^{\approx} \preceq_T \{\langle\alpha\rangle^i\}_T^{\approx}.$$

Remark 4. For any $\{\langle\alpha\rangle^i\}_T^{\approx}$, $\{\bot\}_T^{\approx} \preceq_T \{\langle\alpha\rangle^i\}_T^{\approx}$. Thus $C(\{\langle\alpha\rangle^i\}_T^{\approx})$ always exists. Since T is closed under $\vee$, for any $I(\{\langle\alpha\rangle^i\}_T^{\approx})$ there exists a $\{\beta\}_T^{\approx}$ such that $I(\{\langle\alpha\rangle^i\}_T^{\approx}) = \{\beta\}_T^{\approx}$. Let $I(T^s/_{\approx_T})$ denoted the sets of all $I(\{\langle\alpha\rangle^i\}_T^{\approx})$. Clearly $C(T^s/_{\approx_T}) \subseteq T/_{\approx_T}$. Since T is qb-closed, $T/_{\approx_T}$ is finite. Thus $I(T^s/_{\approx_T})$ is finite.

Lemma 4. *For any* $\{\langle\alpha\rangle^i\}_T^{\approx}, \{\langle\beta\rangle^j\}_T^{\approx} \in T^{\approx}$, *the following hold:*

(1) $I(\{\langle\alpha\rangle^i\}_T^{\approx}) \preceq_T \{\langle\alpha\rangle^i\}_T^{\approx}$.
(2) if $\{\langle\alpha\rangle^i\}_T^{\approx} \preceq_T \{\langle\beta\rangle^j\}_T^{\approx}$, *then* $I(\{\langle\alpha\rangle^i\}_T^{\approx}) \preceq_T I(\{\langle\beta\rangle^j\}_T^{\approx})$.
(3) $I(\{\langle\alpha\rangle^i\}_T^{\approx}) \subseteq I(I(\{\langle\alpha\rangle^i\}_T^{\approx}))$

Proof. (1) Let $I(\{\langle\alpha\rangle^i\}_T^{\approx}) = \{\varphi_1 \vee \ldots \vee \varphi_n\}_T^{\approx}$ such that $\{\varphi_j\}_T^{\approx} \preceq_T \{\langle\alpha\rangle^i\}_T^{\approx}$ where $1 \leq j \leq n$. We suffice to show that $\varphi_1 \vee \ldots \vee \varphi_n \leq_T \langle\alpha\rangle^i$. Since $\{\varphi_j\}_T^{\approx} \preceq_T \{\langle\alpha\rangle^i\}_T^{\approx}$ where $1 \leq j \leq n$, $\varphi_j \leq_T \langle\alpha\rangle^i$ for all $1 \leq j \leq n$. Assume that $\Phi \vdash_{\text{G5}} \langle\langle\alpha\rangle^i\rangle^t \Rightarrow_T \psi$. Then $\Phi \vdash_{\text{G5}} \langle\varphi_j\rangle^t \Rightarrow_T \psi$ for all $1 \leq j \leq n$. Further by $(\vee L)$, one gets $\Phi \vdash_{\text{G5}} \langle\varphi_1 \vee \ldots \vee \varphi_n\rangle^t \Rightarrow_T \psi$. Thus $\varphi_1 \vee \ldots \vee \varphi_n \leq_T \langle\alpha\rangle^i$. Consequently $I(\{\langle\alpha\rangle^i\}_T^{\approx}) \preceq_T \{\langle\alpha\rangle^i\}_T^{\approx}$.

(2) Let $I(\{\langle\alpha\rangle^i\}_T^{\approx}) = \{\varphi\}_T^{\approx}$. By (1) $\{\varphi\}_T^{\approx} \preceq_T \{\langle\alpha\rangle^i\}_T^{\approx}$. Since $\{\langle\alpha\rangle^i\}_T^{\approx} \preceq_T \{\langle\beta\rangle^j\}_T^{\approx}$, $\{\varphi\}_T^{\approx} \preceq_T \{\langle\beta\rangle^j\}_T^{\approx}$. Hence by definition $I(\{\langle\beta\rangle^j\}^{\approx}) = \{\varphi\vee\chi\}_T^{\approx}$ for some $\chi \in T$. We suffice to show that $\varphi \leq_T \varphi \vee \chi$. Assume that $\Phi \vdash_{\text{G5}} \langle\varphi \vee \chi\rangle^t \Rightarrow_T \psi$. Since $\Phi \vdash_{\text{G5}} \varphi \Rightarrow_T \varphi\vee\chi$, by (Cut), one gets $\Phi \vdash_{\text{G5}} \langle\varphi\rangle^t \Rightarrow_T \psi$. Thus $\varphi \leq_T \varphi\vee\chi$. Consequent $I(\{\langle\alpha\rangle^i\}_T^{\approx}) = \{\varphi\}_T^{\approx} \preceq_T \{\varphi \vee \chi\}_T^{\approx} = I(\{\langle\beta\rangle^j\}_T^{\approx})$.

(3) First we show that $\{\varphi\}_T^{\approx} \preceq_T I(\{\varphi\}_T^{\approx})$. Since $\{\varphi\}_T^{\approx} \preceq_T \{\varphi\}_T^{\approx}$, by definition $I(\{\varphi\}_T^{\approx}) = \{\varphi\vee\chi\}_T^{\approx}$ for some $\chi \in T$. Obviously $\varphi \leq_T \varphi\vee\chi$. Thus $\{\varphi \wedge \chi\}_T^{\approx} \preceq_T$. Hence $\{\varphi\}_T^{\approx} \preceq_T I(\{\varphi\}_T^{\approx})T$. Clearly for any $I(\{\langle\alpha\rangle^i\}_T^{\approx})$, there is a $\{\varphi\}_T^{\approx}$ where $\varphi \in T$ such that $I(\{\langle\alpha\rangle^i\}_T^{\approx}) = \{\varphi\}_T^{\approx}$. Consequently $I(\{\langle\alpha\rangle^i\}_T^{\approx}) \subseteq I(I(\{\langle\alpha\rangle^i\}_T^{\approx}))$.

We define a unary operations $\Diamond$ and $\square$ on $T^{\approx}$:

$$\Diamond\{\langle\alpha\rangle^i\}_T^{\approx} = \{\langle\alpha\rangle^{i+1}\}_T^{\approx}$$

$$\square\{\langle\alpha\rangle^i\}_T^{\approx} = \{\langle\varphi\rangle^j\}_T^{\approx} \quad \text{s.t} \quad (bc1) \quad \text{and} \quad (bc2) \quad \text{holds}$$

where (bc1): $\langle\varphi\rangle^{j+1} \leq_T \langle\alpha\rangle^i$ and (bc2): for any $\langle\delta\rangle^k \leq_T \langle\alpha\rangle^i$, $\langle\delta\rangle^k \leq_T \langle\varphi\rangle^{i+1}$ $(k \geq 1)$. Notice that such $\{\langle\varphi\rangle^j\}_T^{\approx}$ always exists. Since for any $\langle\delta_1\rangle_1^t \leq_T \langle\alpha\rangle^i$ and $\langle\delta_2\rangle_2^k \leq_T \langle\alpha\rangle^i$ where $k_1, k_2 \geq 1$, we have $\langle\delta_1\rangle_1^k \leq \langle\delta_1 \vee \delta_2\rangle$, $\langle\delta_1\rangle_1^k \leq_T \langle\delta_1 \vee \delta_2\rangle$ and $\langle\delta_1 \vee \delta_2\rangle \leq_T \langle\alpha\rangle^i$.

Lemma 5. *If* $\{\langle\varphi\rangle^i\}_T^{\approx} \preceq_T \{\langle\psi\rangle^j\}_T^{\approx}$ *where* $i, j \geq 0$, *then* $\Diamond\{\langle\varphi\rangle^i\}_T^{\approx} \preceq_T \Diamond\{\langle\psi\rangle^j\}_T^{\approx}$.

Proof. Assume that $\{\langle\varphi\rangle^i\}_T^{\approx} \preceq_T \{\langle\psi\rangle^j\}_T^{\approx}$. Then $\langle\varphi\rangle^i \leq_T \langle\psi\rangle^j$. Let $\vdash_{G5} \langle\langle\langle\psi\rangle^i\rangle\rangle^t \Rightarrow_T \chi$. We get $\vdash_{G5} \langle\langle\langle\varphi\rangle^j\rangle\rangle^t \Rightarrow_T \chi$ for any $\langle-\rangle^t$ and χ. Hence $\langle\varphi\rangle^{i+1} \leq_T \langle\psi\rangle^{j+1}$. Consequently $\Diamond\{\langle\varphi\rangle^i\}_T^{\approx} \preceq_T \Diamond\{\langle\psi\rangle^j\}_T^{\approx}$.

Lemma 6. *For any* $\{\langle\alpha\rangle^i\}_T^{\approx}, \{\langle\beta\rangle^j\}_T^{\approx} \in T^{\approx}$, $\Diamond C(\{\langle\alpha\rangle^i\}_T^{\approx}) \preceq_T C(\Diamond\{\langle\alpha\rangle^i\}_T^{\approx})$.

Proof. Let $C(\Diamond\{\langle\alpha\rangle^i\}_T^{\approx}) = \{\delta\}_T^{\approx}$. Then Lemma 3 (1) $\Diamond\{\langle\alpha\rangle^i\}_T^{\approx} \leq_T \{\delta\}_T^{\approx}$. Thus $\langle\alpha\rangle^{i+1} \leq_T \delta$. Hence $\vdash_{G5} \langle\alpha\rangle^{i+1} \Rightarrow_T \delta$. By Lemma 2 there is a γ such that $\vdash_{G5} \langle\alpha\rangle^i \Rightarrow_T \gamma$ and $\vdash_{G5} \langle\gamma\rangle \Rightarrow_T \delta$. Consequently $\{\langle\alpha\rangle^i\}_T^{\approx} \preceq_T \{\gamma\}_T^{\approx}$ and $\Diamond\{\gamma\}_T^{\approx} \preceq_T \{\delta\}_T^{\approx}$. By definition, $C(\{\langle\alpha\rangle^i\}_T^{\approx}) \preceq_T \{\gamma\}_T^{\approx}$. By Lemma 5 one gets $\Diamond C(\{\langle\alpha\rangle^i\}_T^{\approx}) \preceq_T \Diamond\{\gamma\}_T^{\approx}$. Hence $\Diamond C(\{\langle\alpha\rangle^i\}_T^{\approx}) \preceq_T \{\delta\}_T^{\approx}$. Therefore $\Diamond C(\{\langle\alpha\rangle^i\}_T^{\approx}) \preceq_T C(\Diamond\{\langle\alpha\rangle^i\}_T^{\approx})$.

We define two unary operation on $T/_{\approx_T}$ as follows:

$$\blacklozenge(\{\varphi\}_T^{\approx}) = C(\Diamond(\{\varphi\}_T^{\approx}))$$

$$\blacksquare(\{\varphi\}_T^{\approx}) = I(\Box(\{\varphi\}_T^{\approx}))$$

Lemma 7. *For any* $\{\varphi\}_T^{\approx} \in T/_{\approx_T}$, *the following hold:*

(1) $\blacklozenge\blacklozenge(\{\varphi\}_T^{\approx}) \preceq_T \blacklozenge(\{\varphi\}_T^{\approx})$
(2) $(\{\varphi\}_T^{\approx}) \preceq_T \blacklozenge(\{\varphi\}_T^{\approx})$
(3) *If* $\blacklozenge((\{\varphi\}_T^{\approx})) \preceq_T \{\psi\}_T^{\approx}$, *then* $\blacklozenge((\{\neg\psi\}_T^{\approx})) \preceq_T \{\neg\varphi\}_T^{\approx}$

Proof. (1) Let $\vdash_{G5} \langle\langle\varphi\rangle\rangle^t \Rightarrow_T \psi$ for some context $\langle-\rangle^t$ and formula $\psi \in T$. By rule (4), one gets $\vdash_{G5} \langle\langle\varphi\rangle^2\rangle^t \Rightarrow_T \psi$. Thus $\langle\varphi\rangle^2 \leq_T \langle\varphi\rangle$. Hence $\Diamond\Diamond\{\varphi\}_T^{\approx} \preceq \Diamond\{\varphi\}_T^{\approx}$. By Lemma 3 (2), one gets $C(\Diamond\Diamond\{\varphi\}_T^{\approx}) \preceq_T C(\Diamond\{\varphi\}_T^{\approx})$. By Lemma 6, one gets $\Diamond C(\Diamond\{\varphi\}_T^{\approx}) \preceq_T C(\Diamond\Diamond\{\varphi\}_T^{\approx})$. Therefore $\Diamond C(\Diamond\{\varphi\}_T^{\approx}) \preceq_T C(\Diamond\{\varphi\}_T^{\approx})$. By Lemma 3 (2) and (3), one gets $C(\Diamond C(\Diamond\{\varphi\}_T^{\approx})) \preceq_T C(\Diamond\{\varphi\}_T^{\approx})$. Hence $\blacklozenge\blacklozenge(\{\varphi\}_T^{\approx}) \preceq_T \blacklozenge(\{\varphi\}_T^{\approx})$.

(2) Let $\vdash_{G5} \langle\langle\varphi\rangle\rangle^t \Rightarrow_T \psi$ for some context $\langle-\rangle^t$ and formula $\psi \in T$. By rule (T), one gets $\vdash_{G5} \langle\varphi\rangle^t \Rightarrow_T \psi$. Thus $\varphi \leq_T \langle\varphi\rangle$. Hence $\{\varphi\}_T^{\approx} \preceq_T \Diamond\{\varphi\}_T^{\approx}$. By Lemma 3 (1), $\Diamond\{\varphi\}_T^{\approx} \preceq_T C(\Diamond\{\varphi\}_T^{\approx})$. Thus $\{\varphi\}_T^{\approx} \preceq_T C(\Diamond\{\varphi\}_T^{\approx})$.

(3) Assume that $\blacklozenge(\{\varphi\}_T^{\approx}) \preceq_T \{\psi\}_T^{\approx}$. By Lemma 6, one can get $\Diamond(\{\varphi\}_T^{\approx}) \preceq_T \blacklozenge(\{\varphi\}_T^{\approx})$. Thus $\Diamond(\{\varphi\}_T^{\approx}) \preceq_T \{\psi\}_T^{\approx}$. Hence $\langle\varphi\rangle \leq_T \psi$. Therefore $\vdash_{G5} \langle\varphi\rangle \Rightarrow_T \psi$. By rule $(\Diamond\Box)$, one gets $\vdash_{G5} \langle\neg\psi\rangle \Rightarrow_T \neg\varphi$. Hence $\Diamond(\{\neg\psi\}_T^{\approx}) \preceq_T \{\neg\varphi\}_T^{\approx}$. By definition $\blacklozenge((\{\neg\psi\}_T^{\approx})) \preceq_T \{\neg\varphi\}_T^{\approx}$

Lemma 8. *For any* $\{\varphi\}_T^{\approx}, \{\psi\}_T^{\approx} \in T/_{\approx_T}$, $\blacklozenge(\{\varphi\}_T^{\approx}) \preceq_T \{\psi\}_T^{\approx}$ *iff* $\{\varphi\}_T^{\approx} \preceq_T \blacksquare\{\psi\}_T^{\approx}$.

Proof. Assume that $\blacklozenge\{\varphi\}_T^{\approx} \preceq_T \{\psi\}_T^{\approx}$. Then $C(\Diamond\{\varphi\}_T^{\approx}) \preceq_T \{\psi\}_T^{\approx}$. Thus $C(\{\langle\varphi\rangle\}_T^{\approx}) \preceq_T \{\psi\}_T^{\approx}$. By Lemma 3 (1), $\{\langle\varphi\rangle\}_T^{\approx} \preceq_T \{\psi\}_T^{\approx}$. Hence $\langle\varphi\rangle \leq_T \psi$. So $\{\varphi\}_T^{\approx} \preceq_T \Box\{\psi\}_T^{\approx}$. By definition $\{\varphi\}_T^{\approx} \preceq_T I(\Box\{\psi\}_T^{\approx}) = \blacksquare\{\psi\}_T^{\approx}$. Conversely assume that $\{\varphi\}_T^{\approx} \preceq_T \blacksquare\{\psi\}_T^{\approx}$. Then $\{\varphi\}_T^{\approx} \preceq_T I(\Box\{\psi\}_T^{\approx}])$. By Lemma 4 (1), one gets $\{\varphi\}_T^{\approx} \preceq_T \Box\{\psi\}_T^{\approx}$. Let $\Box\{\psi\}_T^{\approx} = \{\langle\gamma\rangle^i\}_T^{\approx}$ such that $\langle\gamma\rangle^{i+1} \leq_T \psi$. Hence $\{\langle\gamma\rangle^{i+1}\}_T^{\approx} \preceq_T \{\psi\}_T^{\approx}$. By Lemma 5 $\Diamond\{\varphi\}_T^{\approx} \preceq_T \Diamond\{\langle\gamma\rangle^i\}_T^{\approx}$. Thus $\{\langle\varphi\rangle\}_T^{\approx} \preceq_T \{\langle\gamma\rangle^{i+1}\}_T^{\approx}$. Consequently $\{\langle\varphi\rangle\}_T^{\approx} \preceq_T \{\psi\}_T^{\approx}$. By definition, $C(\Diamond\{\varphi\}_T^{\approx}) = C(\{\langle\varphi\rangle\}_T^{\approx}) \preceq_T \{\psi\}_T^{\approx}$. Thus $\blacklozenge\{\varphi\}_T^{\approx} \preceq_T \{\psi\}_T^{\approx}$.

Now we construct a finite tqBa5 on $T/_{\approx_T}$. For any $\{\varphi\}_T^{\approx}, \{\psi\}_T^{\approx} \in T/_{\approx_T}$ one defines:

- $\{\varphi\}_T^{\approx} \wedge \{\psi\}_T^{\approx} = \{\varphi \wedge \psi\}_T^{\approx}$,
- $\{\varphi\}_T^{\approx} \vee \{\psi\}_T^{\approx} = \{\varphi \vee \psi\}_T^{\approx}$
- $\neg\{\varphi\}_T^{\approx} = \{\neg\varphi\}_T^{\approx}$

Let $\blacklozenge$ and $\blacksquare$ be defined as above. Let $\mathbb{A}(T,\Phi) = (T/_{\approx_T}, \wedge, \vee, \neg, \blacklozenge, \blacksquare)$. Clearly $\mathbb{A}(T,\Phi)$ is a qBa. By Lemma 6 and 7, $\mathbb{A}(T,\Phi)$ is a tqBa5. Since $T/_{\approx_T}$ is finite, $\mathbb{A}(T,\Phi)$ is finite.

We define a assignment σ from T-formulae to $\mathbb{A}(T,\Phi)$ as follows: $\sigma(p) = \{p\}_T^{\approx}$ for any $p \in T$. σ can be extended to formulae and formula structures naturally. By induction on the complexity of formulae, one obtains the following fact

Proposition 3. $\sigma(\varphi) = \{\varphi\}_T^{\approx}$

Lemma 9. $\Diamond^i\{\varphi\}_T^{\approx} \preceq_T \Diamond_c^i\{\varphi\}_T^{\approx}$

Proof. By induction on the number i. If $i = 1$, then the claim holds by Lemma 5. Otherwise by induction hypothesis, one gets $\Diamond^{i-1}\{\varphi\}_T^{\approx} \preceq_T \Diamond_c^{i-1}\{\varphi\}_T^{\approx}$. Then by Lemma 4 and Lemma 3 (2), one gets $C(\Diamond\Diamond^{i-1}\{\varphi\}_T^{\approx}) \preceq_T \Diamond_c^i\{\varphi\}_T^{\approx}$. By Lemma 5, $\Diamond^i\{\varphi\}_T^{\approx} \preceq_T C(\Diamond\Diamond^{i-1}\{\varphi\}_T^{\approx})$. Hence $\Diamond^i\{\varphi\}_T^{\approx} \preceq_T \Diamond_c^i\{\varphi\}_T^{\approx}$.

Lemma 10. *If* $\Phi \not\vdash_{G5} \langle\varphi\rangle^i \Rightarrow_T \psi$, *then* $\Phi \not\models_{\mathbb{A}(T,\Phi)} \sigma(\langle\varphi\rangle^i) \preceq_T \sigma(\psi)$

Proof. Let $\alpha \Rightarrow \beta \in \Phi$. Then $\Phi \vdash_{G5} \alpha \Rightarrow \beta$. Hence $\alpha \leq_T \beta$. Thus $\{\alpha\}_T^{\approx} \preceq \{\beta\}_T^{\approx}$. Hence $\models_{\mathbb{A}(T,\Phi)} \Phi$. Assume that $\models_{\mathbb{A}(T,\Phi)} \sigma(\langle\varphi\rangle^i) \preceq_T \sigma(\psi)$. Since $\sigma(\langle\varphi\rangle^i) = \Diamond_c^i\{\varphi\}_T^{\approx}$ and $\sigma(\psi) = \{\psi\}_T^{\approx}$. Hence $\Diamond_c^i\{\varphi\}_T^{\approx} \preceq_T \{\psi\}_T^{\approx}$. Further by Lemma 8 $\Diamond^i\{\varphi\}_T^{\approx} \preceq_T \Diamond_c^i\{\varphi\}_T^{\approx}$. Thus $\Diamond^i\{\varphi\}_T^{\approx} \preceq_T \{\psi\}_T^{\approx}$ which yields $\langle\varphi\rangle^i \leq_T \psi$. Therefore $\Phi \vdash_{G5} \langle\varphi\rangle^i \Rightarrow \psi$. Contradiction.

Theorem 2. *If* $\Phi \not\vdash_{G5} \langle\varphi\rangle^i \Rightarrow \psi$ *then there exists a model* $(\mathbb{G}, \sigma)$ *s.t.* $\mathbb{G}$ *is finite tqBa5 such that all sequents in* Φ *is true while* $\langle\varphi\rangle^i \Rightarrow \psi$ *is not.*

Proof. Let $\Phi \not\vdash_{G5} \langle\varphi\rangle^i \Rightarrow \psi$. Then $\Phi \not\vdash_{G5} \langle\varphi\rangle^i \Rightarrow_T \psi$. Therefore by Lemma 9, the claim holds.

Theorem 4 means that $G5$ has SFMP so from Remark 1, we get

Theorem 3. *The variety tqBa5 has SFMP*

Theorem 4. *The variety tqBa5 has FEP.*

5   Concluding Remarks

In this paper we have proved FEP for the class of tqBa5s. Indeed the class of Pre-rough algebras also has FEP.

A Pre-rough algebra is a tqBa5 enriched with the following conditions:

(IA1) $\Box a \vee \neg \Box a \leq 1$
(IA2) $\Box(a \vee b) \leq \Box a \vee \Box b$
(IA3) $\Diamond a \leq \Diamond b$ and $\Box a \leq \Box b$ implies $a \leq b$

Clearly in pre-rough one gets (i) $\Diamond(a \wedge b) = \Diamond a \wedge \Diamond b$ and (ii) $\Diamond(a \vee b) = \Diamond a \vee \Diamond b$. Let T be a set of formula closed under $\neg, \wedge, \vee, \Diamond$. By the standard Lindenbaum Tarski method, one can construct a Pre-rough algebra from a sequent calculus for pre-rough with respect to a set of formulae T whose universe is the set of equivalence classes of formulae in T. Clearly by the De-morgan rules and (i), (ii), it is finite. Consequently the class of Pre-rough algebras has SFMP whence has FEP. By similar arguments the classes of intermediate pre-rough algebras those containing (IA2) have FEP. Since tqBa5 does not admit (IA2), the FEP for the class of tqBa5s can not be easily established by standard Lindenbaum Tarski method. The FEP for other classes of intermediate pre-rough algebras without (IA2) including IA1, IA3 remain open. The results in the present paper can also be extended to quasi Boolean algebra enriched with modal logic axioms (K) and (B) and its extensions. Further research can be finding more intermediate pre-rough algebras or modal quasi boolean algebras.

References

1. Banerjee, M., Chakraborty, M.: Rough algebra. Bull. Pol. Acad. Sci. (Math.) **41**(4), 293–297 (1993)
2. Banerjee, M., Chakraborty, M.: Rough sets through algebraic logic. Fundamenta Informaticae **28**(3–4), 211–221 (1996)
3. Pawlak, Z.: Rough sets. Int. J. Comput. Inf. Sci. **11**(5), 341–356 (1982)
4. Rasiowa, H.: An Algebraic Approach to Non-Classical Logics. North-Holland Publishing
5. Banerjee, M.: Rough sets and 3-valued lukasiewicz logic. Fundamenta Informatica **31**, 213–220 (1997)
6. Saha, A., Sen, J., Chakraborty, M.: Algebraic structures in the vicinity of pre-rough algebra and their logics. Inf. Sci. **282**, 296–320 (2014)
7. Saha, A., Sen, J., Chakraborty, M.: Algebraic structures in the vicinity of pre-rough algebra and their logics II. Inf. Sci. **333**, 44–60 (2016)
8. Block, W., van Alten, C.: The finite embeddability property for residuated lattices pocrims and bckcalgebras. Algebra Universalis **48**, 253–271 (2002)
9. Block, W., van Alten, C.: On the finite embeddability property for residuated ordered groupoids. Trans. Am. Math. Soc. **357**(10), 4141–4157 (2004)
10. Farulewski, M.: Finite embeddability property for residuated groupoids. Rep. Math. Logic **43**, 25–42 (2008)
11. Buszkowski, W.: Interpolation and FEP for logic of residuated algebras. Logic J. IGPL **19**(3), 437–454 (2011)

12. Zhe, L.: Non-associative Lambek calculus with modalities: interpolation, complexity and FEP Zhe Lin. Logic J. IGPL **22**(3), 494–512 (2014)
13. Lin, Z., Chakraborty, M.K., Ma, M.: Decidability in pre-rough algebras: extended abstract. In: Nguyen, H.S., Ha, Q.-T., Li, T., Przybyła-Kasperek, M. (eds.) IJCRS 2018. LNCS (LNAI), vol. 11103, pp. 511–521. Springer, Cham (2018). https://doi.org/10.1007/978-3-319-99368-3_40
14. Chakraborty, M., Sen, J.: A study of interconnections between rough and 3-valued lukasiewicz logics. Fundam. Inf. **51**, 311–324 (2002)
15. van Benthem, J., Bezhanishvili, G.: Modal logic of spaces. In: Aliello, M., Pratt-Hartmann, I., van Benthem, J. (eds.) Handbook of Spatial Logics, pp. 217–298. Springer, Dordrecht (2007). https://doi.org/10.1007/978-1-4020-5587-4_5
16. Galatos, N., Jipsen, P., Kowalski, T., Ono, H.: Residuated Lattices: An Algebraic Glimpse at Substructural Logics. Springer (2007)
17. Lambek, J.: On the calculus of syntactic types. Am. Math. Soc. **XII**, C178 (1961)

Specifying Program Properties Using Modal Fixpoint Logics: A Survey of Results

Martin Lange[✉]

University of Kassel, Kassel, Germany
`martin.lange@uni-kassel.de`

Abstract. The modal μ-calculus is a well-known program specification language with desirable properties like decidability of satisfiability and model checking, axiomatisability etc. Its expressive power is limited by Monadic Second-Order Logic or parity tree automata. Hence, it can only express regular properties.

In this talk I will argue in favour of specification languages whose expressiveness reaches beyond regularity. I will present Viswanathan and Viswanathan's Higher-Order Fixpoint Logic as a natural extension of the modal μ-calculus with highly increased expressive power. We will see how this logic can be used to specify some interesting non-regular properties and then survey results on it with a focus on open questions in this area.

1 The Modal μ-Calculus

The modal μ-calculus $\mathcal{L}_\mu$ [23] is a well-known specification formalism for concurrent, reactive systems. Its formulas are interpreted in states of labeled transition systems. It extends multi-modal logic with restricted second-order quantification in the form of least and greatest fixpoints. This makes it a reasonably expressive temporal logic. It can express properties that are built recursively from basic ones like *"there is a successor s.t. ..."* ($\Diamond$) or *"all successors ..."* ($\Box$) using the usual Boolean operations. Least fixpoints (μ) intuitively correspond to terminating recursion, greatest fixpoints (ν) to not necessarily terminating recursion.

Examples of such recursive properties are *"every path ends in some state without a successor"* or *"there is a path on which infinitely many states satisfy p"*. The former is expressed by $\varphi_{\mathsf{end}} := \mu X.\Box X$. A helpful tool for understanding such formulas is the fixpoint unfolding principle $\kappa X.\varphi \equiv \varphi[\kappa X.\varphi/X]$ for $\kappa \in \{\mu, \nu\}$ stating that the set of states satisfying $\kappa X.\varphi$ is indeed a fixpoint of the mapping that takes a set X and returns those satisfying $\varphi(X)$. With this principle we get that

$$\mu X.\Box X \equiv \Box\mu X.\Box X \equiv \Box\Box\mu X.\Box X \equiv \ldots$$

Knowing that a state s that has no successors satisfies $\Box\psi$ for any ψ, one can see that any state that is at the source of finite paths only, satisfies $\mu X.\Box X$.

© Springer-Verlag GmbH Germany, part of Springer Nature 2019
Md. A. Khan and A. Manuel (Eds.): ICLA 2019, LNCS 11600, pp. 42–51, 2019.
https://doi.org/10.1007/978-3-662-58771-3_5

A second helpful tool for understanding formulas that comes on handy at this point is the characterisation of $\mathcal{L}_\mu$'s model checking problem in terms of parity games [33]. Here, two players play with a token on a state s in a transition system and another on φ's syntax tree in order to find out whether s satisfies φ or not. VERIFIER chooses disjuncts whenever a disjunction is reached, and successor states whenever a $\Diamond$ is reached, likewise REFUTER chooses at conjunctions and $\Box$-formulas. When reaching a fixpoint variable the game simply continues with the defining fixpoint formula. VERIFIER wins in a situation where the currently selected state blatantly satisfies the currently selected formula. In case of $\mu X.\Box X$ above, only REFUTER makes choices by continuously selecting successor states. If he follows a (maximal) finite path $s \ldots t$ then this will ultimately end in a situation with t and $\Box X$ being selected, and since t is assumed to have no successors, VERIFIER wins, indicating that the original formula holds in s.

However, if there is an infinite path starting in s, then REFUTER can traverse this and the resulting play of the game is infinite. Then the winner is determined by the type of the outermost fixpoint that gets traversed infinitely often. In the case of $\mu X.\Box X$ there is only one candidate – a least fixpoint – which makes REFUTER the winner. As an exercise one may check that the second property named above about the existence of an infinite path is expressed by the formula $\nu X.\mu Y.(p \wedge \Diamond X) \vee \Diamond Y$. The key is to see that the game rules make VERIFIER select a path, and she can only traverse through the outer greatest fixpoint if this path has infinitely many states satisfying p. Otherwise any play will eventually only traverse through the inner least fixpoint which would make REFUTER the winner again.

$\mathcal{L}_\mu$ is well understood in terms of its expressivity and the computational complexity of the standard decision problems associated with a (temporal) logic. We quickly recall the most important results on $\mathcal{L}_\mu$, for further and more detailed overviews see also [4–6] and [14, Chap. 8].

- $\mathcal{L}_\mu$ respects, like multi-modal logic, bisimulation-invariance, i.e. it cannot distinguish bisimilar models. Despite sounding negatively, this is a good property to have since program specification formalisms should not distinguish states of transition systems that exhibit the same temporal behaviour.
- Many standard temporal and other logics can be embedded into $\mathcal{L}_\mu$, for instance CTL and PDL with simple linear translations, but also CTL* (and therefore LTL) with exponential translations [13], [14, Theorem 10.2.7].
- On trees, $\mathcal{L}_\mu$ is equi-expressive to alternating parity tree automata (APT) and, since alternation can be eliminated, therefore also to nondeterministic tree automata [15]. Bearing the first point in mind, this statement is not quite accurate since APT in their usual form are aware of directions in trees and can therefore specify non-bisimulation-invariant properties. To be precise, $\mathcal{L}_\mu$ is in fact only equi-expressive to APT over classes of ranked trees of bounded branching-degree where it has access to a specific successor, not just some. Over the class of all trees, $\mathcal{L}_\mu$ is equi-expressive to so-called symmetric APT [17,38].

There is of course also a well-known connection between tree automata and Monadic Second-Order Logic (MSO) [32]. Even without automata at hand it is easy to see that $\mathcal{L}_\mu$ can be embedded into MSO. The cannot hold as MSO is not bisimulation-invariant. However, it turns out that $\mathcal{L}_\mu$ is as expressive as MSO when restricted to bisimulation-invariant properties [20].

- Satisfiability is decidable and ExpTime-complete. The upper bound is a consequence of the linear translation into APT and an exponential emptiness test there [15]. The lower bound is inherited from PDL for instance [18].

- There are relatively simple sound and complete axiomatic systems for $\mathcal{L}_\mu$ [1,23], but establishing completeness is typically a challenging task [17,36].

- Model checking over finite transition systems is trivially decidable. It is in fact computationally equivalent to the problem of solving a parity game [33,35]. The best lower bound is known to date is P-hardness since $\mathcal{L}_\mu$ can express winning in a reachability game. The currently best known upper bounds – found only recently after a long time of research in this area – are quasi-polynomial [12,21,27].
  Model checking is even decidable over richer classes of infinite transition systems: for pushdown systems it is ExpTime-complete [37], for higher-order recursion schemes it is of non-elementary complexity [31].

- An interesting source of computational and pragmatic complexity in $\mathcal{L}_\mu$ formulas is the *alternation depth* [16,30], measuring the degree to which recursion is defined by entangling least and greatest fixpoints. It is the determining element in the asymptotic complexity of many algorithms, being exponential in it. It is also a major source of obfuscation when trying to understand the property expressed by a given $\mathcal{L}_\mu$ formula. It is therefore interesting to know how much fixpoint alternation is necessary for writing down all definable properties. It turned out that the fixpoint alternation hierarchy is strict [2,7]: for any alternation depth there are definable properties that cannot be specified using this depth only.

A consequence of $\mathcal{L}_\mu$'s connection to APT and MSO is the fact that it can only define *regular* properties. There are, however, many non-regular properties which are more or less interesting, depending on potential application areas. Typical examples include *"all executions of a program terminate at the same moment"*, *"no two executions can be distinguished from the outside"*, *"there is no underflow in an unbounded buffer"*, *"there is a maximal path of length n^2 for some n"*, etc.

There are a few proposals for modal logics that are capable of expressing non-regular properties, for instance PDL[CFL] [19], FLC [29] and HFL [34]. FLC extends $\mathcal{L}_\mu$, and HFL (vastly) extends FLC. PDL[CFL] is orthogonal to $\mathcal{L}_\mu$ in terms of expressive power but is already captured by FLC. In the following we will turn our attention to *Higher-Order Fixpoint Logic* (HFL), the most expressive among these. We compare it to $\mathcal{L}_\mu$ and its properties as laid out above. We will explain how the increase in expressive power comes at a very high price, not just computationally but also in terms of the number of questions on certain aspects of HFL that remain unanswered to date.

2 Higher-Order Fixpoint Logic

We refrain from giving a detailed definition of the syntax and semantics of the logic HFL. Instead we concentrate on the presentation of those principles that are used there, especially for the semantics. The goal of this exposition is not detailed mathematical completeness but the intuition behind the constructs in a modal fixpoint logic that achieves high expressive power. For a formal definition see [34].

HFL results from a merger between the modal μ-calculus with a simply typed λ-calculus. Its formulas are typed in a simple type system that inductively builds types from a single base type $\bullet$ using three function type constructors:

$$\sigma, \tau ::= \bullet \mid \sigma^v \to \tau \qquad v ::= + \mid - \mid 0$$

Formulas of base type are predicates as in $\mathcal{L}_\mu$; formally the type represents the powerset lattice of the set of states of a given transition system. The type $\sigma^v \to \tau$ then represents functions from objects of type σ to objects of type τ which are monotone (if $v = +$), antitone (if $v = -$) resp. unrestricted (if $v = 0$).

Formulas are given by the following grammar.

$$\varphi ::= p \mid X \mid \varphi \vee \varphi \mid \varphi \wedge \varphi \mid \neg\varphi \mid \langle a \rangle \varphi \mid [a]\varphi \mid \mu X^\tau.\varphi \mid \nu X^\tau.\varphi \mid \lambda X^\tau.\varphi \mid \varphi\,\varphi$$

where X is a variable, p is an atomic proposition interpreted by a set of states in a transition system, a is an action interpreted as a set of edges in a transition system, and τ is a type. However, not every object formed in this way is a formula. The type system guarantees well-formedness of formulas; it mainly ensures that

- in an application of the form $\varphi\,\psi$ the formula ψ has some type σ, and φ has a type $\sigma \to \tau$, and
- in a fixpoint formula $\kappa X.\varphi$, the mapping $X \mapsto \varphi(X)$ is monotone in order to guarantee the existence of least and greatest fixpoints.

The *order* of a type is defined via $ord(\bullet) = 0$ and $ord(\sigma^v \to \tau) = \max\{ord(\tau), 1 + ord(\sigma)\}$. The fragment HFL^k, $k \geq 0$, consists of all formulas of type $\bullet$ which use types of order at most k.

Consider the formula

$$\varphi := \lambda f^{\bullet^0 \to \bullet}.\lambda g^{\bullet^0 \to \bullet}.\lambda X^\bullet.f(g(X)) .$$

Its type is

$$(\bullet^0 \to \bullet)^+ \to (\bullet^0 \to \bullet)^+ \to \bullet^+ \to \bullet$$

and φ is therefore a formula of order 2.

The semantics of a formula with type τ is a function of type τ in a transition system. Its definition is straightforward given that each type induces a complete lattice of pointwise ordered (monotone/antitone/unrestricted) functions in a transition system. Fixpoint formulas can therefore be given meaning through the Knaster-Tarski Theorem. Instead of listing the formal definitions here we present some examples of formulas with the aim of giving some intuition on how to specify complex program properties in HFL. The important concepts for this are the fixpoint unfolding principle and β-reduction: $(\lambda X.\varphi) \, \psi \equiv \varphi[\psi/X]$.

We will use the following abbreviations with appropriate type annotation which are left out for brevity here.

$$g \circ f \; := \; \lambda X.g \, (f \, X), \quad f^i \; := \; \underbrace{f \circ \ldots \circ f}_{i \text{ times}}, \quad \Diamond \; := \; \lambda X.\Diamond X$$

Example 1. Consider the formula

$$\varphi_{\mathsf{qpath}} \; := \; \mu F.\lambda g.\lambda f.(g \, \Box\mathsf{ff}) \vee F \, (g \circ f^2 \circ \Diamond) \, (f \circ \Diamond)$$

Using fixpoint unfolding and β-reduction we see that $\varphi_{\mathsf{qpath}} \, \Diamond \, \Diamond$ unfolds to

$$\Diamond\Box\mathsf{ff} \vee (\varphi_{\mathsf{qpath}} \, \Diamond^4 \, \Diamond^2) \; \equiv \; \Diamond\Box\mathsf{ff} \vee \Diamond^4\Box\mathsf{ff} \vee (\varphi_{\mathsf{qpath}} \, \Diamond^9 \, \Diamond^3)$$

and so on. In fact, after unfolding n times and β-reducing appropriately we obtain $\bigvee_{i=1}^{n} \Diamond^{i^2}\Box\mathsf{ff} \vee (\varphi_{\mathsf{qpath}} \, \Diamond^{(n+1)^2} \, \Diamond^{n+1})$. This uses the fact that $(n+1)^2 = n^2+2n+1$.

Hence, in HFL^2 it is possible to define the property of having a maximal path of quadratic length.

Example 2. The property of a tree being balanced can be defined in HFL^1 already. Note that being balanced means there is some n such that every path of length n ends in a state without successors, and that no path of shorter length does so. This is defined by

$$\left(\mu F.\lambda X.X \vee (F \, (\Diamond\mathsf{tt} \wedge \Box X))\right) \Box\mathsf{ff}$$

which, again, can be unfolded and reduced to yield

$$\bigvee_{i \geq 0} \underbrace{\Diamond\mathsf{tt} \wedge \Box(\Diamond\mathsf{tt} \wedge \Box(\ldots \wedge \Diamond\mathsf{tt} \wedge \Box\Box\mathsf{ff}))}_{i \text{ times}}$$

Example 3. A similar construction principle is used in $\varphi_{\mathsf{unb}} := (\nu F.\lambda X.X \wedge (F \, \Diamond X)) \, \mathsf{tt}$. It unfolds to $\bigwedge_{i=0} \Diamond^i\mathsf{tt}$ and therefore states that are paths of unbounded length. Note that this is not the same as stating there is an infinite path.

Example 4. Note that the context-free grammar $S \to \mathsf{out} \mid \mathsf{in}\, S \, S$ generates the language of all words that have one more out than in's but no prefix does so. It represents the runs of potentially unbounded buffers that see an underflow. This grammar can immediately be transferred into an HFL^1 formula:

$$\neg\left(\left(\mu S.\lambda X.\langle\mathsf{out}\rangle X \vee \langle\mathsf{in}\rangle(S \, (S \, X))\right) \mathsf{tt}\right)$$

states that no execution is of a form that falls into this grammar. Hence, it states that all runs of a buffer do not underflow.

2.1   Results on HFL

We survey results on HFL that are known and problems that are still open, comparing this in particular to the situation with $\mathcal{L}_\mu$.

Embeddings. HFL^s ubsumes $\mathcal{L}_\mu$ in the simple sense that $\mathcal{L}_\mu$ is HFL^0, even syntactically. HFL^1 also subsumes the aforementioned FLC [34] with, in turn, subsumes PDL[CFL] [26].

Model Properties. HFL retains bisimulation-invariance [34]. However, HFL^1 already does not possess the finite model property anymore. Consider the formula $\varphi_{\mathsf{unb}} \wedge \varphi_{\mathsf{end}}$. It requires paths of unbounded length to exist but every path to be finite. This is satisfiable but not in a finite model.

Satisfiability. Strongly connected to the loss of the finite model property is the high undecidability of satisfiability checking, even for HFL^1. It is at least Σ_1^1-hard: this is proved originally for PDL[CFL] [19] and then transferred to stronger logics.

So far, no non-trivial fragments of HFL^w ith a decidable satisfiability problem have been found.

Proof Systems. The situation on the proof-theoretic side of a theory of higher-order modal fixpoint logics is even more bleak. It is not known whether there are fragments of HFL^o r even some HFL^k which can be axiomatised in a sound and complete way, not even when giving up on completeness (looking for non-trivial fragments in that case of course).

Model Checking. The model checking problem for HFL over finite transition systems is decidable and, roughly speaking, k-ExpTime-complete for formulas of order k.[1] The upper bound is obtained in a more or less straightforward way by computing the semantics of a formula bottom-up, the lower bound can be obtained using standard reductions for k-ExpTime-complete problems, for instance tiling game problems [3].

Given that HFL has complete model checking problems for every level of the exponential time hierarchy, it is a fair question to ask whether something similar holds for the exponential space hierarchy. The answer is positive: it is possible to identify a syntactic criterion on formulas called *tail recursion* such that the model checking problem for HFL^k formulas restricted in this way becomes $(k-1)$-ExpSpace-complete [10].

There seems to be no chance to extend the decidability result to any meaningful class of infinite-state systems. One can show undecidability of model checking

---

[1] This does not hold for $k = 0$, i.e. $\mathcal{L}_\mu$. It also requires an assumption of a bound on the number of arguments a function can take. Otherwise the upper bound is one exponential higher.

for FLC, resp. HFL[1] formulas over BPA processes already [29]. It remains to be seen whether there is in fact a – necessarily very small – class of infinite-state transition systems for which HFL[1] model checking is decidable.

On the other hand, there is a connection between model checking higher-order formulas and higher-order model checking: the problems of model checking a $\mathcal{L}_\mu$ formula over a higher-order recursion scheme is computationally equivalent to the problem of model checking an HFL formula over a finite transition system [22]. This can be seen as a trade-off between higher order on the formula side and higher order on the model side. It is worth noting that the translations preserve maximal order.

Automata for HFL. There is a counterpart to HFL in the world of automata. Bruse has been able to come up with an automaton model that captures HFL in the sense that every formula is equivalent to an automaton and vice-versa [8]. The model is called *Alternating Parity Krivine Automata* (APKA) and is an extension of APT that uses the mechanisms of the Krivine machine to handle higher-order functions (using a call-by-name technique). The main combinatorial difficulty in designing such an automaton model is the correct capturing of the interplay of fixpoints in the presence of higher-order features by an appropriate acceptance condition. Bruse has shown [9] that in the case of HFL[1], one can use a neater acceptance condition which is closer to the stair-parity condition [24] used in visibly pushdown games [28].

It remains to be seen whether this neater condition can be extended to fragments beyond first-order functions. We also suspect that Boolean alternation cannot be eliminated from APKA as it can be for APT. There is, however, no proof of this or the contrary.

Fixpoint Alternation. The richness of HFL as opposed to $\mathcal{L}_\mu$ opens up a variety of questions regarding the strictness or collapse of fixpoint alternation hierarchies. Besides the obvious restriction to particular classes of models one can now also ask whether the fixpoint alternation hierarchy in some HFL[k], say $\mathcal{L}_\mu$ for instance, despite being strict in itself, collapses in some HFL[k] for $k > 0$. I.e. it is conceivable that one may be able to reduce fixpoint alternation when one is willing to pay with higher function orders. This is indeed true in some case, namely finite models. One can express the Kleene iteration of length at most ω of a greatest fixpoint at order 0 using a least fixpoint at order 1 and an embedded but non-alternating greatest fixpoint of order 0. Hence, over finite models, every $\mathcal{L}_\mu$ formula is equivalent to an alternation-free HFL[1] formula.

One has to admit, though, that fixpoint alternation is not easy to define syntactically. Using β-expansion it is always possible to decouple nested fixpoints so that syntactically they look like they are not dependent on each other. This, however, only shows that the definition of fixpoint alternation that is used for $\mathcal{L}_\mu$, is too coarse for HFL. Bruse has suggested to define fixpoint alternation via the minimal number of priorities used in equivalent APKA. This way he has managed to show that the fixpoint alternation hierarchy is strict within HFL[1] [9], resembling similar proofs for $\mathcal{L}_\mu$ [2] and FLC [25].

The trick of trading in fixpoint alternation for higher order can be extended slightly beyond order 0 [11]. Here, simulating the Kleene iteration of a greatest fixpoint is more difficult because one has to test two first-order functions for equality, rather than two sets. This would in principle require the enumeration of all possible sets which HFL[2] cannot do due to bisimulation-invariance. It turns out, though, that it suffices to enumerate all modal formulas as possible arguments to such first-order functions.

In summary, results on fixpoint alternation in HFL are sparse. In particular, it is currently open whether general strictness results or, equivalently, strictness over trees, can be extended to order higher than 1. On the other hand, it is equally open whether collapse results based on the trade-in of alternation against higher orders can be extended beyond low orders.

References

1. Afshari, B., Leigh, G.E.: Cut-free completeness for modal mu-calculus. In: Proceedings of the 32nd ACM/IEEE Symposium on Logic in Computer Science, LICS 2017, pp. 1–12. IEEE (2017)
2. Arnold, A.: The modal μ-calculus alternation hierarchy is strict on binary trees. RAIRO Theor. Inform. Appl. **33**, 329–339 (1999)
3. Axelsson, R., Lange, M., Somla, R.: The complexity of model checking higher-order fixpoint logic. Log. Methods Comput. Sci. **3**, 1–33 (2007)
4. Bradfield, J., Stirling, C.: Modal logics and μ-calculi: an introduction. In: Bergstra, J., Ponse, A., Smolka, S. (eds.) Handbook of Process Algebra, pp. 293–330. Elsevier, New York (2001)
5. Bradfield, J., Stirling, C.: Modal mu-calculi. In: Blackburn, P., van Benthem, J., Wolter, F. (eds.) Handbook of Modal Logic: Studies in Logic and Practical Reasoning, vol. 3, pp. 721–756. Elsevier, New York (2007)
6. Bradfield, J., Walukiewicz, I.: The mu-calculus and model checking. In: Clarke, E., Henzinger, T., Veith, H., Bloem, R. (eds.) Handbook of Model Checking, pp. 871–919. Springer, Cham (2018). https://doi.org/10.1007/978-3-319-10575-8_26
7. Bradfield, J.C.: The modal mu-calculus alternation hierarchy is strict. In: Montanari, U., Sassone, V. (eds.) CONCUR 1996. LNCS, vol. 1119, pp. 233–246. Springer, Heidelberg (1996). https://doi.org/10.1007/3-540-61604-7_58
8. Bruse, F.: Alternating Parity Krivine Automata. In: Csuhaj-Varjú, E., Dietzfelbinger, M., Ésik, Z. (eds.) MFCS 2014. LNCS, vol. 8634, pp. 111–122. Springer, Heidelberg (2014). https://doi.org/10.1007/978-3-662-44522-8_10
9. Bruse, F.: Alternation is strict for higher-order modal fixpoint logic. In: Proceedings of the 7th International Symposium on Games, Automata, Logics and Formal Verification, GandALF 2016. EPTCS, vol. 226, pp. 105–119 (2016)
10. Bruse, F., Lange, M., Lozes, E.: Space-efficient fragments of higher-order fixpoint logic. In: Hague, M., Potapov, I. (eds.) RP 2017. LNCS, vol. 10506, pp. 26–41. Springer, Cham (2017). https://doi.org/10.1007/978-3-319-67089-8_3
11. Bruse, F., Lange, M., Lozes, E.: Collapses of fixpoint alternation hierarchies in low type-levels of higher-order fixpoint logic. In: Proceedings Workshop on Programming and Reasoning on Infinite Structures, PARIS 2014 (2018)

12. Calude, C.S., Jain, S., Khoussainov, B., Li, W., Stephan, F.: Deciding parity games in quasipolynomial time. In: Proceedings of the 49th Annual ACM SIGACT Symposium on Theory of Computing, STOC 2017, pp. 252–263. ACM (2017)
13. Dam, M.: CTL* and ECTL* as fragments of the modal μ-calculus. TCS **126**(1), 77–96 (1994)
14. Demri, S., Goranko, V., Lange, M.: Temporal Logics in Computer Science. Cambridge Tracts in Theoretical Computer Science. Cambridge University Press, Cambridge (2016)
15. Emerson, E.A., Jutla, C.S.: Tree automata, μ-calculus and determinacy. In: Proceedings of the 32nd Symposium on Foundations of Computer Science, San Juan, pp. 368–377. IEEE (1991)
16. Emerson, E.A., Lei, C.L.: Efficient model checking in fragments of the propositional μ-calculus. In: Symposon on Logic in Computer Science, Washington, D.C., pp. 267–278. IEEE (1986)
17. Enqvist, S., Seifan, F., Venema, Y.: Completeness for the modal μ-calculus: separating the combinatorics from the dynamics. Theor. Comput. Sci. **727**, 37–100 (2018)
18. Fischer, M.J., Ladner, R.E.: Propositional dynamic logic of regular programs. J. Comput. Syst. Sci. **18**(2), 194–211 (1979)
19. Harel, D., Pnueli, A., Stavi, J.: Propositional dynamic logic of nonregular programs. J. Comput. Syst. Sci. **26**(2), 222–243 (1983)
20. Janin, D., Walukiewicz, I.: On the expressive completeness of the propositional mu-calculus with respect to monadic second order logic. In: Montanari, U., Sassone, V. (eds.) CONCUR 1996. LNCS, vol. 1119, pp. 263–277. Springer, Heidelberg (1996). https://doi.org/10.1007/3-540-61604-7_60
21. Jurdzinski, M., Lazic, R.: Succinct progress measures for solving parity games. In: Proceedings of the 32nd ACM/IEEE Symposium on Logic in Computer Science, LICS 2017, pp. 1–9. IEEE (2017)
22. Kobayashi, N., Lozes, É., Bruse, F.: On the relationship between higher-order recursion schemes and higher-order fixpoint logic. In Proceedings of POPL 2017, pp. 246–259. ACM (2017)
23. Kozen, D.: Results on the propositional μ-calculus. TCS **27**, 333–354 (1983)
24. Lange, M.: Local model checking games for fixed point logic with chop. In: Brim, L., Křetínský, M., Kučera, A., Jančar, P. (eds.) CONCUR 2002. LNCS, vol. 2421, pp. 240–254. Springer, Heidelberg (2002). https://doi.org/10.1007/3-540-45694-5_17
25. Lange, M.: The alternation hierarchy in fixpoint logic with chop is strict too. Inf. Comput. **204**(9), 1346–1367 (2006)
26. Lange, M., Somla, R.: Propositional dynamic logic of context-free programs and fixpoint logic with chop. Inf. Process. Lett. **100**(2), 72–75 (2006)
27. Lehtinen, K.: A modal μ perspective on solving parity games in quasi-polynomial time. In: Proceedings of the 33rd Annual ACM/IEEE Symposium on Logic in Computer Science, LICS 2018, pp. 639–648. ACM (2018)
28. Löding, C., Madhusudan, P., Serre, O.: Visibly pushdown games. In: Lodaya, K., Mahajan, M. (eds.) FSTTCS 2004. LNCS, vol. 3328, pp. 408–420. Springer, Heidelberg (2004). https://doi.org/10.1007/978-3-540-30538-5_34
29. Müller-Olm, M.: A modal fixpoint logic with chop. In: Meinel, C., Tison, S. (eds.) STACS 1999. LNCS, vol. 1563, pp. 510–520. Springer, Heidelberg (1999). https://doi.org/10.1007/3-540-49116-3_48
30. Niwiński, D.: Fixed point characterization of infinite behavior of finite-state systems. Theor. Comput. Sci. **189**(1–2), 1–69 (1997)

31. Ong, C.-H.L.: On model-checking trees generated by higher-order recursion schemes. In: Proceedings of the 21st IEEE Symposium on Logic in Computer Science, LICS 2006, pp. 81–90. IEEE Computer Society (2006)
32. Rabin, M.O.: Decidability of second-order theories and automata on infinite trees. Trans. Am. Math. Soc. **141**, 1–35 (1969)
33. Stirling, C.: Local model checking games (extended abstract). In: Lee, I., Smolka, S.A. (eds.) CONCUR 1995. LNCS, vol. 962, pp. 1–11. Springer, Heidelberg (1995). https://doi.org/10.1007/3-540-60218-6_1
34. Viswanathan, M., Viswanathan, R.: A higher order modal fixed point logic. In: Gardner, P., Yoshida, N. (eds.) CONCUR 2004. LNCS, vol. 3170, pp. 512–528. Springer, Heidelberg (2004). https://doi.org/10.1007/978-3-540-28644-8_33
35. Walukiewicz, I.: Monadic second order logic on tree-like structures. In: Puech, C., Reischuk, R. (eds.) STACS 1996. LNCS, vol. 1046, pp. 399–413. Springer, Heidelberg (1996). https://doi.org/10.1007/3-540-60922-9_33
36. Walukiewicz, I.: Completeness of Kozen's axiomatisation of the propositional μ-calculus. Inf. Comput. **157**(1–2), 142–182 (2000)
37. Walukiewicz, I.: Pushdown processes: games and model-checking. Inf. Comput. **164**(2), 234–263 (2001)
38. Wilke, T.: Alternating tree automata, parity games, and modal μ-calculus. Bull. Belgian Math. Soc. **8**(2), 359–391 (2001)

A Modal Aleatoric Calculus
for Probabilistic Reasoning

Tim French[(⊠)], Andrew Gozzard, and Mark Reynolds

The University of Western Australia, Perth, Western Australia
{tim.french,mark.reynolds}@uwa.edu.au,andrew.gozzard@research.uwa.edu.au

Abstract. We consider multi-agent systems where agents actions and beliefs are determined aleatorically, or "by the throw of dice". This system consists of possible worlds that assign distributions to independent random variables, and agents who assign probabilities to these possible worlds. We present a novel syntax and semantics for such system, and show that they generalise Modal Logic. We also give a sound and complete calculus for reasoning in the base semantics, and a sound calculus for the full modal semantics, that we conjecture to be complete. Finally we discuss some application to reasoning about game playing agents.

Keywords: Probabilistic modal logic · Proof theory · Multi-agent systems

1 Introduction

This paper proposes a probabilistic generalisation of modal logic for reasoning about probabilistic multi-agent systems. There has been substantial work in this direction before [1,6,13]. However, here, rather than extending a propositional modal logic with the capability to represent and reason about probabilities, we revise all logical operators so that they are interpreted probabilistically. Thus we differentiate between *reasoning about probabilities* and *reasoning probabilistically*. Interpreting probabilities as epistemic entities suggests a Bayesian approach [2], where agents assess the likelihood of propositions based on a combination of prior assumptions and observations.

We provide a lightweight logic, the *aleatoric calculus*, for reasoning about systems of independent random variables, and give an extension, the *modal aleatoric calculus* for reasoning about multi-agent systems of random variables. We show that this is a true generalisation of modal logic and provide some initial proof theoretic results. The modal aleatoric calculus allows agents to express strategies in games or theories of how other agents will act, and we present a basic demonstration of this.

2 Related Work

There has been significant and long-standing interest in reasoning about probability and uncertainty, to apply the precision of logical deduction in uncertain

© Springer-Verlag GmbH Germany, part of Springer Nature 2019
Md. A. Khan and A. Manuel (Eds.): ICLA 2019, LNCS 11600, pp. 52–63, 2019.
https://doi.org/10.1007/978-3-662-58771-3_6

and random environments. Hailperin's probability logic [9] and Nilsson's probabilistic logic [16] seek to generalise propositional, so the semantics of true and false are replaced by probability measures. These approaches in turn are generalised in fuzzy logics [19] where real numbers are used to model degrees of truth via T-norms. In [18] Williamson provide an inference system based on Bayesian epistemology.

These approaches lose the simplicity of Boolean logics, as deductive systems must deal with propositions that are not independent. This limits their practicality as well defined semantics require the conditional probabilities of all atoms to be known. However, these approaches have been successfully combined with logic programming [12] and machine learning [3]. Feldman and Harel [7] and Kozen [14] gave a probabilistic variation of propositional dynamic logic for reasoning about the correctness of programs with random variables. Importantly, this work generalises a modal logic (PDL) as a many valued logic.

More general foundational work on reasoning probabilistically was done by de Finetti [4] who established an epistemic notion of probability based on what an agent would consider to be a rational wager (the *Dutch book* argument). In [15], Milne incorporates these ideas into the logic of conditional events. Stalnaker has also considered conditional events and has presented conditional logic [17]. Here, conditional refers to the interpretation of one proposition being contingent on another, although this is not quantified nor assigned a probability.

The other approach to reasoning about uncertainty is to extend traditional Boolean and modal logics with operators for reasoning about probabilities. Modal and epistemic logics have a long history for reasoning about uncertainty, going back to Hintikka's work on possible worlds [11]. More recent work on dynamic epistemic logic [5] has looked at how agents incorporate new information into their belief structures. There are explicit probabilistic extensions of these logics, that maintain the Boolean interpretation of formulae, but include probabilistic terms [6,10]. Probabilistic terms are converted into Boolean terms through arithmetic comparisons. For example, "It is more likely to snow than it is to rain" is a Boolean statement, whereas the likelihood of snow is a probabilistic statement.

3   Syntax and Semantics

We take a many-valued approach here. Rather than presenting a logic that describes what is *true* about a probabilistic scenario, we present the *Modal Aleatoric Calculus* (**MAC**) for determining what is likely. The different is subtle: In probabilistic dynamic epistemic logic [13] it is possible to express that the statement "Alice thinks X has probability 0.5" is true; whereas the calculus here simply has a term "Alice's expectation of X" which may have a value that is greater than 0.5. We present a syntax for constructing complex terms in this calculus, and a semantics for assignment values to terms, given a particular interpretation or model.

3.1   Syntax

The syntax is given for a set of random variables X, and a set of agents N. We also include constants $\top$ and $\bot$. The syntax of the dynamic aleatoric calculus, **MAC**, is as follows:

$$\alpha ::= \ x \mid \top \mid \bot \mid (\alpha?\alpha{:}\alpha) \mid (\alpha\mid\alpha)_i$$

where $x \in X$ is a variable and $i \in N$ is a modality. We typically take an epistemic perspective, so the modality corresponds to an agent's beliefs. As usual, we let $v(\alpha)$ refer to the set of variables that appear in α. We refer to $\top$ as *always* and $\bot$ as *never*. The *if-then-else* operator $(\alpha?\beta{:}\gamma)$ is read *if* α *then* β *else* γ and uses the ternary conditional syntax of programming languages such as C. The *conditional expectation* operator $(\alpha\mid\beta)_i$ is *modality* i's *expectation of* α *given* β (the conditional probability i assigns to α given β).

3.2   Semantics

The modal aleatoric calculus is interpreted over *probability models* similar to the probability structures defined in [10], although they have random variables in place of propositional assignments.

Definition 1. *Given a set S, we use the notation $PD(S)$ to notate the set of probability distributions over S, where $\mu \in PD(S)$ implies: $\mu : S \longrightarrow [0,1]$; and either $\Sigma_{s \in S}\mu(s) = 1$, or $\Sigma_{s \in S}\mu(s) = 0$. In the latter case, we say μ is the* empty *distribution.*

Definition 2. *Given a set of variables X and a set of modalities N, a* probability model *is specified by the tuple $P = (W, \pi, f)$, where:*

- *W is a set of possible worlds.*
- *$\pi : N \longrightarrow W \longrightarrow PD(W)$ assigns for each world $w \in W$ and each modality $i \in N$, a probability distribution $\pi_i(w)$ over W. We will write $\pi_i(w, v)$ in place of $\pi(i)(w)(v)$.*
- *$f : W \longrightarrow X \longrightarrow [0,1]$ is a probability assignment so for each world w, for each variable x, $f_w(x)$ is the probability of x being true.*

Given a model P we identify the corresponding tuple as (W^P, π^P, f^P). A pointed probability model, $P_w = (W, \pi, f, w)$, specifies a world in the model as the point of evaluation.

We note that we have not placed any restraints on the function π. If π were to model agent belief we might expect all worlds in the probability distribution $\pi_i(w)$ to share the same probability distribution of worlds. However, at this stage we have chosen to focus on the unconstrained case.

Given a pointed model P_w, the semantic interpretation of a **MAC** formula α is $P_w(\alpha) \in [0,1]$ which is the expectation of the formula being supported by a sampling of the model, where the sampling is done with respect to the distributions specified by π and f.

Definition 3. *The* semantics of the modal aleatoric calculus *take a pointed probability model, f_w, and a proposition defined in* **MAC**, α, *and calculate the* expectation of α holding at P_w. *Given an agent i, a world w and a $\mathcal{MAC}$ formula α, we define i's expectation of α at w as*

$$E_w^i(\alpha) = \sum_{u \in W} \pi_i(w, u) \cdot P_u(\alpha).$$

Then the semantics of **MAC** *are as follows:*

$$P_w(\top) = 1 \qquad P_w(\bot) = 0 \qquad P_w(x) = f_w(x)$$
$$P_w((\alpha?\beta:\gamma)) = P_w(\alpha) \cdot P_w(\beta) + (1 - P_w(\alpha)) \cdot P_w(\gamma)$$
$$P_w((\alpha \mid \beta)_i) = \frac{E_w^i(\alpha \wedge \beta)}{E_w^i(\beta)} \text{ if } E_w^i(\beta) > 0 \text{ and } 1 \text{ otherwise}$$

We say two formulae, α and β, a semantically equivalent (written $\alpha \cong \beta$) if for all pointed probability models P_w we have $P_w(\alpha) = P_w(\beta)$.

The concept of *sampling* is intrinsic in the rational of these semantics. The word *aleatoric* has its origins in the Latin for dice-player (*aleator*), and the semantics are essentially aleatoric, in that they use dice (or sample probability distributions) for everything. If we ask whether a variable is true at a world, the variable is sampled according to the probability distribution at that world. Likewise, to interpret a modality the corresponding distribution of worlds is sampled, and the formula is evaluated at the selected world. However, we are not interested in the result of any one single sampling activity, but in the *expectation* derived from the sampling activity.

This gives us an interesting combination approaches for understanding probability. Aleatoric approaches appeal to frequentist interpretations of probability, where likelihoods are fixed and assume arbitrarily large sample sizes. This contrasts the Bayesian approach where probability is intrinsically epistemic, where we consider what likelihood an agent would assign to an event, given the evidence they have observed. Our approach can be seen as an aleatoric implementation of a Bayesian system. By this we mean that: *random variables are aleatoric*, always sampled from a fixed distribution, and *modalities are Bayesian*, always conditioned on a set of possible worlds.

The if-then-else operator, $(\alpha?\beta:\gamma)$, can be imagined as a sampling protocol. We first sample α, and if α is true, we proceed to sample β and otherwise we sample γ. We imagine an evaluator interpreting the formula by flipping a coin: if it lands heads, we evaluate β; if it lands tails, we evaluate γ. This corresponds to the additive property of Bayesian epistemology: *if A and B are disjoint events, then P(A or B) = P(A) + P(B)* [2]. Here the two disjoint events are α *and* β and $\neg\alpha$ *and* γ, but disjointedness is only guaranteed if α and $\neg\alpha$ are evaluated from the same sampling.

The conditional expectation operator $(\alpha \mid \beta)_i$ expresses modality i's expectation of α *marginalised* by the expectation of β. This is, as in the Kolmogorov definition of conditional probability, i's expectation of $\alpha \wedge \beta$ divided by i's expectation of β. The intuition for these semantics corresponds to a sampling protocol.

The modality i samples a world from the probability distribution and samples β at that world. If β is true, then i samples α at that world and returns the result. Otherwise agent i resamples a world from their probability distribution, and repeats the process. In the case that β is never true, we assign $(\alpha \mid \beta)_i$ probability 1, as being vacuously true.

Abbreviations: Some abbreviations we can define in $\mathcal{MAC}$ are as follows:

$$\alpha \wedge \beta = (\alpha?\beta:\bot) \qquad \alpha \vee \beta = (\alpha?\top:\beta) \qquad \alpha \to \beta = (\alpha?\beta:\top) \qquad \neg\alpha = (\alpha?\bot:\top)$$
$$E_i\alpha = (\alpha \mid \top)_i \qquad \Box_i\alpha = (\bot \mid \neg\alpha)_i$$
$$\alpha^{\frac{0}{b}} = \top \qquad \alpha^{\frac{a}{b}} = \bot \text{ if } b < a \qquad \alpha^{\frac{a}{b}} = (\alpha?\alpha^{\frac{a-1}{b-1}}:\alpha^{\frac{a}{b-1}}) \text{ if } a \leq b$$

where a and b are natural numbers. We will show later that under certain circumstances these operators do correspond with their Boolean counterparts. However, this is not true in the general case. The formula $\alpha \wedge \beta$ does not interpret directly as α is true and β is true. Rather it is the likelihood of α being sampled as true, followed by β being sampled as true. For this reason $\alpha \wedge \alpha$ is *not* the same as α. Similarly $\alpha \vee \beta$ is the likelihood of α being sampled as true, or in the instance that it was not true, that β was sampled as true.

The modality $E_i\alpha$ is agent i's expectation of α being true, which is just α conditioned on the uniformly true $\top$. The operator $\Box_i\alpha$ corresponds to the necessity operator of standard modal logic, and uses a property of the conditional operator: it evaluates $(\alpha \mid \beta)_i$ as vacuously true if and only if there is no expectation that β can ever be true. Therefore, $(\bot \mid \neg\alpha)_i$ can only be true if modality i always expects $\neg\alpha$ to be false, and thus for the modality i, α is necessarily true. The formula $\alpha^{\frac{a}{b}}$ allows us to explicitly represent degrees of belief in the language. It is interpreted as α *is true at least a times out of b*. Note that this is not a statement saying what the frequency of α is. Rather it describes the event of α being true a times out of b. Therefore, if α was unlikely (say true 5% of the time) then $\alpha^{\frac{9}{9}}$ describes an incredibly unlikely event.

3.3 Example

We will give simple example of reasoning in **MAC**. Suppose we have an aleator (dice player), considering the outcome of a role of a die. While the dice is fair, our aleator does not know whether it is a four sided die or a six sided die. We consider a single proposition: p_1 if the result of throw of the die is 1. The aleator considers two worlds equally possible: w_4 where the die has four sides, and w_6 where the die has 6 sides. The probability model $P = (W, \pi, f)$ is depicted in Fig. 1: We can formulate properties such as "at least one of the next two throws will be a 1": $p_1^{\frac{1}{2}} = (p_1?\top:p_1)$. We can calculate $P_{w_4}(p_1^{\frac{1}{2}}) = \frac{7}{16}$, while $P_{w_6}(p_1^{\frac{1}{2}}) = \frac{11}{36}$. Now if we asked our aleator what are the odds of rolling a second 1, given the first roll was 1, we would evaluate the formula $(p_1 \mid p_1)_a$ (where a is our aleator), and in either world this evaluates to $\frac{5}{24}$. Note that this involves some speculation from the aleator.

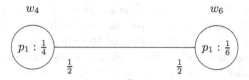

Fig. 1. A probability model for an aleator who does not know whether the die is four sided (w_4) or six sided (w_6).

4   Axioms for the Modal Aleatoric Calculus

Having seen the potential for representing stochastic games, we will now look at some reasoning techniques. First we will consider some axioms to derive constraints on the expectations of propositions, as an analogue of a Hilbert-style proof system for modal logic. In the following section we will briefly analyse the model checking problem, as a direct application of the semantic definitions.

Our approach here is to seek a derivation system that can generate equalities that are always valid in **MAC**. For example, $\alpha \wedge \beta \simeq \beta \wedge \alpha$ will be satisfied by every world of every model. We use the relation $\simeq$ to indicate that two formulae are equivalent in the calculus, and the operator $\cong$ to indicate the expectation assigned to each formula will be equal in all probability models. We show that the calculus is sound, and sketch a proof of completeness in the most basic case.

4.1   The Aleatoric Calculus

The *aleatoric calculus*, AC, is the language of $\top$, $\bot$, x and $(\alpha?\beta:\gamma)$, where $x \in X$. The interpretation of this fragment only depends on a single world and it is the analogue of propositional logic in the non-probabilistic setting. The axioms of the calculus are:

id	$x \simeq x$	**vacuous**	$(x?\top:\bot) \simeq x$
ignore	$(x?y:y) \simeq y$	**tree**	$((x?y:z)?p:q) \simeq (x?(y?p:q):(z?p:q))$
always	$(\top?x:y) \simeq x$	**swap**	$(x?(y?p:q):(y?r:s)) \simeq (y?(x?p:r):(x?q:s))$
never	$(\bot?x:y) \simeq y$		

We also have the rules of transitivity, symmetry and substitution for $\simeq$:

> **Trans :**   If $\alpha \simeq \beta$ and $\beta \simeq \gamma$ then $\alpha \simeq \gamma$
> **Sym :**   If $\alpha \simeq \beta$ then $\beta \simeq \alpha$
> **Subs :**   If $\alpha \simeq \beta$ and $\gamma \simeq \delta$ then $\alpha[x\backslash\gamma] \simeq \beta[x\backslash\delta]$

where $\alpha[x\backslash\gamma]$ is α with every occurrence of the variable x replaced by γ. We let this system of axioms and rules be referred to as $\mathfrak{AC}$.

As an example of reasoning in this system, we will show that the commutativity of $\wedge$ holds:

$$(x?y:\bot) \simeq (x?(y?\top:\bot):(y?\bot:\bot)) \quad \textbf{vacuous, ignore}$$
$$\simeq (y?(x?\top:\bot):(x?\bot:\bot)) \quad \textbf{swap}$$
$$\simeq (y?x:\bot) \quad\quad\quad\quad \textbf{vacuous, ignore}$$

The axiom system $\mathfrak{AC}$ is sound. The majority of these axioms are simple to derive from Definition 3, and all proofs essentially show that the semantic interpretation of the left and right side of the equation are equal. The rules **Trans** and **Sym** come naturally with equality, and the rule **Subs** follows because at any world, all formulae are probabilistically independent.

We present arguments for the soundness of the less obvious **tree** and **swap** in the long version of the paper [8].

Lemma 1. *The axiom system $\mathfrak{AC}$ is sound for AC.*

Proof. This follows from Lemmas 2 and 3 presented in [8], and Definition 3.2. Also, from the semantics we can see that the interpretation of subformulae are independent of one another, so the substitution rule holds, and the remaining rules follow directly from the definition of $\simeq$.

To show that $\mathfrak{AC}$ complete for the aleatoric calculus, we aim to show that any aleatoric calculus formula that are semantically equivalent can be transformed into a common form. As the axioms of $\mathfrak{AC}$ are equivalences this is sufficient to show that the formulae are provably equivalent. The proofs are presented in the long version of the paper.

A *tree form* Aleatoric Calculus formula is either atomic, or it has an atomic random variable condition and both its left and right subformulae are in tree form.

Definition 4. *The set of all* tree form *Aleatoric Calculus formulae $T \subset \Phi$ is generated by the following grammar:*

$$\varphi :: = \top \mid \bot \mid (x?\varphi : \varphi)$$

Lemma 2. *For any Aleatoric Calculus formula there exists an equivalent (by $\simeq$) tree form formula.*

Definition 5. *A* path *in a tree form aleatoric calculus formula is a sequence of tokens from the set $\{x, \overline{x} \mid x \in X\}$ corresponding to the outcomes of random trials involved in reaching a terminal node in the tree. We define the functions $\mathbb{T}$ and $\perp$ to be the set of paths that terminate in a $\top$ or a $\bot$ respectively:*

$$\mathbb{T}(\top) = \perp(\bot) = \{()\}, \qquad \mathbb{T}(\bot) = \perp(\top) = \emptyset,$$

$$\mathbb{T}((x?\alpha : \beta)) = \{(x)^\frown a \mid a \in \mathbb{T}(\alpha)\} \cup \{(\overline{x})^\frown b \mid b \in \mathbb{T}(\beta)\}$$

$$\perp((x?\alpha : \beta)) = \{(x)^\frown a \mid a \in \perp(\alpha)\} \cup \{(\overline{x})^\frown b \mid b \in \perp(\beta)\}$$

where $^\frown$ is the sequence concatenation operator:

$$(a_1, \ldots, a_n)^\frown (b_1, \ldots, b_n) = (a_1, \ldots, a_n, b_1, \ldots, b_n)$$

We say two paths are multi-set equivalent *if the multiset of tokens that appear in each path are equivalent, and define $\boldsymbol{P}(\phi) = \mathbb{T}(\phi) \cup \perp(\phi)$ to be the set of all paths through a formula.*

Lemma 3. *For any tree form aleatoric calculus formula* α:

$$P(\alpha) = \sum_{t \in \mathbb{T}(\alpha)} \prod_{x \in t} P(x)$$

where $P(\overline{x}) = (1 - P(x))$.

Proof. This follows immediately from the Definition 3.

Lemma 4. *Suppose that* ϕ *is a formula in tree form such that* $a = (a_0, \ldots, a_n) \in \mathbb{T}(\phi)$ *(resp.* $\perp\!\!\!\perp(\phi)$*). Then, for any* $i < n$ *there is some formula* ϕ_a^i *such that:*

1. $\phi \simeq \phi_a^i$
2. $(a_0, \ldots, a_{i-1}, a_{i+1}, a_i, a_{i+2}, \ldots, a_n) \in \mathbb{T}(\phi_a^i)$ *(resp.* $\perp\!\!\!\perp(\phi)$*)*
3. ϕ *and* ϕ_a^i *agree on all paths that do not have the prefix* $(a_0, \ldots, a_{i-1}$*. That is, for all* $b \in \boldsymbol{P}(\phi) \cup \boldsymbol{P}(\phi_a^i)$*, where for some* $j < i$*,* $b_j \neq a_j$*, we have* $b \in \mathbb{T}(\phi)$ *if and only if* $b \in \mathbb{T}(\phi_a^i)$ *and* $b \in \perp\!\!\!\perp(\phi)$ *if and only if* $b \in \perp\!\!\!\perp(\phi_a^i)$*.*

Lemma 5. *Given a pair of multi-set equivalent paths* a *and* b *in a tree form aleatoric calculus formula,* ϕ*, such that* $a \in \mathbb{T}(\phi)$ *and* $b \in \perp\!\!\!\perp(\phi)$*, we can find a formula* $\phi_a^b \simeq \phi$ *where*

1. $a \in \perp\!\!\!\perp(\phi_a^b)$ *and* $b \in \mathbb{T}(\phi_a^b)$
2. $\mathbb{T}(\phi_a^b) - \{b\} = \mathbb{T}(\phi) - \{a\}$,
3. $\perp\!\!\!\perp(\phi_a^b) - \{a\} = \perp\!\!\!\perp(\phi) - \{b\}$.

Lemma 6. *For any pair of tree form aleatoric calculus formulae,* ϕ *and* ψ*, there exists a pair of tree forms* $\phi' \simeq \phi$ *and* $\psi' \simeq \psi$*, such that* $\boldsymbol{P}(\phi') = \boldsymbol{P}(\psi')$*.*

Theorem 1. *For any pair of semantically equivalent aleatoric calculus formulae* ϕ *and* ψ*, we can show* $\phi \simeq \psi$*.*

Proof. By Lemma 2 it is possible to convert both formulae in question, ϕ and ψ, into tree form, respectively ϕ^τ and ψ^τ. By Lemma 6 it is then possible to convert ϕ^τ and ψ^τ to a pair of equivalent formulae, respectively Φ and Ψ, with the same structure (so $\mathbb{T}(\Phi) \cup \perp\!\!\!\perp(\Phi) = \mathbb{T}(\Psi) \cup \perp\!\!\!\perp(\Psi)$), but possibly different leaves (so $\mathbb{T}(\Phi)$ is possibly not the same as $\mathbb{T}(\Psi)$). By Lemma 5 it is possible to swap any multiset-equivalent paths between $\mathbb{T}(\Phi)$ and $\perp\!\!\!\perp(\Phi)$. By Lemma 3 two formula, Φ and Ψ, with the same structure are semantically equivalent if and only if there is a one-to-one correspondence between paths of Φ and Ψ such that corresponding paths a and b are multi-set equivalent, and $a \in \mathbb{T}(\Phi)$ if and only if $b \in \mathbb{T}(\Psi)$. Therefore, if and only if the two formulae are equivalent we are able to define Φ' by swapping paths between $\mathbb{T}(\Phi)$ and $\perp\!\!\!\perp(\Phi)$ such that $\Phi' = \Psi$. As all steps are performed using the axioms and are reversible, this is sufficient to show $\phi \simeq \psi$.

4.2 The Modal Aleatoric Calculus

The modal aleatoric calculus includes the propositional fragment, as well as the conditional expectation operator $(\alpha \mid \beta)_i$ that depends on the modality i's probability distribution over the set of worlds.

The axioms we have for the conditional expectation operator are as follows:

$$
\begin{array}{ll}
\mathbf{A0}: & ((x?y\!:\!z)\,|\,c)_i \simeq ((x\,|\,c)_i?(y\,|\,(x?c\!:\!\bot))_i\!:\!(z\,|\,(x?\bot\!:\!c))_i). \\
\mathbf{A1}: & (\bot\,|\,x)_i \wedge (x\,|\,y)_i \simeq (\bot\,|\,x \vee y)_i \\
\mathbf{A2}: & (\bot\,|\,x)_i \simeq ((\bot\,|\,x)_i?(\bot\,|\,x)_i\!:\!\neg(\bot\,|\,x)_i) \\
\mathbf{A3}: & (\top\,|\,x)_i \simeq \top \\
\mathbf{A4}: & (x\,|\,\bot)_i \simeq \top
\end{array}
$$

We let the axiom system $\mathfrak{MAC}$ be the axiom system $\mathfrak{AC}$ along with the axioms **A0-A5**.

We note that the conditional expectation operator $(x\,|\,y)_i$ is its own dual, but only in the case that agent i does not consider x and y to be mutually exclusive: $(\neg x\,|\,y)_i \simeq (x\,|\,y)_i \rightarrow \Box_i(\neg(x \wedge y))$. We can see this in the following derivation:

$$
\begin{array}{lll}
(\neg x\,|\,y)_i & \simeq ((x?\bot\!:\!\top)\,|\,y)_i & \text{abb.} \\
& \simeq ((x\,|\,y)_i?(\bot\,|\,x \wedge y)_i\!:\!(\top\,|\,(x?\bot\!:\!\bot))_i) & \text{A0} \\
& \simeq (x\,|\,y)_i \rightarrow \Box_i\neg(x \wedge y) & \text{abb.}
\end{array}
$$

The main axiom in $\mathfrak{MAC}$ is the axiom **A0** which is a rough analogue of the **K** axiom in modal logic. We note that in this axiom:

$$
((x?y\!:\!z)\,|\,c)_i \simeq ((x\,|\,c)_i?(y\,|\,(x?c\!:\!\bot))_i\!:\!(z\,|\,(x?\bot\!:\!c))_i)
$$

if we substitute $\top$ for y and $\bot$ for w, we have: $E_i x \wedge (y\,|\,x)_i \simeq E_i(x \wedge y)$ whenever agent i considers x possible (so that $(\bot\,|\,\neg x)_i \simeq \bot$). In that case we can "divide" both sides of the semantic equality by $P_w(E_i x)$ which gives the Kolmogorov definition of conditional probability:

$$
P_w((y\,|\,x)_i) = \frac{P_w(E_i(x \wedge y))}{P_w(E_i x)}.
$$

Axioms **A1** and **A2** deal with formulas of the type $(\bot\,|\,\alpha)_i$. The probability associated with such formulas is non-zero if and only if α is impossible in all the successor states, so in these states, we are able to substitute α with $\bot$.

Finally axioms **A3** and **A4** allow us to eliminate conditional expectation operators.

As with the aleatoric calculus, soundness can be shown by deriving equivalence of the semantic evaluation equations, although the proofs are more complex.

Lemma 7. *The system $\mathfrak{MAC}$ is sound for* **MAC**.

Proof. The nontrivial cases of **A0** and **A1** are presented in the long version of the paper [8]. and the axioms **A2**, **A3** and **A4** follow immediately from the semantics.

Given its correspondence to axiomatizations of modal logic and adequacy for proving small equivalences, we conjecture that $\mathfrak{MAC}$ is complete for the given semantics.

5   Expressivity

In this section we show that the modal aleatoric calculus generalises the modal logic K. The syntax and semantics of modal logic are given over a set of atomic propositions Q. The syntax of K_n is given by:

$$\phi ::= \; q \mid \phi \wedge \phi \mid \neg \phi \mid \Box \phi$$

where $q \in Q$, and the operators are respectively *and, not, necessary*. The semantics of K_n are given with respect to an epistemic model $M = (W, R, V)$ where W is the nonempty, countable set of possible worlds, $R \subseteq W \times W$ is the accessibility relation, and $V : Q \longrightarrow 2^W$ is an assignment of propositions to states. We require that:

1 $\forall w, u, v \in W,\; u \in R(w)$ and $v \in R(w)$ implies $v \in R(u)$
2 $\forall w, u, v \in W,\; u \in R(w)$ and $v \in R(u)$ implies $v \in R(w)$
3 $\forall w \in W,\; R(w) \neq \emptyset$.

We describe the set of worlds $\|\alpha\|^M$ in the model $M = (W, R, V)$ that satisfy the formula α by induction as follows:

$$\|q\|^M = V(q) \qquad\qquad \|\alpha \wedge \beta\|^M = \|\alpha\|^M \cap \|\beta\|^M$$
$$\|\neg\alpha\|^M = W - \|\alpha\|^M \qquad \|\Box\alpha\|^M = \{u \in W \mid uR \subseteq \|\alpha\|^M\}$$

where $\forall u \in W$, $uR^\alpha = uR \cap \|\alpha\|^M$, if $uR^\alpha \cap \|\alpha\|^M \neq \emptyset$ and $uR^\alpha = uR$, otherwise.

We say **MAC** *generalises* K if there is some map Λ from pointed epistemic models to pointed probability models, and some map λ from K formulae to **MAC** formulae such that for all pointed epistemic models M_w, for all K formulae ϕ, $w \in \|\phi\|^M$ if and only if $\Lambda(M_w)(\lambda(\phi)) = 1$.

We suppose that for every atomic proposition $q \in Q$, there is a unique atomic variable $x_q \in X$. Then the map Λ is defined as follows: Given $M = (W, R, V)$ and $w \in W$, $\Lambda(M_w) = P_w$ where $P = (W, \pi, f)$ and

- $\forall u, v \in W$, $\pi_i(u, v) > 0$ if and only if $v \in uR^1$.
- $\forall w \in W$, $\forall q \in Q$, $f_w(x_q) = 1$ if $w \in V(q)$ and $f_w(x_q) = 0$ otherwise.

This transformation replaces the atomic propositions with variables that, at each world, are either always true or always false, and replaces the accessibility relation at a world w with a probability distribution that is non-zero for precisely the worlds accessible from w. It is clear that there is a valid probability model that satisfies these properties.

We also define the map λ from K to **MAC** with the following induction:

$$\lambda(q) = x_q \qquad\qquad \lambda(\alpha \wedge \beta) = (\lambda(\alpha)?\lambda(\beta) : \bot)$$
$$\lambda(\neg\alpha) = (\lambda(\alpha)?\bot : \top) \qquad \lambda(\Box\alpha) = (\bot \mid (\lambda(\alpha)?\bot : \top))$$

---

[1] We note this function is not deterministic, but this does not impact the final result.

Lemma 8. *For all epistemic models $M = (W, R, V)$, for all $w \in W$, for all K formula ϕ, we have $w \in \|\phi\|^M$ if and only if $\Lambda(M_w)(\lambda(\phi)) = 1$.*

The proof may be found in the long version of this paper [8].

6   Case Study

We present a case study using some simple actions in a dice game illustrating the potential for reasoning in AI applications. A simple version of the game *pig*[2] uses a four sided dice, and players take turns. Each turn, the player rolls the dice as many times as they like, adding the numbers the roll to their turn total. However, if they roll a 1, their turn total is set to 0, and their turn ends. They can elect to stop at any time, in which case their turn total is added to their score.

To illustrate the aleatoric calculus we suppose that for our dice we have two random variables, odd and gt2 (*greater than 2*). Every roll of the dice can be seen as a sampling of these two variables: 1 is an odd number not greater than 2, and so on. Now we suppose that there is some uncertainty to the fairness of the dice, so it is possible that there is a 70% chance of the dice rolling a number greater than 2. However, we consider this unlikely and only attach a 10% likelihood to this scenario. Finally, we suppose there is an additional random variable called risk which can be used to define a policy. For example, we might roll again if the risk variable is sampled as true. This scenario if visualised in Fig. 2, and the formalization of relevant properties is given in Fig. 3.

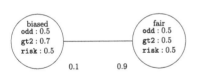

Name	Formula	Description	Prob
bust	(gt2?⊥:odd)	prob of rolling 1	0.15
four	(odd?⊥:gt2)	prob of rolling 4	0.35
ifBust	(bust|⊤)	expectation of a 1	0.265
if-4-1	(bust|four)	expectation of 1 given 4	0.237
roll	(ifBust$^{1/2}$?risk:⊤)	whether to roll again	0.77

Fig. 2. A simple two world representation of the game pig, where the dice is possibly biased.

Fig. 3. Formulas describing different event in the model at the left world.

These formulas show the different types of information that can be represented: bust and four are true random variables (aleatoric information), whereas ifBust and if-4-1 are based on an agent's mental model (Bayesian information). Finally roll describes the condition for a policy to roll again. In a dynamic extension of this calculus, given prior assumptions about policies, agents may apply Bayesian conditioning to learn probability distributions from observations.

---

[2] https://en.wikipedia.org/wiki/Pig_(dice_game).

7 Conclusion

The modal aleatoric calculus is shown to be a true generalisation of modal logic, but gives a much richer language that encapsulates probabilistic reasoning and degrees of belief. We have shown that the modal aleatoric calculus is able to describe probabilistic strategies for agents. We have provided a sound axiomatization for the calculus, shown it is complete for the aleatoric calculus and we are working to show that the axiomatization is complete for the modal aleatoric calculus. Future work will consider dynamic variations of the logic, where agents apply Bayesian conditioning based on their observations to learn the probability distribution of worlds.

References

1. Baltag, A., Smets, S.: Group belief dynamics under iterated revision: fixed points and cycles of joint upgrades. In: Proceedings of 12th TARK, pp. 41–50 (2009)
2. Bovens, L., Hartmann, S.: Bayesian Epistemology. Oxford University Press, Oxford (2003)
3. Cohen, W.W., Yang, F., Mazaitis, K.R.: TensorLog: deep learning meets probabilistic DBs. arXiv preprint arXiv:1707.05390 (2017)
4. De Finetti, B.: Theory of Probability: A Critical Introductory Treatment. Wiley, Chichester (1970)
5. van Ditmarsch, H., van der Hoek, W., Kooi, B.: Dynamic Epistemic Logic. Synthese Library, vol. 337. Springer, Dordrecht (2007)
6. Fagin, R., Halpern, J.Y., Megiddo, N.: A logic for reasoning about probabilities. Inf. Comput. **87**(1–2), 78–128 (1990)
7. Feldman, Y.A., Harel, D.: A probabilistic dynamic logic. In: Proceedings of the Fourteenth Annual ACM Symposium on Theory of Computing, pp. 181–195 (1982)
8. French, T., Gozzard, A., Reynolds, M.: A modal aleatoric calculus for probabalistic reasoning: extended version (2019). http://arxiv.org/abs/1812.11741
9. Hailperin, T.: Boole's Logic and Probability: A Critical Exposition from the Standpoint of Contemporary Algebra, Logic and Probability Theory. North-Holland Pub. Co., Amsterdam (1976)
10. Halpern, J.Y.: Reasoning About Uncertainty. MIT Press, Cambridge (2017)
11. Hintikka, J.: Knowledge and Belief. Cornell University Press, Ithaca (1962)
12. Hommersom, A., Lucas, P.J.: Generalising the interaction rules in probabilistic logic. In: 22nd International Joint Conference on Artificial Intelligence (2011)
13. Kooi, B.P.: Probabilistic dynamic epistemic logic. J. Log. Lang. Inf. **12**(4), 381–408 (2003)
14. Kozen, D.: A probabilistic PDL. J. Comput. Syst. Sci. **30**(2), 162–178 (1985)
15. Milne, P.: Bruno de Finetti and the logic of conditional events. Br. J. Philos. Sci. **48**(2), 195–232 (1997)
16. Nilsson, N.J.: Probabilistic logic. Artif. Intell. **28**(1), 71–87 (1986)
17. Stalnaker, R.C., Thomason, R.H.: A semantic analysis of conditional logic 1. Theoria **36**(1), 23–42 (1970)
18. Williamson, J.: From Bayesian epistemology to inductive logic. J. Appl. Log. **11**(4), 468–486 (2013)
19. Zadeh, L.A.: Fuzzy sets. In: Fuzzy Sets, Fuzzy Logic, And Fuzzy Systems: Selected Papers by Lotfi A Zadeh, pp. 394–432. World Scientific (1996)

Public Announcements for Epistemic Models and Hypertheories

Nenad Savić$^{(\boxtimes)}$ and Thomas Studer

Institute of Computer Science, University of Bern, Bern, Switzerland
{savic,tstuder}@inf.unibe.ch
http://www.ltg.unibe.ch

Abstract. Artemov has recently proposed a modernization of the semantics and proof theory of epistemic logic. We take up his approach and extend his framework with public announcements and the corresponding belief change operation. We establish a soundness and completeness result and show that our model update operation satisfies the AGM postulate of minimal change. Further, we also show that the standard approach cannot be directly employed to capture knowledge change by truthful announcements.

Keywords: Modal logic · Public announcements · Epistemic models · Hypertheories

1 Introduction

Artemov [3–6] suggests a modernization of the semantics and proof theory of epistemic logic. He proposes new foundations for epistemic logic with

1. a semantics that does not assume models to be common knowledge and
2. a matching framework of hypertheories for reasoning with partial information.

He introduces the class of epistemic models, which includes Kripke models, but can cover many more epistemic situations. The main difference is that in epistemic models, the Kripkean definition of satisfiability of a belief formula

$$u \Vdash \Box_i A \iff R_i(u) \Vdash A, \tag{1}$$

is replaced by a weaker condition

$$u \models \Box_i A \implies R_i(u) \models A,$$

where we write $R_i(u) \Vdash A$ for $\forall v (R_i(u,v) \Rightarrow v \Vdash A)$ and similarly for $R_i(u) \models A$. Hence the fully explanatory property of models is avoided, i.e., we do not have that if a sentence holds at all possible states, then it is believed.

This work was supported by the Swiss National Science Foundation grant 200021_165549.

Md. A. Khan and A. Manuel (Eds.): ICLA 2019, LNCS 11600, pp. 64–75, 2019.
https://doi.org/10.1007/978-3-662-58771-3_7

In this paper, we extend Artemov's epistemic models and hypertheories with public announcements. The idea behind public announcements in the Kripkean case is that a public announcement induces a model change: after the public announcement of a formula A, the model is restricted in a way that preserves only the relations between states where A holds [11,14,16,18]. In the case of epistemic models, however, only restricting the model does not yield new beliefs since (1) does not hold. To model public announcements properly, we also have to explicitly take care of what the new beliefs are. We will make use of public announcements that are total [7,16,17], i.e., new information can always be announced. Moreover, our explicit treatment of belief change is influenced by dynamic epistemic justification logics [2,8,9,12,13,15].

Our approach gives us more control over the belief dynamics that takes place when an announcement occurs. In particular, we can define the updated model such that the minimal change property of the AGM postulates is satisfied [1,10]. This is not possible in the traditional Kripkean setting.

We also show that there is no straightforward adaptation of our approach to the case of knowledge change. Namely, with a help of one example, it is showed that a restricted $S4_n$-models are not well-defined.

The content of this paper is as follows. In Sect. 2 we present epistemic models and hypertheories formally. In Sect. 3 we propose the logic KPA_n of public announcements and prove soundness and completeness of the corresponding hypertheories w.r.t. epistemic models. In Sect. 4 we discuss a problem concerning truthful public announcements over $S4_n$-epistemic models. We conclude the paper in Sect. 5.

2 Epistemic Models and Hypertheories

Recall the multi-agent modal logic K_n. Let $\mathsf{Prop} = \{p, q, r, \dots\}$ be a countable set of propositional letters. The language of the logic K_n consists Prop, the classical propositional connectives $\neg$ and $\wedge$, and modalities $\Box_i$, for $i = 1, \dots, n$. The set of formulas Fml is generated by the following grammar:

$$A := p \mid \neg A \mid (A \wedge A) \mid \Box_i A.$$

The other connectives are defined as usual. An axiomatization of K_n contains, besides the axioms of classical propositional logic and Modus Ponens, the following belief postulates:

Distributivity: $\Box_i(A \to B) \to (\Box_i A \to \Box_i B)$;
Necessitation rule: From A, infer $\Box_i A$.

The semantics of K_n is standard Kripke semantics. Namely, a model is a tuple $\mathcal{K} = (W, R_1, \dots, R_n, \Vdash)$, where

(K1) $W \neq \emptyset$;
(K2) R_i is a binary relation on W, for $i = 1, \dots, n$;
(K3) $\Vdash: \mathsf{Prop} \to 2^W$.

We call $(W, R_1, \ldots, R_n)$ a *frame*.

Truth in Kripke semantics is then defined inductively, starting from atomic propositions, with classical conditions for Boolean connectives and

$$u \Vdash \Box_i A \; :\Leftrightarrow \; R_i(u) \Vdash A. \tag{2}$$

In epistemic models, the situation is quite different since we use belief sets instead of (2) to model the agents' beliefs. Belief sets contain all theorems of a logic and they are closed under Modus Ponens. We use the following closure operation.

Definition 1 (Closure of a Set of Formulas). *Let* L *be a logic and* T *be a set of formulas.*

1. $cl_{\mathsf{L}}^0(T) = T \cup \{A \mid \mathsf{L} \vdash A\}$;
2. $cl_{\mathsf{L}}^{j+1}(T) = cl_{\mathsf{L}}^j(T) \cup \{A \mid B \in cl_{\mathsf{L}}^j(T) \text{ and } B \to A \in cl_{\mathsf{L}}^j(T), \text{ for some } B\}$;
3. $F \in cl_{\mathsf{L}}(T)$ *iff* $F \in cl_{\mathsf{L}}^j(T)$, *for some* j.

Definition 2 (Belief Set). *For a given logic* L, *an* L-*belief set is a set of formulas* T *with* $T = cl_{\mathsf{L}}(T)$.

Remark 1. The set $cl_{\mathsf{L}}(T)$ is a belief set for any set of formulas T.

Remark 2. Instead of belief sets, Artemov uses complete truth assignments in his definition of epistemic model. For the purpose of this paper, however, belief sets are better suited.

For a set $Z \subseteq X \times Y$ and an element $x \in X$, we set $(Z)_x := \{y \mid (x, y) \in Z\}$.

Definition 3 (Pre-epistemic Model). *Let* L *be a logic. A* pre-epistemic L-*model is a tuple* $\mathcal{E} = (W, R_1, \ldots, R_n, \nu, \nu_\mathcal{B}^1, \ldots, \nu_\mathcal{B}^n)$, *where:*

- $W \neq \emptyset$ *is a non-empty set of states;*
- $R_1, \ldots, R_n$ *are binary relations on* W;
- $\nu \subseteq W \times \mathsf{Prop}$;
- $\nu_\mathcal{B}^i \subseteq W \times \mathsf{For}$, *such that for every* $u \in W$, $(\nu_\mathcal{B}^i)_u$ *is an* L-*belief set.*

Definition 4 (Satisfaction Relation). *Let* $\mathcal{E}$ *be a pre-epistemic* L-*model and* $u \in W$. *The satisfaction relation,* $\models$, *is defined as follows:*

- $\mathcal{E}, u \models p$ *iff* $(u, p) \in \nu$;
- $\mathcal{E}, u \models A \wedge B$ *iff* $\mathcal{E}, u \models A$ *and* $\mathcal{E}, u \models B$;
- $\mathcal{E}, u \models \neg A$ *iff* $\mathcal{E}, u \not\models A$;
- $\mathcal{E}, u \models \Box_i A$ *iff* $(u, A) \in \nu_\mathcal{B}^i$.

Definition 5 (Epistemic Model). *Let* L *be a logic. An* epistemic L-*model is a pre-epistemic* L-*model that satisfies*

$$\mathcal{E}, u \models \Box_i A \; \Rightarrow \; \mathcal{E}, R_i(u) \models A. \tag{3}$$

In contrast to Kripke models, the truth value of belief formulas in epistemic models is provided by belief sets and (3) is a set of constraints. Also note that in (3) we only have the implication 'from left to right', while we have an equivalence in (2). Hence the fully explanatory property, which states that

if a sentence is valid at all possible states, then it is believed,

does not hold for epistemic models.

The following theorem shows the relationship between Kripke and epistemic models: for any given epistemic K_n-model, there is a Kripke model that contains it.

Theorem 1 (Embedding Theorem). *For any epistemic K_n-model*

$$\mathcal{E} = (W, R_1, \ldots, R_n, \nu, \nu_B^1, \ldots, \nu_B^n),$$

there exists a Kripke model

$$\mathcal{K} = (\widetilde{W}, \widetilde{R_1}, \ldots, \widetilde{R_n}, \Vdash),$$

such that:

(a) $W \subseteq \widetilde{W}$;
(b) $R_i \subseteq \widetilde{R_i}$;
(c) *for each $u \in W$ and each formula A,*

$$\mathcal{E}, u \models A \quad \textit{iff} \quad \mathcal{K}, u \Vdash A.$$

Theorem 1 tells us that epistemic models are contained in Kripke models, where the containing Kripke model is obtained by adding appropriate states to the epistemic model. The following example illustrates this fact.

Example 1. Consider W consisting of a single state u at which p is true but the agent does not believe that p. The appropriate epistemic model is:

$$u$$
$$\bullet$$
$$p, \neg\Box_i p$$

Note that it is not a Kripke model (since p holds in every possible state but is not believed), but can be extended to one:

$$u \qquad\qquad v$$
$$\bullet \longleftarrow\!\!\longrightarrow \bullet$$
$$p, \neg\Box_i p \qquad\qquad \neg p, \neg\Box_i p$$

This example shows one important difference between epistemic and Kripke models: in epistemic models we do not have to add new states in order to represent situations where an agent does not believe a true fact, as it is the case in Kripke models.

Hypertheories provide the proof-theoretic framework that matches epistemic models.

Definition 6 (Hypertheory). *A hypertheory is a tuple*

$$\mathcal{H} = (W, R_1, \ldots, R_n, \mathcal{T}),$$

where:

- $(W, R_1, \ldots, R_n)$ *is a frame;*
- $\mathcal{T}$ *assigns a set of formulas T_u to each $u \in W$.*

Note that T_u need not be maximal, i.e., we may have neither $p \in T_u$ nor $\neg p \in T_u$. This reflects the fact, mentioned in the introduction, that hypertheories represent a tool for dealing with partial information.

Definition 7. *An epistemic L-model $\mathcal{E} = (W, R_1, \ldots, R_n, \nu, \nu_{\mathcal{B}}^1, \ldots, \nu_{\mathcal{B}}^n)$ is a model of $\mathcal{H} = (W, R_1, \ldots, R_n, \mathcal{T})$ if for each $u \in W$,*

$$u \models T_u \quad (i.e.\ u \models A,\ for\ each\ A \in T_u).$$

A formula A logically L-follows from a hypertheory $\mathcal{H} = (W, R_1, \ldots, R_n, \mathcal{T})$ at state $u \in W$, denoted by

$$\mathcal{H}, u \models_\mathsf{L} A,$$

if $\mathcal{E}, u \models A$ for each epistemic L-model $\mathcal{E}$ of $\mathcal{H}$.

Hyperderivations provide the syntactic consequence relation for hypertheories.

Definition 8 (Hyperderivation). *Let $\mathcal{H}$ be a hypertheory. A formula A is L-hyperderivable at $u \in W$ (write $\mathcal{H}, u \Vdash_\mathsf{L} A$) if A can be obtained by the rules:*

(1) classical inference[1]:
  (a) $u \Vdash_\mathsf{L} A$,   if $A \in T_u$;
  (b) $u \Vdash_\mathsf{L} A$,   if $\mathsf{L} \vdash A$;
  (c) $u \Vdash_\mathsf{L} A$,   if $u \Vdash_\mathsf{L} B \to A$ and $u \Vdash_\mathsf{L} B$ for some formula B;
(2) transition: $u \Vdash_\mathsf{L} \Box_i A \Rightarrow R_i(u) \Vdash_\mathsf{L} A$;
(3) deduction: $u \cup A \Vdash_\mathsf{L} B \Rightarrow u \Vdash_\mathsf{L} A \to B$, where for a hypertheory $\mathcal{H}$, $\mathcal{H}^{v \cup A}$ is defined as a hypertheory $\mathcal{H}$ where T_v is replaced by $T_v \cup \{A\}$ and $u \cup A \Vdash_\mathsf{L} B$ stands for $\mathcal{H}^{u \cup A}, u \Vdash_\mathsf{L} B$;
(4) consistency: if $u R_i v$, then $v \Vdash_\mathsf{L} \bot \Rightarrow u \Vdash_\mathsf{L} \bot$.

---

[1] As usual, we write $u \Vdash_\mathsf{L} A$ instead of $\mathcal{H}, u \Vdash_\mathsf{L} A$ when $\mathcal{H}$ is clear from the context.

Note that the *transition* rule "goes only in one direction", which corresponds to (3).

Artemov established the following soundness and completeness result for K_n.

Theorem 2. *For a hypertheory $\mathcal{H}$ and any* Fml*-formula A,*

$$\mathcal{H}, u \Vdash_{\mathsf{K}_n} A \quad \textit{iff} \quad \mathcal{H}, u \models_{\mathsf{K}_n} A.$$

3   Public Announcements

In this section we discuss how to model public announcements in epistemic models. For simplicity we follow an approach where the agents believe any formula that is announced—no matter whether the announcement is truthful or whether it is consistent with the current beliefs.

The logic KPA_n is an extension of the logic K_n with an announcement operator $[\cdot]$. The set of formulas $\mathsf{Fml}_{[\cdot]}$ is generated by the following grammar:

$$A := p \mid \neg A \mid (A \wedge A) \mid \square_i A \mid [A]A.$$

Read $[A]B$ as: "after the announcement of A, it holds that B". The logic KPA_n is given by the following axioms and rules:

Axiom schemes:

(B1) all instantiations of classical propositional tautologies
(B2) $\square_i(A \rightarrow B) \rightarrow (\square_i A \rightarrow \square_i B)$
(B3) $[A]p \leftrightarrow p$
(B4) $[A]\neg B \leftrightarrow \neg[A]B$
(B5) $[A](B \wedge C) \leftrightarrow ([A]B \wedge [A]C)$
(B6) $[A]\square_i B \leftrightarrow \square_i(A \rightarrow B)$
(B7) $[A][B]C \leftrightarrow [A \wedge B]C$

Inference rules:

(IR1) Modus Ponens
(IR2) From A, infer $\square_i A$.

Since atomic propositions represent facts, axiom $B3$ says that announcements of formulas do not change facts (only agents' beliefs). Axioms $B4$ and $B5$ state that negation and conjunction behave as expected, while axiom $B6$ explains what it means that after an announcements of a formula, an agent beliefs that B. Finally, axiom $B7$ says that announcing first a formula A and then B is the same as announcing $A \wedge B$.

We already have the definitions of KPA_n-belief sets and pre-epistemic KPA_n-models. To define the corresponding satisfaction relation, we add the following clause to Definition 4:

$- \ \mathcal{E}, u \models [A]B$ iff $\mathcal{E}|_A, u \models B$, where the restricted model $\mathcal{E}|_A = (W', R'_1, \ldots, R'_n, \nu', \nu'^1_B, \ldots, \nu'^n_B)$ is given by:

$$W' = W,$$
$$R'_i = R_i \cap (\llbracket A \rrbracket_\mathcal{E} \times \llbracket A \rrbracket_\mathcal{E}),$$
$$\nu' = \nu,$$
$$\nu'^i_B = \{(w, B) \mid w \in W \text{ and } B \in cl_{\mathsf{KPA}_n}((\nu^i_B)_w \cup \{A\})\},$$

with $\llbracket A \rrbracket_\mathcal{E} = \{w \in W \mid \mathcal{E}, w \models A\}$.

Epistemic KPA_n-models are now given by Definition 5. We show that the restriction $\mathcal{E}|_A$ of an epistemic KPA_n-model is well-defined.

Lemma 1. *For any formula A, if $\mathcal{E}$ is an epistemic KPA_n-model, then the restricted model $\mathcal{E}|_A$ is an epistemic KPA_n-model, too.*

Proof. Directly from the definition, we have that W' is non-empty, R'_i are binary relations on W', $\nu' \subseteq W' \times \mathsf{Prop}$, and for any $u \in W'$, $(\nu'^i_B)_u$ is a belief set and therefore $\mathcal{E}|_A$ is a pre-epistemic KPA_n-model. We need to prove that for any formula B, the constraint

$$\mathcal{E}|_A, u \models \square_i B \Rightarrow \mathcal{E}|_A, R_i(u) \models B \tag{4}$$

holds as well. Suppose that $\mathcal{E}|_A, u \models \square_i B$, i.e.,

$$B \in cl_{\mathsf{KPA}_n}((\nu^i_B)_u \cup \{A\}).$$

By induction on the buildup of $cl_{\mathsf{KPA}_n}((\nu^i_B)_u \cup \{A\})$, we prove $\mathcal{E}|_A, R_i(u) \models B$.

(1) (i) If $B \in (\nu^i_B)_u$, from the assumption that $\mathcal{E}$ is an epistemic model we get that $\mathcal{E}|_A, R_i(u) \models B$.
   (ii) If $B = A$ or $\mathsf{KPA}_n \vdash B$, the claim follows from the definition of an restricted model.
(2) There exists a formula C, such that both

$$C, C \rightarrow B \in cl_{\mathsf{KPA}_n}((\nu^i_B)_u \cup \{A\}).$$

By induction hypothesis, $\mathcal{E}|_A, R_i(u) \models C$ and $\mathcal{E}|_A, R_i(u) \models C \rightarrow B$. Thus $\mathcal{E}|_A, R_i(u) \models B$. $\qquad\square$

Note that KPA_n is a conservative extension of the modal logic K_n with respect to announcement-free formulas. We have for each Fml-formula A,

$$\mathsf{KPA}_n \vdash A \quad \text{iff} \quad \mathsf{K}_n \vdash A.$$

This conservativity result can be transfered to logical consequence.

Lemma 2. *Let $\mathcal{H}$ be a hypertheory consisting of Fml-formulas and let u be a state of $\mathcal{H}$. We have for each Fml-formula A,*

$$\mathcal{H}, u \models_{\mathsf{KPA}_n} A \quad \text{iff} \quad \mathcal{H}, u \models_{\mathsf{K}_n} A.$$

We say that a formula A is KPA_n-*valid*, if for any epistemic KPA_n-model $\mathcal{E}$ and state u, we have that $\mathcal{E}, u \models A$. We have the following result.

Lemma 3. *Axioms B3–B7 are* KPA_n-*valid.*

Proof

(B3)
$$\mathcal{E}, u \models [A]p \quad \text{iff} \quad \mathcal{E}|_A, u \models p \quad \text{iff} \quad \mathcal{E}, u \models p.$$

(B4)
$$\begin{aligned}
\mathcal{E}, u \models \neg[A]B \quad &\text{iff} \quad \mathcal{E}, u \not\models [A]B \\
&\text{iff} \quad \mathcal{E}|_A, u \not\models B \\
&\text{iff} \quad \mathcal{E}|_A, u \models \neg B \\
&\text{iff} \quad \mathcal{E}, u \models [A]\neg B.
\end{aligned}$$

(B5)
$$\begin{aligned}
\mathcal{E}, u \models [A](B \wedge C) \quad &\text{iff} \quad \mathcal{E}|_A, u \models B \wedge C \\
&\text{iff} \quad \mathcal{E}|_A, u \models B \text{ and } \mathcal{E}|_A, u \models C \\
&\text{iff} \quad \mathcal{E}, u \models [A]B \text{ and } \mathcal{E}, u \models [A]C \\
&\text{iff} \quad \mathcal{E}, u \models [A]B \wedge [A]C.
\end{aligned}$$

(B6)

$(\leftarrow)$ $(u, A \to B) \in \nu_{\mathcal{B}}^i$ implies $(u, A \to B) \in \nu_{\mathcal{B}}'^i$, where
$$\nu_{\mathcal{B}}'^i = \{(w, B) \mid w \in W \text{ and } B \in cl_{\mathsf{KPA}_n}((\nu_{\mathcal{B}}^i)_w \cup \{A\})\}.$$

Obviously, $(u, A) \in \nu_{\mathcal{B}}'^i$ as well and hence $(u, B) \in \nu_{\mathcal{B}}'^i$, i.e., $\mathcal{E}, u \models [A]\Box_i B$.

$(\to)$
$$\begin{aligned}
\mathcal{E}, u \models [A]\Box_i B \quad &\text{iff} \quad \mathcal{E}|_A, u \models \Box_i B \\
&\text{iff} \quad B \in cl_{\mathsf{KPA}_n}((\nu_{\mathcal{B}}^i)_u \cup \{A\}).
\end{aligned}$$

We prove that $\mathcal{E}, u \models \Box_i(A \to B)$ by induction on the construction of $cl_{\mathsf{KPA}_n}((\nu_{\mathcal{B}}^i)_u \cup \{A\})$.

(1) (*i*) If $(u, B) \in \nu_{\mathcal{B}}^i$, since $(u, B \to (A \to B)) \in \nu_{\mathcal{B}}^i$ as well, we obtain that $(u, A \to B) \in \nu_{\mathcal{B}}^i$, i.e., $\mathcal{E}, u \models \Box_i(A \to B)$.
(*ii*) If $B = A$, since $\mathsf{KPA}_n \vdash A \to A$, we get $(u, A \to A) \in \nu_{\mathcal{B}}^i$.
(*iii*) If $\mathsf{KPA}_n \vdash B$, the claim follows from the same reasoning as in (*i*).
(2) Suppose that there exists a formula C, such that both
$$C, C \to B \in cl_{\mathsf{KPA}_n}((\nu_{\mathcal{B}}^i)_u \cup \{A\}).$$

By induction hypothesis,
$$\mathcal{E}, u \models \Box_i(A \to C) \text{ and } \mathcal{E}, u \models \Box_i(A \to (C \to B)).$$

Since belief sets are closed under classical propositional reasoning, we finally obtain $\mathcal{E}, u \models \Box_i(A \to B)$.

(B7) Directly from the fact that $(\mathcal{E}|_A)|_B = \mathcal{E}|_{A \wedge B}$.   □

From the axiomatization of the logic KPA_n, it is clear that we have the usual "rewriting" property for public announcements, which makes it possible to remove all announcements from an arbitrary formula (see, e.g., [18]). Namely, it can be proved that for any $\mathsf{Fml}_{[\cdot]}$-formula A, there exists an Fml-formula A^K such that

$$\mathsf{KPA}_n \vdash A \leftrightarrow A^\mathsf{K}. \tag{5}$$

Hence we can do the usual completeness by reduction proof for KPA_n.

Theorem 3 (Soundness and Completeness Theorem). *Let $\mathcal{H}$ be a hyper-theory containing Fml-formulas. For each $\mathsf{Fml}_{[\cdot]}$-formula A, we have*

$$\mathcal{H}, u \Vdash_{\mathsf{KPA}_n} A \quad \textit{iff} \quad \mathcal{H}, u \models_{\mathsf{KPA}_n} A.$$

Proof. From Lemma 3 we know that axioms B3–B7 are sound, while from the definition of an epistemic KPA_n-model follows that the axioms B1–B2, both inference rules, as well as transition, deduction and consistency constraints from KPA_n-hyperderivations are sound as well, i.e., the direction from left to right is established.

Completeness is obtained from the following observation:

$$\begin{aligned}
\mathcal{H}, u \models_{\mathsf{KPA}_n} A \quad &\text{implies} \quad \mathcal{H}, u \models_{\mathsf{KPA}_n} A^\mathsf{K} \quad \text{(by (5) and soundness)} \\
&\text{implies} \quad \mathcal{H}, u \models_{\mathsf{K}_n} A^\mathsf{K} \quad \text{(Lemma 2)} \\
&\text{implies} \quad \mathcal{H}, u \Vdash_{\mathsf{K}_n} A^\mathsf{K} \quad \text{(Theorem 2)} \\
&\text{implies} \quad \mathcal{H}, u \Vdash_{\mathsf{KPA}_n} A.
\end{aligned}$$

                    □

Our belief change operation satisfies the AGM postulates [1,10] for belief expansion. First of all, it is obvious from our semantics that all announcements are *successful*, i.e., after any announcement of A, each agent beliefs A. Formally, we have that

$$[A]\Box_i A$$

is KPA_n-valid.

Further, we have *persistence of beliefs*, i.e, no announcement will change existing beliefs. The formula

$$\Box_i A \rightarrow [B]\Box_i A$$

is KPA_n-valid. Indeed, for an arbitrary epistemic KPA_n-model $\mathcal{E}$ and formula B:

$$\begin{aligned}
\mathcal{E}, u \models \Box_i A \quad &\text{iff} \quad (u, A) \in \nu_B^i \\
&\text{then} \quad A \in cl_{\mathsf{KPA}_n}((\nu_B^i)_u \cup \{B\}) \\
&\text{iff} \quad \mathcal{E}|_B, u \models \Box_i A \\
&\text{iff} \quad \mathcal{E}, u \models [B]\Box_i A.
\end{aligned}$$

The belief sets of the restricted model are given as a least fixed point of a monotone operator. It is an immediate consequence of this definition that our model update satisfies the requirement of *minimal change*. We have the following lemma.

Lemma 4 (Minimal Change). *Let $\mathcal{E}$ be an epistemic KPA_n-model with a state w. Let $\mathcal{F}$ be any epistemic KPA_n-model with a state v such that*

1. $\mathcal{F}, v \models \Box_i A$
2. $\mathcal{E}, w \models \Box_i B$ *implies* $\mathcal{F}, v \models \Box_i B$ *for all formulas B.*

Then we find that for all formulas B,

$$\mathcal{E}|_A, w \models \Box_i B \text{ implies } \mathcal{F}, v \models \Box_i B.$$

4 The Case of Knowledge Change in Epistemic Models

In this section we show that there is no straightforward adaptation of our approach to the case of knowledge change. Let us investigate public announcements over $\mathsf{S4}_n$. In order to model them, we consider so-called *truthful* announcements. For a given Kripke model $\mathcal{M} = (W, R_1, \ldots, R_n, \Vdash)$, we define satisfiability of announcement formulas by

$$\mathcal{M}, s \Vdash [A]B \quad \text{iff} \quad \mathcal{M}, s \Vdash A \text{ implies } \mathcal{M}|_A, s \Vdash B, \tag{6}$$

where $\mathcal{M}|_A = (W', R_1', \ldots, R_n', \Vdash')$ is the restriction of the model defined as

$$W' = [\![A]\!]_{\mathcal{M}},$$
$$R_i' = R_i \cap ([\![A]\!]_{\mathcal{M}} \times [\![A]\!]_{\mathcal{M}}),$$
$$\Vdash' = \Vdash \cap [\![A]\!]_{\mathcal{M}},$$

for $[\![A]\!]_{\mathcal{M}} = \{w \in W \mid w \Vdash A\}$.

Adapting this strategy for an epistemic $\mathsf{S4}_n$-model

$$\mathcal{E} = (W, R_1, \ldots, R_n, \nu, \nu_{\mathcal{B}}^1, \ldots, \nu_{\mathcal{B}}^n)$$

yields the following definition of satisfiability:

$$\mathcal{E}, s \models [A]B \quad \text{iff} \quad \mathcal{E}, s \models A \text{ implies } \mathcal{E}|_A, s \models B,$$

where $\mathcal{E}|_A$ is given by

$$W' = [\![A]\!]_{\mathcal{E}},$$
$$R_i' = R_i \cap ([\![A]\!]_{\mathcal{E}} \times [\![A]\!]_{\mathcal{E}}),$$
$$\nu' = \nu,$$
$$\nu_{\mathcal{B}}'^i = \{(w, B) \mid w \in [\![A]\!]_{\mathcal{E}} \text{ and } B \in cl_{\mathsf{S4}_n}((\nu_{\mathcal{B}}^i)_w \cup \{A\})\},$$

where cl_{S4_n} is given as in Definition 1 with the addition of

- if $A \in cl^j_{S4_n}(T)$, then $\Box_i A \in cl^{j+1}_{S4_n}(T)$,
- if $\Box_i A \in cl^j_{S4_n}(T)$, then $A \in cl^{j+1}_{S4_n}(T)$.

Unfortunately, restricted $S4_n$-models are not well-defined. There exists an epistemic $S4_n$-model $\mathcal{E}$ and a formula A such that the restriction $\mathcal{E}|_A$ is not an epistemic $S4_n$-model. Consider $\mathcal{E} = (W, R, \nu, \nu_{\mathcal{B}})$ with

$$W = \{w\},$$
$$R = \{(w, w)\},$$
$$\nu = \{(w, p)\},$$
$$\nu_{\mathcal{B}} = \{(w, A) \mid A \in cl_{S4_n}(\{\Box p \to \Box q\})\}.$$

This is an epistemic $S4_n$-model. We can depict it as follows:

$$p, \neg q, \neg \Box p, \neg \Box q, \Box(\Box p \to \Box q)$$

Since p holds at the state w, the restriction of $\mathcal{E}$ to the formula p yields

$$\mathcal{E}|_p = (W, R, \nu, \nu'_{\mathcal{B}}) \text{ with } \nu'_{\mathcal{B}} = \{(w, A) \mid A \in cl_{S4_n}(\{\Box p \to \Box q, p\})\}.$$

Thus, by the closure conditions of cl_{S4_n}, we get $\Box p \in (\nu'_{\mathcal{B}})_w$, thus $\Box q \in (\nu'_{\mathcal{B}})_w$ and finally $q \in (\nu'_{\mathcal{B}})_w$. However, now we have the situation that $\mathcal{E}|_p, w \models \Box q$ but also $\mathcal{E}|_p, w \not\models q$. Since we also have $R(w, w)$, we find that condition (3) in the definition of an epistemic model is not satisfied.

5   Conclusion

We introduced public announcements for epistemic models and studied the corresponding belief dynamics. We showed that our model update operation satisfies the AGM postulate of minimal change. We also adapted hypertheories to support public announcements and established soundness and completeness.

In the case of knowledge change and truthful announcements, the situation gets more complicated. We presented an example showing that the standard approach cannot be used in a straightforward way to capture public announcements over epistemic $S4_n$-models. This remains a topic for future research.

Moreover, it will be interesting to see how other belief change operations can be implemented in the framework of epistemic models. The fact that minimal change is satisfied for public announcements is a strong hint that using epistemic models is a promising approach to dealing with belief change.

References

1. Alchourrón, C.E., Gärdenfors, P., Makinson, D.: On the logic of theory change: partial meet contraction and revision functions. J. Symb. Log. **50**(2), 510–530 (1985)
2. Artemov, S.N.: The logic of justification. RSL **1**(4), 477–513 (2008)
3. Artemov, S.N.: Knowing the model. ArXiv e-prints https://arxiv.org/abs/1610. 04955 (2016)
4. Artemov, S.N.: New foundations of epistemic logic. Talk given at OST 2018, Bern (2018). http://ost18.inf.unibe.ch/
5. Artemov, S.N.: Rebuilding epistemic logic. Talk given at Trends in Logic XVIII, Milan (2018). https://www.unicatt.it/meetings/trends-home
6. Artemov, S.N.: Revising epistemic logic. Talk given at LFCS 2018, Deerfield Beach (2018). http://lfcs.ws.gc.cuny.edu/lfcs-2018/
7. Brünnler, K., Flumini, D., Studer, T.: A logic of blockchain updates. In: Artemov, S., Nerode, A. (eds.) LFCS 2018. LNCS, vol. 10703, pp. 107–119. Springer, Cham (2018). https://doi.org/10.1007/978-3-319-72056-2_7
8. Bucheli, S., Kuznets, R., Renne, B., Sack, J., Studer, T.: Justified belief change. In: Arrazola, X., Ponte, M. (eds.) LogKCA-2010, pp. 135–155. University of the Basque Country Press (2010)
9. Bucheli, S., Kuznets, R., Studer, T.: Realizing public announcements by justifications. J. Comput. Syst. Sci. **80**(6), 1046–1066 (2014)
10. Gärdenfors, P.: Knowledge in Flux. The MIT Press, Cambridge (1988)
11. Gerbrandy, J., Groeneveld, W.: Reasoning about information change. J. Log. Lang. Inf. **6**(2), 147–169 (1997)
12. Kuznets, R., Studer, T.: Update as evidence: belief expansion. In: Artemov, S., Nerode, A. (eds.) LFCS 2013. LNCS, vol. 7734, pp. 266–279. Springer, Heidelberg (2013). https://doi.org/10.1007/978-3-642-35722-0_19
13. Kuznets, R., Studer, T.: Logics of Proofs and Justifications. College Publications. (in preparation)
14. Plaza, J.: Logics of public communications. Synthese **158**(2), 165–179 (2007). Reprinted from Emrich, M.L., et al. (eds.) Proceedings of the 4th International Symposium on Methodologies for Intelligent Systems (ISMIS 1989), pp. 201–216. Oak Ridge National Laboratory, ORNL/DSRD-24 (1989)
15. Renne, B.: Public communication in justification logic. J. Log. Comput. **21**(6), 1005–1034 (2011). Published online July 2010
16. Steiner, D.: Belief change functions for multi-agent systems. Ph.D. thesis, University of Bern (2009)
17. Steiner, D., Studer, T.: Total public announcements. In: Artemov, S.N., Nerode, A. (eds.) LFCS 2007. LNCS, vol. 4514, pp. 498–511. Springer, Heidelberg (2007). https://doi.org/10.1007/978-3-540-72734-7_35
18. van Ditmarsch, H., van der Hoek, W., Kooi, B.: Dynamic Epistemic Logic. Springer, Dordrecht (2008). https://doi.org/10.1007/978-1-4020-5839-4

Revisiting the
Generalized Łoś-Tarski Theorem

Abhisekh Sankaran[(⊠)]

Department of Computer Science and Technology,
University of Cambridge, Cambridge, UK
as2269@cam.ac.uk

Abstract. We present a new proof of the generalized Łoś-Tarski theorem ($\mathsf{GLT}(k)$) from [6], over arbitrary structures. Instead of using λ-saturation as in [6], we construct just the "required saturation" directly using ascending chains of structures. We also strengthen the failure of $\mathsf{GLT}(k)$ in the finite shown in [7], by strengthening the failure of the Łoś-Tarski theorem in this context. In particular, we prove that not just universal sentences, but for each fixed k, even Σ_2^0 sentences containing k existential quantifiers fail to capture hereditariness in the finite. We conclude with two problems as future directions, concerning the Łoś-Tarski theorem and $\mathsf{GLT}(k)$, both in the context of all finite structures.

Keywords: Łoś-Tarski theorem · k-hereditary · k-ary cover · Chain

1 Introduction

Preservation theorems are a class of results from classical model theory that provide syntactic characterizations of first order (FO) definable classes of arbitrary structures (structures that could be finite or infinite), that are closed under given model-theoretic operations. One of the earliest such results is the Łoś-Tarski theorem that states that a class of arbitrary structures defined by an FO sentence is hereditary (closed under substructures) if, and only if, it is definable by a universal sentence (an FO sentence that contains only universal quantifiers) [2]. The theorem in "dual" form characterizes extension closed FO definable classes of arbitrary structures in terms of existential sentences. The theorem extends to theories (sets of sentences) as well. The Łoś-Tarski theorem is historically important for classical model theory since its proof constituted the earliest applications of the FO Compactness theorem (a central result of model theory), and since it triggered off an extensive study of preservation theorems for various other model-theoretic operations (homomorphisms, unions of chains, direct products, etc.), also for logics beyond FO (such as infinitary logics) [3].

Recently [6], a generalization of the Łoś-Tarski theorem was proven by introducing and characterizing a new semantic property that generalizes hereditariness in a parameterized manner. We refer to this property, called *preservation under substructures modulo k-cruxes* in [6], as *k-hereditariness* in this paper.

© Springer-Verlag GmbH Germany, part of Springer Nature 2019
Md. A. Khan and A. Manuel (Eds.): ICLA 2019, LNCS 11600, pp. 76–88, 2019.
https://doi.org/10.1007/978-3-662-58771-3_8

A class of structures is said to be k-hereditary if every structure in the class contains a set of at most k elements, called a k-*crux* of the structure, such that all substructures (of the mentioned structure) *containing the k-crux* are also in the class. For instance, consider the class of arbitrary graphs that contain a dominating set of size at most k. (A dominating set in graph is a set S of vertices such that every vertex in the graph is either in S or adjacent to a vertex in S.) This class can be described by the FO sentence $\varphi := \exists x_1 \ldots \exists x_k \forall y (\bigvee_{i=1}^{i=k} ((y = x_i) \vee E(y, x_i)))$. In any model of φ, any witnesses to the existential quantifiers of φ form a dominating set, and any such set is a k-crux of the model; then φ defines a k-hereditary class. Observe that φ is an $\exists^k \forall^*$ sentence, i.e. a sentence in prenex normal form whose quantifier prefix is a string of k existential quantifiers followed by universal quantifiers[1]. By a similar reasoning as above, it can be shown that any $\exists^k \forall^*$ sentence defines a k-hereditary class. The authors of [6] proved that the converse is true as well, that any FO definable k-hereditary class of arbitrary structures is always definable by an $\exists^k \forall^*$ sentence, thus proving a *generalized Łoś-Tarski theorem*, that we denote GLT(k) (following [6]). Observe that the Łoś-Tarski theorem is a special case of GLT(k) when k is 0.

The proof of GLT(k) from [6] goes via first showing GLT(k) over a special class of structures called λ-*saturated* structures where λ is an infinite cardinal. These structures, intuitively speaking, realize many types (maximal consistent sets of formulae in a given number of free variables); in particular, such a structure $\mathfrak{A}$ realizes all the types that are realized in all structures elementarily equivalent to $\mathfrak{A}$, i.e. structures which satisfy the same FO sentences as $\mathfrak{A}$. Then using the fact that every structure has an elementarily equivalent extension that is λ-saturated for some λ, the truth of GLT(k) is "transferred" to all structures. To show GLT(k) over λ-saturated structures, a notion dual to k-hereditariness is introduced, called *preservation under k-ary covered extensions*, that we call k-*extension closure* in this paper. Given a structure $\mathfrak{A}$, define a set $\mathcal{R}$ of substructures of $\mathfrak{A}$ to be a k-*ary cover of* $\mathfrak{A}$ if every set of k elements of $\mathfrak{A}$ is contained in some structure of $\mathcal{R}$. We then say $\mathfrak{A}$ is a k-extension of $\mathcal{R}$. A class is k-*extension closed* if every k-extension of a set of structures of the class, is also in the class. One sees that a class is k-extension closed if, and only if, its complement is k-hereditary. Then GLT(k) is shown by proving its dual form that characterizes k-extension closure in terms of $\forall^k \exists^*$ sentences. The heart of this proof – Lemma 4.2 of [6] – shows that if Γ is the theory of the $\forall^k \exists^*$ implications of a sentence φ that defines a k-extension closed class, then every λ-saturated model of Γ has a k-ary cover consisting of the models of φ. It follows that the λ-saturated model then itself models φ, showing that φ and Γ are equivalent; then one application of the Compactness theorem shows φ to be equivalent to a single sentence of Γ.

The first result of this paper is motivated by the above proof of [6]. In particular, we give a new proof of GLT(k) that completely avoids using λ-saturated structures, by making the key observation that the full power of λ-saturation is hardly used in the proof of the mentioned Lemma 4.2 of [6]. The formulae

---

[1] See [5] for a variety of graph properties of interest in parameterized algorithms and finite model theory, that are k-hereditary and expressible as $\exists^k \forall^*$ sentences.

that play a central role in the proof are not arbitrary FO formulae, but are in fact formulae that have only one quantifier alternation at best. We therefore construct just the "required saturation" as is needed for our proof, by showing a "weaker" version of the mentioned Lemma 4.2, that states that for Γ and φ as above, every model of Γ has an elementarily equivalent extension that might not be λ-saturated for any λ, but still contains a k-ary cover consisting of models of φ; see (1) $\rightarrow$ (3) of Lemma 3 of this paper. Then showing (the dual form of) GLT(k) over the class of the mentioned elementary extensions is sufficient to transfer GLT(k) out to all structures. The aforementioned implication is in turn shown by defining in the natural way, the more general notion of a k-ary cover of a structure *in* a superstructure of it, and then using (transfinite) induction over the k-tuples of elements of a given model $\mathfrak{A}$ of Γ, to construct an elementary extension $\mathfrak{A}'$ of $\mathfrak{A}$ such that $\mathfrak{A}$ has a k-ary cover consisting of models of φ *in* $\mathfrak{A}'$; see (1) $\rightarrow$ (2) of Lemma 3. Applying this implication iteratively to the elementary extensions it gives, we get a chain of structures whose union is an elementary extension of $\mathfrak{A}$ that has a (self-contained) k-ary cover of models of φ; see (2) $\rightarrow$ (3) of Lemma 3. Our new proof is therefore much "from the scratch" as opposed to the proof in [6] which uses established notions of model theory.

The second result of this paper is a strengthening of the failure of the Łoś-Tarski theorem in the finite. In the research programme of investigating classical model theoretic results over all finite structures, which is amongst the major themes of finite model theory, one of the first results identified to fail was the Łoś-Tarski theorem [1]. (In fact, Tait had already shown this failure in 1959 [8].) Specifically, there is an FO sentence that is hereditary over the class of all finite structures, but that is not equivalent over this class to any universal sentence. In the spirit of [1], one can ask if there is a different syntactic characterization of hereditariness in the finite, or even a syntactic (proper) subfragment of FO that is expressive enough to contain (up to equivalence) all FO sentences that are hereditary when restricted to the finite. We show in Theorem 4 that for no fixed k, is the class of $\exists^k \forall^*$ sentences such a subfragment. Specifically, for each k, we construct a sentence φ_k whose finite models form a hereditary class, and yet φ_k is not equivalent over all finite structures to any $\exists^k \forall^*$ sentence.

This result also strengthens the failure of GLT(k) in the finite as shown in [7]. For every k, the authors of [7] present a counterexample to GLT(k) (over all finite structures) that is k-hereditary but not $(k-1)$-hereditary. The sentence φ_k given by our Theorem 4 provides a counterexample to GLT(k), that is l-hereditary for all l. The proof of Theorem 4 proceeds by constructing for each $\exists^k \forall^n$ sentence γ, a model $\mathfrak{A}$ of φ_k such that, if $\mathfrak{A}$ models γ, then one can "edit" $\mathfrak{A}$ depending on the witnesses (in $\mathfrak{A}$) of the existential quantifiers of γ, to obtain a non-model $\mathfrak{B}$ of φ_k, that also models γ. This proof can be seen as being based on an Ehrenfeucht-Fraïssé game in which the Spoiler picks k elements from $\mathfrak{A}$ in the first move, in response to which the Duplicator first constructs the structure $\mathfrak{B}$ and then picks k elements from it, and in the next move, the Spoiler picks n elements from $\mathfrak{B}$ to which the Duplicator responds by picking n elements from $\mathfrak{A}$. Interestingly, the sentence φ_k itself turns out to be equivalent to an $\exists^{k+1} \forall^*$ sentence.

Paper Organization: In Sect. 2, we introduce terminology and notation used in the paper, and formally state GLT(k). In Sect. 3, we present our new proof of GLT(k) and in Sect. 4, we prove the strengthened failure of the Łoś-Tarski theorem in the finite. We conclude in Sect. 5 by presenting two problems for future investigation, one concerning the Łoś-Tarski theorem and the other concerning GLT(k), both in the context of all finite structures.

2 Preliminaries and Background

We assume the reader is familiar with standard notation and terminology used in the syntax and semantics of FO [2]. A *vocabulary* τ is a set of predicate, function and constant symbols. In this paper, we will always be concerned with arbitrary *finite* vocabularies, unless explicitly stated otherwise. We denote by FO(τ) the set of all FO formulae over vocabulary τ. A sequence $(x_1,\ldots,x_k)$ of variables is denoted by $\bar{x}$. A formula ψ whose free variables are among $\bar{x}$, is denoted by $\psi(\bar{x})$. A formula with no free variables is called a *sentence*. An FO(τ) *theory* is a set of FO(τ) sentences. An FO(τ) theory *with free variables* $\bar{x}$ is a set of FO(τ) formulae, all of whose free variables are among $\bar{x}$. When τ is clear from context, we call an FO(τ) theory, a theory simply. We denote by $\mathbb{N}$, the natural numbers *including zero*. We abbreviate a block of quantifiers of the form $Qx_1 \ldots Qx_k$ by $Q^k\bar{x}$ or $Q\bar{x}$ (depending on what is better suited for understanding), where $Q \in \{\forall, \exists\}$ and $k \in \mathbb{N}$. By Q^*, we mean a block of k Q quantifiers, for some $k \in \mathbb{N}$. For every non-zero $n \in \mathbb{N}$, we denote by Σ_n^0 and Π_n^0, the classes of all FO formulae in prenex normal form, whose quantifier prefixes begin with $\exists$ and $\forall$ respectively, and consist of $n-1$ alternations of quantifiers. We call Σ_1^0 formulae *existential* and Π_1^0 formulae *universal*. We call Σ_2^0 formulae with k existential quantifiers $\exists^k\forall^*$ *formulae*, and Π_2^0 formulae with k universal quantifiers $\forall^k\exists^*$ *formulae*.

We use standard notions of τ-structures (denoted $\mathfrak{A}, \mathfrak{B}$ etc.; we refer to these simply as structures when τ is clear from context), substructures (denoted $\mathfrak{A} \subseteq \mathfrak{B}$), extensions, isomorphisms (denoted $\mathfrak{A} \cong \mathfrak{B}$), isomorphic embeddings (denoted $\mathfrak{A} \hookrightarrow \mathfrak{B}$), elementary equivalence (denoted $\mathfrak{A} \equiv \mathfrak{B}$), elementary substructures (denoted $\mathfrak{A} \preceq \mathfrak{B}$) and elementary extensions, as defined in [2]. Given a structure $\mathfrak{A}$, we use $U_\mathfrak{A}$ to denote the universe of $\mathfrak{A}$, and $|\mathfrak{A}|$ to denote the size (or *power*) of $\mathfrak{A}$ which is the cardinality of $U_\mathfrak{A}$. For an FO sentence φ and an FO theory T, we denote by $\mathfrak{A} \models \varphi$ and $\mathfrak{A} \models T$ that $\mathfrak{A}$ is a model of φ and T respectively. In Sect. 3 of the paper, we consider structures that could be finite or infinite, whereas in Sect. 4 we restrict ourselves to only finite structures.

Finally, we use standard abbreviations of English phrases that commonly appear in mathematical literature. Specifically, 'w.l.o.g' stands for 'without loss of generality', 'iff' stands for 'if and only if', and 'resp.' stands for 'respectively'.

2.1 The Generalized Łoś-Tarski Theorem

We recall the notions of preservation under substructures modulo k-cruxes, k-ary covered extensions and preservation under k-ary covered extensions introduced in [6], that we resp. call in this paper k-hereditariness, k-extensions and k-extension closure. These notions for $k = 0$ correspond exactly to hereditariness, extensions and extension closure resp.

Definition 1 (Definition 3.1 [6]).

a. *Let $\mathcal{U}$ be a class of arbitrary structures and $k \in \mathbb{N}$. A subclass $\mathcal{S}$ of $\mathcal{U}$ is said to be k-hereditary over $\mathcal{U}$, if for every structure $\mathfrak{A}$ of $\mathcal{S}$, there is a set $C \subseteq \mathsf{U}_{\mathfrak{A}}$ of size $\leq k$ such that if $\mathfrak{B} \subseteq \mathfrak{A}$, $\mathfrak{B}$ contains C and $\mathfrak{B} \in \mathcal{U}$, then $\mathfrak{B} \in \mathcal{S}$. The set C is called a k-crux of $\mathfrak{A}$ w.r.t. $\mathcal{S}$ over $\mathcal{U}$.*

b. *Given theories T and V, we say T is k-hereditary modulo V, if the class of models of $T \cup V$ is k-hereditary over the class of models of V. A sentence φ is k-hereditary modulo V if the theory $\{\varphi\}$ is k-hereditary modulo V.*

Definition 2 (Definitions 3.5 and 3.8 [6]).

a. *Given a structure $\mathfrak{A}$, a non-empty collection $\mathcal{R}$ of substructures of $\mathfrak{A}$ is said to be a k-ary cover of $\mathfrak{A}$ if for every set $C \subseteq \mathsf{U}_{\mathfrak{A}}$ of size $\leq k$, there is a structure in $\mathcal{R}$ that contains C. We call $\mathfrak{A}$ a k-extension of $\mathcal{R}$.*

b. *For a class $\mathcal{U}$ of arbitrary structures and $k \in \mathbb{N}$, a subclass $\mathcal{S}$ of $\mathcal{U}$ is said to be k-extension closed over $\mathcal{U}$ if for every collection $\mathcal{R}$ of structures of $\mathcal{S}$, if $\mathfrak{A}$ is a k-extension of $\mathcal{R}$ and $\mathfrak{A} \in \mathcal{U}$, then $\mathfrak{A} \in \mathcal{S}$.*

c. *Given theories V and T, we say T is k-extension closed modulo V if the class of models of $T \cup V$ is k-extension closed over the class of models of V. A sentence φ is k-extension closed modulo V if the theory $\{\varphi\}$ is k-extension closed modulo V.*

We extend the above definitions slightly to formulae and theories with free variables. Given a vocabulary τ, let τ_n denote the vocabulary obtained by expanding τ with n fresh and distinct constant symbols $c_1, \ldots, c_n$. For a given FO(τ) theory $T(x_1, \ldots, x_n)$, let T' denote the FO(τ_n) theory (without free variables) obtained by substituting c_i for x_i in $T(x_1, \ldots, x_n)$ for each $i \in \{1, \ldots, n\}$. Then we say $T(x_1, \ldots, x_n)$ is k-hereditary, resp. k-extension closed, modulo an FO(τ) theory V (without free variables) if T' is k-hereditary, resp. k-extension closed, modulo V where V is seen as an FO(τ_n) theory. A formula $\varphi(x_1, \ldots, x_n)$ is k-hereditary, resp. k-extension closed, modulo V if the theory $\{\varphi(x_1, \ldots, x_n)\}$ is k-hereditary, resp. k-extension closed, modulo V. The following lemma establishes the duality of the introduced preservation properties.

Lemma 1 (Lemma 3.9 [6]). *Let $\mathcal{U}$ be a class of arbitrary structures, $\mathcal{S}$ be a subclass of $\mathcal{U}$ and $\overline{\mathcal{S}}$ be the complement of $\mathcal{S}$ in $\mathcal{U}$. Then $\mathcal{S}$ is k-hereditary over $\mathcal{U}$ iff $\overline{\mathcal{S}}$ is k-extension closed over $\mathcal{U}$, for each $k \in \mathbb{N}$. In particular, if $\mathcal{U}$ is defined by a theory V, then a formula $\varphi(\bar{x})$ is k-hereditary modulo V iff $\neg\varphi(\bar{x})$ is k-extension closed modulo V.*

We now recall $\mathsf{GLT}(k)$ as proved in [6]. This theorem gives syntactic characterizations of FO definable k-hereditary and k-extension closed classes of structures. Observe that the case of $k = 0$ gives exactly the Łoś-Tarski theorem. Below, for

FO(τ) formulae $\varphi(\bar{x})$ and $\psi(\bar{x})$ where $\bar{x} = (x_1, \ldots, x_n)$, we say $\varphi(\bar{x})$ is equivalent to $\psi(\bar{x})$ *modulo* V if for every τ-structure $\mathfrak{A}$ and every n-tuple $\bar{a}$ from $\mathfrak{A}$, we have $(\mathfrak{A}, \bar{a})$ is a model of $\{\varphi(\bar{x})\} \cup V$ iff it is a model of $\{\psi(\bar{x})\} \cup V$.

Theorem 1 (Generalized Łoś-Tarski theorem: GLT(k); Corollaries 4.4 and 4.6 [6]). *Let $\varphi(\bar{x})$ and V be a given formula and theory respectively, and $k \in \mathbb{N}$. Then the following are true:*

1. *The formula $\varphi(\bar{x})$ is k-hereditary modulo V iff it is equivalent modulo V to an $\exists^k \forall^*$ formula whose free variables are among $\bar{x}$.*
2. *The formula $\varphi(\bar{x})$ is k-extension closed modulo V iff it is equivalent modulo V to a $\forall^k \exists^*$ formula whose free variables are among $\bar{x}$.*

3 A New Proof of GLT(k)

We give a new proof to a more general result than Theorem 1, from [6]. This result is a generalization of the "extensional" version of GLT(k) to theories. We extend in the natural way the aforestated notion of equivalence modulo a theory, of formulae, to theories with free variables.

Theorem 2 (Theorem 4.1 [6]). *A theory $T(\bar{x})$ is k-extension closed modulo a theory V if, and only if, $T(\bar{x})$ is equivalent modulo V to a theory (consisting) of $\forall^k \exists^*$ formulae all of whose free variables are among $\bar{x}$.*

Using the above result, Theorem 1 can be proved as below.

Proof (of Theorem 1). We prove part (2) of Theorem 1. Part (1) of Theorem 1 easily follows from part (2) and Lemma 1.

The 'If' direction is straightforward. Let $\varphi(\bar{x})$ be equivalent modulo V to the $\forall^k \exists^*$ formula $\psi(\bar{x})$. Then the theory $\{\varphi(\bar{x})\}$ is equivalent modulo V to the theory $\{\psi(\bar{x})\}$. Then $\{\varphi(\bar{x})\}$, and hence $\varphi(\bar{x})$, is k-extension closed modulo V by Theorem 2. For the 'Only if' direction, let $\varphi(\bar{x})$ be k-extension closed modulo V; then so is the theory $\{\varphi(\bar{x})\}$. By Theorem 2, $\{\varphi(\bar{x})\}$ is equivalent to a theory $Z(\bar{x})$ of $\forall^k \exists^*$ formulae whose free variables are among $\bar{x}$. By Compactness theorem, $\{\varphi(\bar{x})\}$ is equivalent modulo V to a finite subset $Y(\bar{x})$ of $Z(\bar{x})$. Then $\varphi(\bar{x})$ is equivalent modulo V to the conjunction of the formulae of $Y(\bar{x})$. Since any conjunction of $\forall^k \exists^*$ formulae is equivalent (modulo any theory) to a single $\forall^k \exists^*$ formula, the result follows. $\square$

Towards Theorem 2, we first recall some important notions and results from the classical model theory literature [2] that are needed for our proof.

Lemma 2 (Corollary 5.4.2, Chap. 5 [2]). *Let $\mathfrak{A}$ and $\mathfrak{B}$ be structures such that every existential sentence that is true in $\mathfrak{B}$ is true in $\mathfrak{A}$. Then $\mathfrak{B}$ is isomorphically embeddable in an elementary extension of $\mathfrak{A}$.*

Given a cardinal λ, an *ascending chain*, or simply a chain, $(\mathfrak{A}_\eta)_{\eta<\lambda}$ of structures is a sequence $\mathfrak{A}_0, \mathfrak{A}_1, \ldots$ of structures such that $\mathfrak{A}_0 \subseteq \mathfrak{A}_1 \subseteq \ldots$. The *union* of this chain is a structure $\mathfrak{A}$ defined as follows: (i) $U_{\mathfrak{A}} = \bigcup_{\eta<\lambda} U_{\mathfrak{A}_\eta}$, (ii) $c^{\mathfrak{A}} = c^{\mathfrak{A}_\eta}$ for every constant symbol $c \in \tau$ and every $\eta < \lambda$, (iii) $R^{\mathfrak{A}} = \bigcup_{\eta<\lambda} R^{\mathfrak{A}_\eta}$ for every relation symbol $R \in \tau$, and (iv) $f^{\mathfrak{A}} = \bigcup_{\eta<\lambda} f^{\mathfrak{A}_\eta}$ for every function symbol $f \in \tau$ (here, in taking the union of functions, we view an n-ary function as its corresponding $(n+1)$-ary relation). Observe that $\mathfrak{A}$ is well-defined. We denote $\mathfrak{A}$ as $\bigcup_{\eta<\lambda} \mathfrak{A}_\eta$. If it is additionally the case that $\mathfrak{A}_0 \preceq \mathfrak{A}_1 \preceq \ldots$ above, then we say $(\mathfrak{A}_\eta)_{\eta<\lambda}$ is an *elementary chain*. We now have the following result.

Theorem 3 (Tarski-Vaught elementary chain theorem, Theorem 3.1.9, Chap. 3 [2]). *Let* $(\mathfrak{A}_\eta)_{\eta<\lambda}$ *be an elementary chain of structures. Then* $\bigcup_{\eta<\lambda} \mathfrak{A}_\eta$ *is an elementary extension of* $\mathfrak{A}_\eta$ *for each* $\eta < \lambda$.

The key element of our proof of Theorem 2 is the notion of a k-ary cover of a structure $\mathfrak{A}$ *in an extension of* $\mathfrak{A}$. Below is the definition. Observe that this notion generalizes the notion of k-ary cover seen in Definition 2 – the latter corresponds to the notion in Definition 3, with $\mathfrak{A}^+$ being the same as $\mathfrak{A}$.

Definition 3. *Let* $\mathfrak{A}$ *be a structure and* $\mathfrak{A}^+$ *be an extension of* $\mathfrak{A}$. *A non-empty collection* $\mathcal{R}$ *of substructures of* $\mathfrak{A}^+$ *is said to be a* k-ary *cover of* $\mathfrak{A}$ *in* $\mathfrak{A}^+$ *if for every* k-tuple $\bar{a}$ *of elements of* $\mathfrak{A}$, *there exists a structure in* $\mathcal{R}$ *that contains (the elements of)* $\bar{a}$.

The following lemma is at the heart of our proof. It (along with its application in proving Theorem 2) shows why "full" λ-saturation as is used in a similar result (Lemma 4.2) in [6], is not needed for Theorem 2. Below, a *consistent* theory is one that has a model, and Γ is the set $\{\varphi \mid (V \cup T) \to \varphi \text{ where } \varphi \text{ is a } \forall^k \exists^* \text{ sentence}\}$.

Lemma 3. *Let* V *and* T *be consistent theories and* $k \in \mathbb{N}$. *Let* Γ *be the set of* $\forall^k \exists^*$ *consequences of* T *modulo* V. *Then for every model* $\mathfrak{A}$ *of* V, *the following are equivalent:*

1. $\mathfrak{A}$ *is a model of* $V \cup \Gamma$.
2. $\mathfrak{A}$ *is a model of* $V \cup \Gamma$, *and there exists an elementary extension* $\mathfrak{A}^+$ *of* $\mathfrak{A}$ *and a* k-ary *cover* $\mathcal{R}$ *of* $\mathfrak{A}$ *in* $\mathfrak{A}^+$ *such that* $\mathfrak{B} \models (V \cup T)$ *for every* $\mathfrak{B} \in \mathcal{R}$.
3. *There exists an elementary extension* $\mathfrak{A}^+$ *of* $\mathfrak{A}$ *and a* k-ary *cover* $\mathcal{R}$ *of* $\mathfrak{A}^+$ *(in* $\mathfrak{A}^+$*) such that* $\mathfrak{B} \models (V \cup T)$ *for every* $\mathfrak{B} \in \mathcal{R}$.

Using the above lemma, Theorem 2 can be proved as follows.

Proof (of Theorem 2). We prove the theorem for theories without free variables; the proof for theories with free variables follows from definitions.

If: Suppose T is equivalent modulo V to a theory Z of $\forall^k \exists^*$ sentences. Let $\mathfrak{A} \models V$ and let $\mathcal{R}$ be a k-ary cover of $\mathfrak{A}$ consisting of models of $V \cup T$. We show that $\mathfrak{A} \models T$. Consider a sentence $\varphi := \forall^k \bar{x} \psi(\bar{x}) \in Z$ where $\psi(\bar{x})$ is an existential formula. Let $\bar{a}$ be a k-tuple from $\mathfrak{A}$. Since $\mathcal{R}$ is a k-ary cover of $\mathfrak{A}$, there exists $\mathfrak{B} \in \mathcal{R}$ such that $\mathfrak{B} \subseteq \mathfrak{A}$ and $\mathfrak{B}$ contains $\bar{a}$. Since $\mathfrak{B} \models V \cup T$, we have $\mathfrak{B} \models Z$ (since Z and T are equivalent modulo V); then $\mathfrak{B} \models \varphi$ and hence $(\mathfrak{B}, \bar{a}) \models \psi(\bar{x})$. Since existential formulae are preserved under extensions by Łoś-Tarski theorem, we have $(\mathfrak{A}, \bar{a}) \models \psi(\bar{x})$. Since $\bar{a}$ is an arbitrary k-tuple of $\mathfrak{A}$, we have $\mathfrak{A} \models \varphi$. Finally, since φ is an arbitrary sentence of Z, we have $\mathfrak{A} \models Z$, and hence $\mathfrak{A} \models T$.

Only if: Conversely, suppose T is k-extension closed modulo V. If $V \cup T$ is unsatisfiable, we are trivially done. Else, let Γ be the set of $\forall^k \exists^*$ consequences of T modulo V. Then $(V \cup T) \rightarrow (V \cup \Gamma)$. Conversely, suppose $\mathfrak{A} \models (V \cup \Gamma)$. By Lemma 3, there exists an elementary extension $\mathfrak{A}^+$ of $\mathfrak{A}$ (hence $\mathfrak{A}^+ \models V$) for which there is a k-ary cover consisting of models of $V \cup T$. Then $\mathfrak{A}^+ \models T$ since T is k-extension closed modulo V, whereby $\mathfrak{A} \models T$. In other words, $(V \cup \Gamma) \rightarrow (V \cup T)$, so that T is equivalent to Γ modulo V. Then Γ is the desired $\forall^k \exists^*$ theory. $\qquad\square$

Towards the proof of Lemma 3, we would require an auxiliary lemma that we state and prove below.

Lemma 4. *Let V, T and Γ be as in the statement of Lemma 3, and suppose $\mathfrak{A} \models (V \cup \Gamma)$. Given an elementary extension $\mathfrak{A}'$ of $\mathfrak{A}$ and a k-tuple $\bar{a}$ of $\mathfrak{A}$, there exist an elementary extension $\mathfrak{A}''$ of $\mathfrak{A}'$ and a substructure $\mathfrak{B}$ of $\mathfrak{A}''$, such that (i) $\mathfrak{B}$ contains $\bar{a}$ and (ii) $\mathfrak{B} \models (V \cup T)$.*

Proof. Let $\mathsf{tp}_{\Pi,\mathfrak{A},\bar{a}}(\bar{x})$ denote the Π_1^0-type of $\bar{a}$ in $\mathfrak{A}$, that is, the set of all Π_1^0 formulae that are true of $\bar{a}$ in $\mathfrak{A}$ (so $|\bar{x}| = |\bar{a}|$). Let $Z(\bar{x})$ be the theory given by $Z(\bar{x}) := V \cup T \cup \mathsf{tp}_{\Pi,\mathfrak{A},\bar{a}}(\bar{x})$. We show below that $Z(\bar{x})$ is satisfiable. Assuming this, it follows that if $(\mathfrak{D}, \bar{d}) \models Z(\bar{x})$, then every existential sentence that is true in $(\mathfrak{D}, \bar{d})$ is also true in $(\mathfrak{A}, \bar{a})$, and hence in $(\mathfrak{A}', \bar{a})$. Then by Lemma 2, there is an isomorphic embedding f of $(\mathfrak{D}, \bar{d})$ in an elementary extension $(\mathfrak{A}'', \bar{a})$ of $(\mathfrak{A}', \bar{a})$. If the vocabulary of $\mathfrak{A}$ is τ, then taking $\mathfrak{B}$ to be the τ-reduct of the image of $(\mathfrak{D}, \bar{d})$ under f, we see that $\mathfrak{B}$ and $\mathfrak{A}''$ are as desired.

We show $Z(\bar{x})$ is satisfiable by contradiction. Suppose $Z(\bar{x})$ is inconsistent; then by Compactness theorem, there is a finite subset of $Z(\bar{x})$ that is inconsistent. Since $\mathsf{tp}_{\Pi,\mathfrak{A},\bar{a}}(\bar{x})$ is closed under finite conjunctions and since each of $\mathsf{tp}_{\Pi,\mathfrak{A},\bar{a}}(\bar{x})$, V and T is consistent, there exists $\psi(\bar{x}) \in \mathsf{tp}_{\Pi,\mathfrak{A},\bar{a}}(\bar{x})$ such that $V \cup T \cup \{\psi(\bar{x})\}$ is inconsistent. In other words, $(V \cup T) \rightarrow \neg\psi(\bar{x})$. Since $V \cup T$ has no free variables, we have $(V \cup T) \rightarrow \varphi$, where $\varphi := \forall^k \bar{x}\, \neg\psi(\bar{x})$. Observe that $\neg\psi(\bar{x})$ is equivalent to an existential formula; then φ is equivalent to a sentence in Γ, and hence $\mathfrak{A} \models \varphi$. Then $(\mathfrak{A}, \bar{a}) \models \neg\psi(\bar{x})$, contradicting our inference that $\psi(\bar{x}) \in \mathsf{tp}_{\Pi,\mathfrak{A},\bar{a}}(\bar{x})$. $\qquad\square$

Proof (of Lemma 3). $\underline{(3) \rightarrow (1)}$: This implication is established along similar lines as the 'If' direction of Theorem 2. We show that $\mathfrak{A}^+$ models φ for each sentence φ of Γ; then $\mathfrak{A}$ models φ as well since $\mathfrak{A} \preceq \mathfrak{A}^+$, and hence $\mathfrak{A} \models \Gamma$.

$\underline{(1) \rightarrow (2)}$: We have two cases here depending on whether $\mathfrak{A}$ is finite or infinite.

(1) $\mathfrak{A}$ is finite: Given a k-tuple $\bar{a}$ of $\mathfrak{A}$, by Lemma 4 there exists an elementary extension $\mathfrak{A}''$ of $\mathfrak{A}$ and a substructure $\mathfrak{B}_{\bar{a}}$ of $\mathfrak{A}''$ such that (i) $\mathfrak{B}_{\bar{a}}$ contains $\bar{a}$ and (ii) $\mathfrak{B}_{\bar{a}} \models (V \cup T)$. Since $\mathfrak{A}$ is finite, and since elementary equivalence is the

same as isomorphism over finite structures [2], we have $\mathfrak{A}'' = \mathfrak{A}$. Then taking $\mathfrak{A}^+ = \mathfrak{A}$ and $\mathcal{R} = \{\mathfrak{B}_{\bar{a}} \mid \bar{a} \in (U_{\mathfrak{A}})^k\}$, we see that $\mathfrak{A}^+$ and $\mathcal{R}$ are respectively indeed the desired elementary extension of $\mathfrak{A}$ and k-ary cover of $\mathfrak{A}$ in $\mathfrak{A}^+$.

(2) $\mathfrak{A}$ is infinite: The proof for this case is along the lines of the proof of the characterization of Π_2^0 sentences in terms of the property of preservation under unions of chains (see proof of Theorem 3.2.3 in Chap. 3 of [2]). Let λ be the successor cardinal of $|\mathfrak{A}|$ and $(\bar{a}_\kappa)_{\kappa < \lambda}$ be an enumeration of the k-tuples of $\mathfrak{A}$. For $\eta \leq \lambda$, given sequences $(\mathfrak{E}_\kappa)_{\kappa < \eta}$ and $(\mathfrak{F}_\kappa)_{\kappa < \eta}$ of structures, we say that $\mathcal{P}((\mathfrak{E}_\kappa)_{\kappa < \eta}, (\mathfrak{F}_\kappa)_{\kappa < \eta})$ is true iff $(\mathfrak{E}_\kappa)_{\kappa < \eta}$ is an elementary chain and $\mathfrak{A} \preceq \mathfrak{E}_0$, and for each $\kappa < \eta$, we have (i) $\mathfrak{F}_\kappa \subseteq \mathfrak{E}_\kappa$, (ii) $\mathfrak{F}_\kappa$ contains $\bar{a}_\kappa$, and (iii) $\mathfrak{F}_\kappa \models (V \cup T)$. We show below the existence of sequences $(\mathfrak{A}_\kappa)_{\kappa < \lambda}$ and $(\mathfrak{B}_\kappa)_{\kappa < \lambda}$ of structures such that $\mathcal{P}((\mathfrak{A}_\kappa)_{\kappa < \lambda}, (\mathfrak{B}_\kappa)_{\kappa < \lambda})$ is true. Then taking $\mathfrak{A}^+ = \bigcup_{\kappa < \lambda} \mathfrak{A}_\kappa$ and $\mathcal{R} = \{\mathfrak{B}_\kappa \mid \kappa < \lambda\}$, we see by Theorem 3 that $\mathfrak{A}^+$ and $\mathcal{R}$ are respectively indeed the elementary extension of $\mathfrak{A}$ and k-ary cover of $\mathfrak{A}$ in $\mathfrak{A}^+$ as desired.

We construct the sequences $(\mathfrak{A}_\kappa)_{\kappa < \lambda}$ and $(\mathfrak{B}_\kappa)_{\kappa < \lambda}$ by constructing for each positive ordinal $\eta < \lambda$, the partial (initial) sequences $(\mathfrak{A}_\kappa)_{\kappa < \eta}$ and $(\mathfrak{B}_\kappa)_{\kappa < \eta}$ and showing that $\mathcal{P}((\mathfrak{A}_\kappa)_{\kappa < \eta}, (\mathfrak{B}_\kappa)_{\kappa < \eta})$ is true. We do this by (transfinite) induction on η. For the base case of $\eta = 1$, we see by Lemma 4 that if $\mathfrak{A}' = \mathfrak{A}$, then there exists an elementary extension $\mathfrak{A}''$ of $\mathfrak{A}$ and a substructure $\mathfrak{B}$ of $\mathfrak{A}''$ such that (i) $\mathfrak{B}$ contains $\bar{a}_0$ and (ii) $\mathfrak{B} \models (V \cup T)$. Then taking $\mathfrak{A}_0 = \mathfrak{A}''$ and $\mathfrak{B}_0 = \mathfrak{B}$, we see that $\mathcal{P}((\mathfrak{A}_0), (\mathfrak{B}_0))$ is true. As the induction hypothesis, assume that we have constructed sequences $(\mathfrak{A}_\kappa)_{\kappa < \eta}$ and $(\mathfrak{B}_\kappa)_{\kappa < \eta}$ such that $\mathcal{P}((\mathfrak{A}_\kappa)_{\kappa < \eta}, (\mathfrak{B}_\kappa)_{\kappa < \eta})$ is true. Then by Theorem 3, the structure $\mathfrak{A}' = \bigcup_{\kappa < \eta} \mathfrak{A}_\kappa$ is such that $\mathfrak{A}_\kappa \preceq \mathfrak{A}'$ for each $\kappa < \eta$. Then for the tuple $\bar{a}_\eta$ of $\mathfrak{A}$, by Lemma 4, there exists an elementary extension $\mathfrak{C}$ of $\mathfrak{A}'$ and a substructure $\mathfrak{D}$ of $\mathfrak{C}$ such that (i) $\mathfrak{D}$ contains $\bar{a}_\eta$ and (ii) $\mathfrak{D} \models (V \cup T)$. Then taking $\mathfrak{A}_\eta = \mathfrak{C}$ and $\mathfrak{B}_\eta = \mathfrak{D}$, and letting μ be the successor ordinal of η, we see that $\mathcal{P}((\mathfrak{A}_\kappa)_{\kappa < \mu}, (\mathfrak{B}_\kappa)_{\kappa < \mu})$ is true, completing the induction.

(2) $\rightarrow$ (3): Any elementary extension of $\mathfrak{A}$ models $V \cup \Gamma$. Then by applying the implication (1) $\rightarrow$ (2) iteratively to the elementary extensions that (2) produces, we get a sequence $(\mathfrak{A}_i)_{i \geq 0}$ of elementary extensions of $\mathfrak{A}$, and a sequence $(\mathcal{R}_i)_{i \geq 0}$ of collections of structures with the following properties:

1. $(\mathfrak{A}_i)_{i \geq 0}$ is an elementary chain such that $\mathfrak{A}_0 = \mathfrak{A}$ (whereby $\mathfrak{A}_i \models V$ for $i \geq 0$).
2. For each $i \geq 0$, $\mathcal{R}_i$ is a k-ary cover of $\mathfrak{A}_i$ in $\mathfrak{A}_{i+1}$ such that $\mathfrak{B} \models (V \cup T)$ for every $\mathfrak{B} \in \mathcal{R}_i$.

Consider the structure $\mathfrak{A}^+ = \bigcup_{i \geq 0} \mathfrak{A}_i$. By Theorem 3, we have $\mathfrak{A}_i \preceq \mathfrak{A}^+$ for each $i \geq 0$, and (hence) that $\mathfrak{A}^+ \models V$. Consider any k-tuple $\bar{a}$ of $\mathfrak{A}^+$; there exists $j \geq 0$ such $\bar{a}$ is contained in $\mathfrak{A}_j$. Then there exists a structure $\mathfrak{B}_{\bar{a}} \in \mathcal{R}_j$ such that (i) $\mathfrak{B}_{\bar{a}}$ contains $\bar{a}$ and (ii) $\mathfrak{B}_{\bar{a}} \models (V \cup T)$. Since $\mathfrak{B}_{\bar{a}} \in \mathcal{R}_j$, we have $\mathfrak{B}_{\bar{a}} \subseteq \mathfrak{A}_{j+1}$ and since $\mathfrak{A}_{j+1} \preceq \mathfrak{A}^+$, we have $\mathfrak{B}_{\bar{a}} \subseteq \mathfrak{A}^+$. Then $\mathcal{R} = \{\mathfrak{B}_{\bar{a}} \mid \bar{a} \text{ is a } k\text{-tuple from } \mathfrak{A}^+\}$ is the desired k-ary cover of $\mathfrak{A}^+$ such that $\mathfrak{B} \models (V \cup T)$ for each $\mathfrak{B} \in \mathcal{R}$. $\square$

4 A Stronger Failure of Łoś-Tarski Theorem in the Finite

In this section, we strengthen the known failure of the Łoś-Tarski theorem in the finite [8]. As a consequence, we get a strengthening of the failure of $\mathsf{GLT}(k)$ in the finite for each k, over the one proved in [7]. Below, by φ_k is $(k\text{-})$hereditary over S we mean that the class of finite models of φ_k is $(k\text{-})$hereditary over S.

Theorem 4. *There exists a vocabulary τ such that if S is the class of all finite τ-structures, then for each $k \geq 0$, there exists an $FO(\tau)$ sentence φ_k that is hereditary over S, but that is not equivalent over S, to any $\exists^k\forall^*$ sentence. It follows that there is a sentence that is k-hereditary over S (φ_k being one such sentence) but that is not equivalent over S to any $\exists^k\forall^*$ sentence.*

Proof. The second part of the theorem follows from the first part since a sentence that is hereditary over S is also k-hereditary over S for each $k \geq 0$. We now prove the first part of the theorem. Consider the vocabulary $\tau = \{\leq, S, P, c, d\}$ where $\leq$ and S are binary relation symbols, P is a unary relation symbol, and c and d are constant symbols. The sentence φ_k is constructed along the lines of the counterxample to the Łoś-Tarski theorem in the finite as given in [1].

$$\varphi_k := (\xi_1 \wedge \xi_2 \wedge \xi_3) \wedge \neg(\xi_4 \wedge \xi_5)$$
$$\xi_1 := \text{``}\leq \text{ is a linear order''}$$
$$\xi_2 := \text{``}c \text{ is minimum under } \leq \text{ and } d \text{ is maximum under } \leq \text{''}$$
$$\xi_3 := \forall x \forall y \, S(x,y) \rightarrow \text{``}y \text{ is the successor of } x \text{ under } \leq \text{''}$$
$$\xi_4 := \forall x \, (x \neq d) \rightarrow \exists y S(x,y)$$
$$\xi_5 := \text{``There exist at most } k \text{ elements in (the set interpreting) } P\text{''}$$

Each of ξ_1, ξ_2, ξ_3 and ξ_5 can be expressed using a universal sentence. In particular, ξ_1 and ξ_3 can be expressed using a $\forall^3$ sentence each, ξ_2 using a $\forall$ sentence, and ξ_5 using a $\forall^{k+1}$ sentence. Then φ_k is equivalent to an $\exists^{k+1}\forall^3$ sentence.

We first show that φ_k is hereditary over S, by showing that $\psi_k := \neg\varphi_k$ is extension closed over S. Let $\mathfrak{A} \models \psi_k$ and $\mathfrak{A} \subseteq \mathfrak{B}$. If $\alpha := (\xi_1 \wedge \xi_2 \wedge \xi_3)$ is such that $\mathfrak{A} \models \neg\alpha$, then since $\neg\alpha$ is equivalent to an existential sentence, we have $\mathfrak{B} \models \neg\alpha$; then $\mathfrak{B} \models \psi_k$. Else, $\mathfrak{A} \models \alpha \wedge \xi_4$. Suppose $\mathfrak{B} \models \alpha$ and b is an element of $\mathfrak{B}$ that is not in $\mathfrak{A}$. Then there are two cases as below based on the position of b in the linear order underlying $\mathfrak{B}$. In both of these cases, we get a contradiction, showing that $\mathfrak{B} \models \neg\alpha$ and hence $\mathfrak{B} \models \psi_k$.

1. $(\mathfrak{B}, a_1, b, a_2) \models ((x \leq y) \wedge (y \leq z))$ for two elements a_1, a_2 of $\mathfrak{A}$ such that $(\mathfrak{A}, a_1, a_2) \models S(x, z)$; then $\mathfrak{B} \models \neg\xi_3$ and hence $\mathfrak{B} \models \neg\alpha$.
2. $(\mathfrak{B}, b) \models ((d \leq x) \vee (x \leq c))$. Since the interpretations of c, d in $\mathfrak{B}$ are resp. the same as those of c, d in $\mathfrak{A}$, we have $\mathfrak{B} \models \neg\xi_2$ and hence $\mathfrak{B} \models \neg\alpha$.

We now show that φ_k is not equivalent over S to any $\exists^k\forall^*$ sentence. Towards a contradiction, suppose φ_k is equivalent over S to the sentence $\gamma := \exists x_1 \ldots \exists x_k$ $\forall^n \bar{y} \beta(x_1, \ldots, x_k, \bar{y})$, where β is a quantifier-free formula. Consider the structure $\mathfrak{A} = (U_\mathfrak{A}, \leq^\mathfrak{A}, S^\mathfrak{A}, P^\mathfrak{A}, c^\mathfrak{A}, d^\mathfrak{A})$, where the universe $U_\mathfrak{A} = \{1, \ldots, (8n+1)\times(k+1)\}$, $\leq^\mathfrak{A}$ and $S^\mathfrak{A}$ are respectively the usual linear order and successor relation on $U_\mathfrak{A}$, $c^\mathfrak{A} = 1, d^\mathfrak{A} = (8n+1)\times(k+1)$ and $P^\mathfrak{A} = \{(4n+1)+i\times(8n+1) \mid i \in \{0, \ldots, k\}\}$. We see that $\mathfrak{A} \models (\xi_1 \wedge \xi_2 \wedge \xi_3 \wedge \xi_4 \wedge \neg\xi_5)$ and hence $\mathfrak{A} \models \varphi_k$. Then $\mathfrak{A} \models \gamma$. Let $a_1, \ldots, a_k$ be the witnesses in $\mathfrak{A}$ to the k existential quantifiers of γ.

It is clear that there exists $i^* \in \{0, \ldots, k\}$ such that a_j does not belong to $\{(8n+1) \times i^*+1, \ldots, (8n+1) \times (i^*+1)\}$ for each $j \in \{1, \ldots, k\}$. Then consider the structure $\mathfrak{B}$ that is identical to $\mathfrak{A}$ except that $P^{\mathfrak{B}} = P^{\mathfrak{A}} \setminus \{(4n+1)+i^* \times (8n+1)\}$. It is clear from the definition of $\mathfrak{B}$ that $\mathfrak{B} \models (\xi_1 \wedge \xi_2 \wedge \xi_3 \wedge \xi_4 \wedge \xi_5)$ and hence $\mathfrak{B} \models \neg\varphi_k$. We now show a contradiction by showing that $\mathfrak{B} \models \gamma$.

We show that $\mathfrak{B} \models \gamma$ by showing that $(\mathfrak{B}, a_1, \ldots, a_k) \models \forall^n \bar{y} \beta(x_1, \ldots, x_k, \bar{y})$. This is in turn done by showing that for any n-tuple $\bar{e} = (e_1, \ldots, e_n)$ from $\mathfrak{B}$, there exists an n-tuple $\bar{f} = (f_1, \ldots, f_n)$ from $\mathfrak{A}$ such that the (partial) map $\rho : \mathfrak{B} \to \mathfrak{A}$ given by $\rho(1) = 1$, $\rho((8n+1) \times (k+1)) = (8n+1) \times (k+1)$, $\rho(a_j) = a_j$ for $j \in \{1, \ldots, k\}$ and $\rho(e_j) = f_j$ for $j \in \{1, \ldots, n\}$ is such that ρ is a partial isomorphism from $\mathfrak{B}$ to $\mathfrak{A}$. Then since $(\mathfrak{A}, a_1, \ldots, a_k) \models \forall^n \bar{y} \beta(x_1, \ldots, x_k, \bar{y})$, we have $(\mathfrak{A}, a_1, \ldots, a_k, \bar{f}) \models \beta(x_1, \ldots, x_k, \bar{y})$ whereby $(\mathfrak{B}, a_1, \ldots, a_k, \bar{e}) \models \beta(x_1, \ldots, x_k, \bar{y})$. As $\bar{e}$ is an arbitrary n-tuple from $\mathfrak{B}$, we have $(\mathfrak{B}, a_1, \ldots, a_k) \models \forall^n \bar{y} \beta(x_1, \ldots, x_k, \bar{y})$.

Define a *contiguous segment in* $\mathfrak{B}$ to be a set of l distinct elements of $\mathfrak{B}$, for some $l \geq 1$, that are contiguous w.r.t. the linear ordering in $\mathfrak{B}$. That is, if $b_1, \ldots, b_l$ are the distinct elements of the aforesaid contiguous segment, then $(b_j, b_{j+1}) \in S^{\mathfrak{B}}$ for $j \in \{1, \ldots, l-1\}$. We represent such a contiguous segment as $[b_1, b_l]$, and view it as an interval in $\mathfrak{B}$. Given an n-tuple $\bar{e}$ from $\mathfrak{B}$, a *contiguous segment of* $\bar{e}$ *in* $\mathfrak{B}$ is a contiguous segment in $\mathfrak{B}$, all of whose elements belong to (the set underlying) $\bar{e}$. A *maximal contiguous segment of* $\bar{e}$ *in* $\mathfrak{B}$ is a contiguous segment of $\bar{e}$ in $\mathfrak{B}$ that is not strictly contained in another contiguous segment of $\bar{e}$ in $\mathfrak{B}$. Let CS be the set of all maximal contiguous segments of $\bar{e}$ in $\mathfrak{B}$. Let $\mathsf{CS}_1 \subseteq \mathsf{CS}$ be the set of all those segments of CS that have an intersection with the set $\{1, \ldots, (8n+1) \times i^*\} \cup \{(8n+1) \times (i^*+1)+1, \ldots, (8n+1) \times (k+1)\}$. Let $\mathsf{CS}_2 = \mathsf{CS} \setminus \mathsf{CS}_1$. Then all intervals in CS_2 are contained in the interval $[(8n+1) \times i^*+1, (8n+1) \times (i^*+1)]$. Let $\mathsf{CS}_2 = \{[i_1, j_1], [i_2, j_2] \ldots, [i_r, j_r]\}$ such that $i_1 \leq j_1 < i_2 \leq j_2 < \ldots < i_r \leq j_r$. Observe that $r \leq n$. Let CS_3 be the set of contiguous segments in $\mathfrak{A}$ defined as $\mathsf{CS}_3 = \{[i'_1, j'_1], [i'_2, j'_2], \ldots, [i'_r, j'_r]\}$ where $i'_1 = (8n+1) \times i^*+n+1, j'_1 = i'_1 + (j_1 - i_1)$, and for $2 \leq l \leq r$, we have $i'_l = j'_{l-1} + 2$ and $j'_l = i'_l + (j_l - i_l)$. Observe that the sum of the lengths of the segments of CS_2 is at most n, so that $j'_r \leq (8n+1) \times i^* + 3n + 1$.

Now consider the tuple $\bar{f} = (f_1, \ldots, f_n)$ defined using $\bar{e} = (e_1, \ldots, e_n)$ as follows. Let Elements(CS_1), resp. Elements(CS_2), denote the elements contained in the segments of CS_1, resp. CS_2. For $1 \leq l \leq n$, if $e_l \in$ Elements(CS_1), then $f_l = e_l$. Else suppose e_l belongs to the segment $[i_s, j_s]$ of CS_2 where $1 \leq s \leq r$, and suppose that $e_l = i_s + t$ for some $t \in \{0, \ldots, (j_s - i_s)\}$. Then choose $f_l = i'_s + t$. We now verify that the (partial) map $\rho : \mathfrak{B} \to \mathfrak{A}$ given by $\rho(1) = 1$, $\rho((8n+1) \times (k+1)) = (8n+1) \times (k+1)$, $\rho(a_j) = a_j$ for $j \in \{1, \ldots, k\}$ and $\rho(e_l) = f_l$ for $l \in \{1, \ldots, n\}$, is indeed a partial isomorphism from $\mathfrak{B}$ to $\mathfrak{A}$. $\square$

5   Conclusion and Future Directions

In this paper, we presented a new proof of the extensional form of the generalized Łoś-Tarski theorem ($\mathsf{GLT}(k)$) for theories, first shown in [6], and thereby obtained a new proof of the theorem for sentences in both its forms substructural and extensional. Our proof avoids using λ-saturation as used in [6], and instead constructs structures with just the "needed saturation" to prove the theorem. As our second result, we presented a strengthening of the failure of the Łoś-Tarski theorem in the finite by showing that not only universal sentences, but even $\exists^k\forall^*$ sentences for any fixed k are not expressive enough to capture the semantic property of hereditariness in the finite.

We now mention two future directions concerning our results. The first is in connection with the Łoś-Tarski theorem in the finite. The counterexample to this theorem in the finite as presented in [1] uses two binary relations and two constants. But what happens if the vocabulary contains only one binary relation and some constants/unary relations? There are positive results shown when the binary relation is constrained to be interpreted as special kinds of posets, specifically linear orders or (more generally) poset-theoretic trees, or special kinds of graphs, specifically subclasses of bounded clique-width graphs such as classes of bounded tree-depth/shrub-depth and m-partite cographs [4]. (In fact, over all these classes, even $\mathsf{GLT}(k)$ is true for all k.) But the case of an unconstrained binary relation remains open, motivating the following question.

Problem 1. *Is the (relativized version of the) Łoś-Tarski theorem true over all finite colored directed graphs? The same question also for undirected graphs.*

Our second future direction concerns $\mathsf{GLT}(k)$ over all finite structures. Theorem 4 exhibits for each k, a sentence φ_k that is hereditary over all finite structures but that is not equivalent over this class to any $\exists^k\forall^*$ sentence. We however observe that φ_k is itself equivalent to an $\exists^{k+1}\forall^*$ sentence. So that this counterexample to $\mathsf{GLT}(k)$ is not a counterexample to $\mathsf{GLT}(k+1)$. This raises the natural question of whether all counterexamples to $\mathsf{GLT}(k)$ in the finite, are simply Σ_2^0 sentences, or sentences equivalent to these. Given that any Σ_2^0 sentence is k-hereditary for some k, we pose the aforesaid question as the following problem.

Problem 2. *Is it the case that over the class of all finite structures, a sentence is k-hereditary for some k if, and only if, it is equivalent to a Σ_2^0 sentence?*

Observe that the version of Problem 2 in which arbitrary structures are considered instead of finite structures, has a positive answer due to Theorem 1 (which is a stronger statement). Much like the Łoś-Tarski theorem, results from classical model theory almost invariably fail in the finite [1]. Resolving Problem 2 in the affirmative would then give us a preservation theorem that survives passage to all finite structures.

Acknowledgments. I would like to thank Anuj Dawar for pointing out the Ehrenfeucht-Fraïssé game perspective to the arguments contained in the proof of Theorem 4. I also thank the anonymous referees for their comments and suggestions.

References

1. Alechina, N., Gurevich, Y.: Syntax vs. semantics on finite structures. In: Mycielski, J., Rozenberg, G., Salomaa, A. (eds.) Structures in Logic and Computer Science. A Selection of Essays in Honor of A. Ehrenfeucht. LNCS, vol. 1261, pp. 14–33. Springer, Heidelberg (1997). https://doi.org/10.1007/3-540-63246-8_2
2. Chang, C.C., Keisler, H.J.: Model Theory, vol. 73. Elsevier, Amsterdam (1990)
3. Hodges, W.: Model Theory (Draft 20 Jul 00) (2000). http://wilfridhodges.co.uk/history07.pdf
4. Sankaran, A.: A generalization of the Łoś-Tarski preservation theorem. Ph.D. thesis, Department of Computer Science and Engineering, Indian Institute of Technology Bombay. CoRR abs/1609.06297 (2016)
5. Sankaran, A.: A generalization of the Łoś-Tarski preservation theorem – dissertation summary. CoRR abs/1811.01014 (2018)
6. Sankaran, A., Adsul, B., Chakraborty, S.: A generalization of the Łoś-Tarski preservation theorem. Ann. Pure Appl. Log. **167**(3), 189–210 (2016)
7. Sankaran, A., Adsul, B., Madan, V., Kamath, P., Chakraborty, S.: Preservation under substructures modulo bounded cores. In: Proceedings of WoLLIC 2012, Buenos Aires, Argentina, 3–6 September, 2012, pp. 291–305 (2012)
8. Tait, W.W.: A counterexample to a conjecture of Scott and Suppes. J. Symb. Log. **24**(1), 15–16 (1959)

Model Theory for Sheaves of Modules

Mike Prest[✉]

School of Mathematics, University of Manchester, Manchester M13 9PL, UK
mprest@manchester.ac.uk

Abstract. We describe how the model theory of modules is adapted to deal with sheaves of modules.

Keywords: Model theory · Sheaves · Multisorted · Modules

1 Introduction

A sheaf may be thought of as a set of structures, indexed by the points of a topological space, which "vary in a continuous way". For example, a sheaf $\mathcal{O}$ of rings over a topological space X is given by a set $\{\mathcal{O}_{X,x} : x \in X\}$ of rings (which we assume to be associative, not necessarily commutative, and each with a 1) together with a certain type of topology on the union of these sets. This is the étalé-space view of a sheaf, which we will point out after approaching the definition of a sheaf through that of a presheaf.

Sheaves arise typically in geometry, topology and analysis. Our, algebraic/ model-theoretic, interest will be in the model theory of sheaves of modules over sheaves of rings.

The model theory of modules is very well-developed and has found many applications. We will describe how to set up model theory for sheaves of modules in a way which naturally generalises how this is done for modules over a fixed ring (that is the case where the space X has just one point). The key change is that we should regard sheaves as multi-sorted structures. The outcome is that, over topological spaces X which satisfy some mild conditions, one can apply all the techniques and results of the model theory of modules.

A great deal of what we say applies to sheaves over sites (where Grothendieck topologies replace topologies in the usual sense) and to sheaves of structures other than modules but our aim is to explain the particularities of the model theory of sheaves of modules in the relatively concrete context of sheaves over topological spaces.

2 Model Theory for Modules

Here we give a very brief overview of some relevant aspects of the model theory of modules.

© Springer-Verlag GmbH Germany, part of Springer Nature 2019
Md. A. Khan and A. Manuel (Eds.): ICLA 2019, LNCS 11600, pp. 89–102, 2019.
https://doi.org/10.1007/978-3-662-58771-3_9

The model theory of modules was originally set up to deal with modules over a fixed ring R (always assumed to be associative and with a 1).[1] By default, by "R-module" we will mean "right R-module", and we will denote the category of these by Mod-R. Since the left R-modules are the right modules over the ring R^{op} with the opposite multiplication, it is immaterial for the general theory whether we deal with right or left modules.

The language $\mathcal{L}_R$ used has a binary operation symbol $+$ for the addition on a module, a constant symbol 0 for the zero element of a module and, for each $r \in R$, a 1-ary function symbol to express multiplication-by-r on a module. In practice we use natural notation, writing the value of multiplication by $r \in R$ on an element a of a module M by ar, rather than introducing a more explicitly functional notation (such as $f_r(a)$). Also, for instance, we would write an atomic formula in variables $x_1, \ldots, x_n$ in the (simplified, using the theory of R-modules) form $\sum_{i=1}^n x_i r_i = 0$. The background theory is that generated by the usual axioms for (right) R-modules.

The key result in the model theory of modules is pp-elimination of quantifiers (see, for example, [8, Sect. 2.4]).

Theorem 1. *Let R be any ring. Modulo the theory of (right) R-modules, every formula is equivalent to the conjunction of a sentence and finite boolean combination of pp formulas. Moreover, every sentence is a finite boolean combination of invariants conditions.*

A **pp** (for **positive primitive**), also called **regular**, formula, is an existentially quantified conjunction of atomic formulas, that is, in our context, an existentially quantified system of R-linear equations (perhaps inhomogeneous equations if the formula contains parameters from a module). The solution set, $\phi(M)$, of a pp formula ϕ, in any module M, is a subgroup of M^n, where n is the number of free variables of ϕ. These subgroups are the **groups pp-definable in** M or, as commonly said more loosely, the **pp-definable subgroups of** M. If ϕ, ψ are pp formulas in the same free variables, then we write $\psi \leq \phi$ if $\psi(M) \leq \phi(M)$ for every module M. In fact (see e.g. [9, 1.2.23]), it is enough to check this for every finitely presented module M, because every module is a direct limit (=directed colimit) $\varinjlim_\lambda M_\lambda$ of finitely presented modules M_λ and pp formulas commute with direct limits in the sense that $\phi(\varinjlim_\lambda M_\lambda) = \varinjlim_\lambda \phi(M_\lambda)$. Recall that a module M is **finitely presented** if it is finitely generated and finitely related, equivalently if the functor $\mathrm{Hom}_R(M, -)$, which we simply denote by $(M, -)$, commutes with direct limits.

An **invariants condition** is a sentence which says, of some **pp-pair** $\psi \leq \phi$, that the index of the subgroup defined by ψ in that defined by ϕ either is less than, equal to, or greater than, n, for some particular integer n.

---

[1] One could let the ring vary by using a two-sorted language: one sort for the ring, one for the module, so that the structures are (ring, module) pairs (R, M_R). The model theory of such pairs is, however, much less well-behaved than that for modules over a fixed ring, and not at all as amenable to useful analysis.

This partial elimination of quantifiers allowed greatly simplified proofs of much that had already been shown about the model theory of modules and it stimulated a fundamental transformation of the subject.

Elimination of quantifiers also partly explained the, already-recognised, importance (see e.g. [3,6,19]) of notions such as purity and pure-injectivity in the model theory of modules, where we say that an inclusion $N \to M$ of modules is **pure** if, for every pp formula ϕ in n free variables[2], we have $\phi(N) = N^n \cap \phi(M)$. The key role of pp formulas is also seen in that they are exactly those whose solution sets are preserved by R-linear maps: if $f : M \to N$ is a homomorphism of R-modules then $f\phi(M) \leq \phi(N)$.

With elimination of quantifiers to hand, Garavaglia introduced (e.g. [5]) new ideas and connections with algebra which inspired the fundamental paper [22] of Ziegler. The area has subsequently seen yet further transformations as well as many algebraic applications, for which one may look at [8] for model theory *per se* and at [9] for the more algebraic/category-theoretic form of the theory and many applications. Since then, there has been further widening in viewpoint, for which one many consult [11,12].

In a short paper one can say little of all that has been done but, for the purpose in hand, we pick out a couple of important aspects.

One is the extension of the theory to apply to multisorted modules, that is, modules over rings with many objects or, said otherwise, additive functors from a skeletally small preadditive category $\mathcal{R}$ to the category **Ab** of abelian groups. This viewpoint is explained in [13] and below, in Sect. 4, we give the details that we will need here. A preadditive category with one object is simply a ring and an additive functor from that to **Ab** is exactly a module over that ring. Essentially everything about modules, and about the model theory of modules, extends, almost without change, to the general case of such "multisorted modules". The requirement that the preadditive category $\mathcal{R}$ be skeletally small, that is, to have just a *set* of objects up to isomorphism, avoids set-theoretic difficulties. Model-theoretically, the change in moving from modules over rings to modules over rings with many objects is that we use a multisorted language with at least one sort for each isomorphism class of object of $\mathcal{R}$; we will see some examples later.

An important notion is that of a definable category. If $\mathcal{M}$ is the category of modules over a ring (or, more generally, over a ring with many objects), then a **definable subcategory** of $\mathcal{M}$ is the full subcategory on a class of modules which is closed under isomorphism, direct products, direct limits and pure submodules. We also use the term for the underlying class of objects.

Theorem 2. *If $\mathcal{M}$ is the category of modules over a ring (or over a skeletally small preadditive category), then the following conditions are equivalent on a class $\mathcal{D}$ of modules:*

(i) $\mathcal{D}$ is a definable subcategory, that is, closed in $\mathcal{M}$ under isomorphism, direct products, direct limits and pure submodules;

---

[2] In fact, [18], see [9, 2.1.6], it is enough to check for $n = 1$.

*(ii) $\mathcal{D}$ is the class of models of a theory which is axiomatised by sentences of
the form $\forall \overline{x}\, (\phi \to \psi)$ where ψ, ϕ are pp formulas (in free variables $\overline{x}$);*

*(iii) $\mathcal{D}$ is an axiomatisable class of modules satisfying $\mathcal{D} = \mathrm{Add}(\mathcal{D})$ (in fact
$\mathcal{D} = \mathrm{add}(\mathcal{D})$ is enough).*

By $\mathrm{Add}(\mathcal{D})$, respectively $\mathrm{add}(\mathcal{D})$, we mean the closure of $\mathcal{D}$ under direct
summands and arbitrary, resp. finite, direct sums.

By a **definable category** we mean one which is equivalent to a definable
subcategory of some module category. It has turned out that definable categories
are the natural context for the model theory of modules, in the sense that the
techniques apply, and the general results which hold for modules also hold (with
minor modifications) in any definable category. Furthermore, the model theory
of any definable category $\mathcal{D}$ is intrinsic, in the sense that an appropriate language
and theory for which $\mathcal{D}$ is the category of models, may be defined just from the
category structure of $\mathcal{D}$ (see [10, Chap. 12]).

3   Presheaves and Sheaves of Modules

Let X be a topological space. We use the notation X also for the underlying set
of the topology and we write Op_X for the poset of open subsets of X, ordered by
inclusion. We can regard Op_X as a category with objects being the open sets and
a (unique) arrow from V to U iff $V \subseteq U$. Let $\mathcal{C}$ be a category of structures - for
example the category of abelian groups, or rings, or commutative rings with 1.

A **presheaf** F of structures from $\mathcal{C}$ over X is given by the following data:

- for each open subset U of X, an object FU of $\mathcal{C}$;
- for each inclusion of open subsets $V \subseteq U$ of X, a morphism $r_{UV}^F = r_{UV}$:
  $FU \to FV$, usually referred to as a **restriction** map, of $\mathcal{C}$,
  such that:
- for every open set U, r_{UU} is the identity map id_{FU} of FU and
- given open sets $W \subseteq V \subseteq U$ we have $r_{UW} = r_{VW} r_{UV}$.

In other words, a presheaf in $\mathcal{C}$ over X is a contravariant functor from the
poset Op_X, regarded as a category, to $\mathcal{C}$. From that point of view, a morphism
$f : F \to G$ of presheaves is defined simply to be a natural transformation, that
is, an Op_X-indexed set $(f_U : FU \to GU)_U$ of morphisms of $\mathcal{C}$ such that, for
every inclusion $V \subseteq U$ the following diagram commutes.

$$
\begin{array}{ccc}
FU & \xrightarrow{\ f_U\ } & GU \\
{\scriptstyle r_{UV}^F} \downarrow & & \downarrow {\scriptstyle r_{UV}^G} \\
FV & \xrightarrow[\ f_V\ ]{} & GV
\end{array}
$$

Thus we obtain the category of presheaves of $\mathcal{C}$-objects over X and morphisms
between them.

In particular, if $\mathcal{C}$ is the category of associative rings with 1 then we obtain the notion of a presheaf of rings.

As a specific example, take X to be the unit circle $S^1 \subseteq \mathbb{R}^2$ in the real plane with its usual topology, and define a presheaf of rings by assigning, to each open subset U of S^1, the ring $\mathcal{C}(U, \mathbb{R})$ of continuous functions from U to $\mathbb{R}$, and by assigning, to an inclusion $V \subseteq U$ of open subsets of S^1, the map from $\mathcal{C}(U, \mathbb{R})$ to $\mathcal{C}(V, \mathbb{R})$ which takes a continuous function on U to its restriction to V. It is easily checked that this is indeed a presheaf of rings. In fact it is a *sheaf* in the sense of the following definition, which we state in a form which applies when the category $\mathcal{C}$ is a category of sets with structure. In that case we refer to the elements of FU, where F is a presheaf and U an open set, as **sections** (of F) over U.

A presheaf F on a space X is a **sheaf** if:

- given an open cover $U = \bigcup_\lambda U_\lambda$ of an open set $U \subseteq X$, and given, for each λ, some section $s_\lambda \in FU_\lambda$, if, for every λ, μ, we have $r_{U_\lambda, U_\lambda \cap U_\mu}(s_\lambda) = r_{U_\mu, U_\lambda \cap U_\mu}(s_\mu)$, then there is a section $s \in FU$ such that, for every λ, the restriction of s to U_λ is s_λ, that is, $r_{U,U_\lambda}(s) = s_\lambda$, and
- given an open cover $U = \bigcup_\lambda U_\lambda$ of an open set in X, and given sections $s, t \in FU$ which agree on each member of the cover - that is, if, for each λ we have $r_{UU_\lambda}(s) = r_{UU_\lambda}(t)$ - then $s = t$.

The second condition says that sections which locally agree (that is, agree on some open cover) must be equal; in the case that the objects of $\mathcal{C}$ have an underlying abelian group structure then it is enough to take $t = 0$. The first condition says that sections on a cover may be glued together to make a section on the set being covered provided that they agree on the intersections; if the second condition also holds, then there is a unique such section.

Given a presheaf F over a space X, with values in a category $\mathcal{C}$ which has direct limits (that is, directed colimits) and given a point $x \in X$, we define the **stalk** of F at x to be $F_x = \varinjlim_{U \ni x} FU$, the direct limit being taken over the directed (by intersection) system of open subsets that contain x. In the example above, of continuous functions on open subsets of the circle S^1, the stalk at a point $x \in S^1$ is the ring of so-called "germs" of continuous functions at x.

Now we come to our main definition, that of a sheaf of modules.

Suppose that $R_X = (X, R)$ is a **ringed space**, that is a sheaf R of rings (associative with 1 under our conventions) over a topological space X. For each open subset $U \subseteq X$ we have the ring RU and the corresponding category, which we denote by Mod-RU, of right modules over RU and for every point $x \in X$ we have the ring R_x and its corresponding category Mod-R_x of right modules. We will use the notations R and R_X fairly interchangeably.

We define a (right) R_X-**premodule** to be a presheaf M which assigns to each open subset $U \subseteq X$ a right RU-module MU such that, for each inclusion $V \subseteq U$ of open subsets, the restriction $r_{UV}^M : MU \to MV$ is a homomorphism of RU-modules, where we regard MV as an RU-module *via* restriction of scalars along the ring homomorphism $r_{UV}^{R_X} : RU \to RV$. Strictly speaking this is not, as defined, a presheaf in the sense of our earlier definition since the category where M takes values varies with U! But there are ways around this - for example we could just let M take values in the category of abelian groups and then add extra conditions concerning the actions of the elements of the various RU. In any case, it is convenient to think of the codomain category as varying. Note that the stalk M_x at a point x will be a right $R_{X,x}$-module.

A morphism $f : M \to N$ of R_X-premodules is, if we regard M and N as functors, a natural transformation, that is, for each open set U, an RU-linear map $f_U : MU \to NU$ such that, if $V \subseteq U$ are open, then the diagram commutes.

$$
\begin{array}{ccc}
MU & \xrightarrow{\ f_U\ } & NU \\
{\scriptstyle r^M_{UV}}\big\downarrow & & \big\downarrow{\scriptstyle r^N_{UV}} \\
MV & \xrightarrow{\ f_V\ } & NV
\end{array}
$$

One may check that such a morphism induces, at each $x \in X$, an $R_{X,x}$-linear map $M_x \to N_x$ of stalks at x. These definitions give us the category PreMod-R_X of R_X-premodules. It is an abelian[3] category and, is in fact, Grothendieck and locally finitely presented. We will examine the reason for the latter since it leads directly to setting up the model theory of such structures. We denote by Mod-R_X the category of **sheaves of modules** - the full subcategory of PreMod-R_X with objects those presheaves of modules which are actually sheaves. This also is abelian and Grothendieck, but is not always locally finitely presented, though in many important cases it is.

Bundling together the stalks of a sheaf gives an alternative view of a sheaf. Given a sheaf F, we form the disjoint union of the stalks F_x, for $x \in X$ of F. This union is then given a topology which locally looks like that of X (see, e.g., [21] for details) and the sections FU of F at an open set U then become the continuous maps s, from U to the resulting *étalé space*, which are such that $\pi s = \mathrm{id}_U$ where π is the projection map from the étalé space to X which takes an element of F_x to the point x (and id_U is the identity map on U). One more piece of notation: if F is a sheaf of additive structures then we define the **support** of a section $s \in FU$ to be the set, $\mathrm{supp}(s) = \{x \in X : s_x \neq 0\}$, of points where the image s_x of s in F_x, under the natural map $FU \to F_x$, is nonzero.

4   Model Theory for Presheaves of Modules

A category $\mathcal{C}$ is said to be **finitely accessible** (see [1]) if it has direct limits, if there is, up to isomorphism, just a set of finitely presented objects of $\mathcal{C}$ and if every object of $\mathcal{C}$ is a direct limit of finitely presented objects, where an object $A \in \mathcal{C}$ is **finitely presented** if the functor $\mathcal{C}(A, -)$, which we abbreviate as $(A, -)$, commutes with direct limits. This is equivalent to the object A being

---

[3] We will not present background on abelian category theory here but there are many suitable references, for example [4,20].

"finitely generated and finitely related" in $\mathcal{C}$ if those terms make sense in $\mathcal{C}$ (for instance, if $\mathcal{C}$ is the category of rings, or of groups, or of modules over a ring).

To expand on the condition that the representable functor $(A, -)$ commute with direct limits, this means that, given any directed system, $((C_\lambda)_{\lambda \in \Lambda}, (f_{\lambda\mu} : C_\lambda \to C_\mu)_{\lambda \leq \mu \in \Lambda})$ with direct limit $(C, (f_{\lambda\infty} : C_\lambda \to C)_\lambda)$ - so $f_{\mu\infty} f_{\lambda\mu} = f_{\lambda\infty}$ for all $\lambda \leq \mu$ - and given any $g : A \to C$, there is λ' and $g'_\cdot : A \to C_{\lambda'}$ such that $f_{\lambda'\infty} g' = g$. Moreover, such a factorisation must be essentially unique in the sense that, if also there is λ'' and $g'' : A \to C_{\lambda''}$ such that $f_{\lambda''\infty} g'' = g$, then there is $\mu \geq \lambda', \lambda''$ such that $f_{\lambda'\mu} g' = f_{\lambda''\mu} g''$.

We denote by $\mathcal{C}^{\mathrm{fp}}$ the full subcategory of finitely presented objects of $\mathcal{C}$. If $\mathcal{C}$ is finitely accessible and both complete and cocomplete then it is a **locally finitely presented** category. Both the category all presheaves and of all sheaves of modules over a ringed spaces are complete and cocomplete, so we will use the terms finitely accessible and locally finitely presented interchangeably for these.

The fact that, for any ringed space R_X, PreMod-R_X is locally finitely presented is a special case of a general fact for functor categories and, as in that general case, it is the representable functors which provide a generating collection (a set up to isomorphism since Op_X has just a set of objects) of finitely presented presheaves. We describe these representable functors in specific presheaf terms.

Let $U \in \mathrm{Op}_X$ be any open subset of X and let $j : U \to X$ denote the inclusion map. We define the presheaf $j_0 R_U$ as follows (we will explain the notation after that):

$$j_0 R_U(V) = \begin{cases} R(V) & \text{if } V \subseteq U \\ 0 & \text{otherwise} \end{cases}.$$

Here j_0 denotes a functor from PreMod-R_U to PreMod-R_X and R_U denotes the restriction, $R_X|_U$, of R_X to U. For any presheaf F on a space X and open subset U of X, the **restriction** of F to U is the presheaf $F|_U$ on U which is given by $F|_U V = FV$ for V an open subset of U.[4] It is direct from the definition that the restriction of a sheaf is again a sheaf. For any $G \in$ PreMod-R_U the presheaf j_0 defined by $j_0 G \cdot V = \begin{cases} GV & \text{if } V \subseteq U \\ 0 & \text{otherwise} \end{cases}$ is the **extension by** 0 of G (to a presheaf on X).

Proposition 1 *(e.g. [2, p. 7 Proposition 6]). If R_X is any ringed space, then the category* PreMod-R_X *of presheaves of R_X-modules is abelian Grothendieck and locally finitely presented, with the $j_0 R_U$, for $U \in \mathrm{Op}_X$ being a generating set of finitely presented objects.*

By a **generating set** $\mathcal{G}$ of objects of an Grothendieck abelian category $\mathcal{C}$ we mean that for every object $C \in \mathcal{C}$ there is an exact sequence $H' \to H \to C \to 0$ where H', H are direct sums (possibly infinite) of copies of objects in $\mathcal{G}$. We do not mean that it is $\underrightarrow{\lim}$-generating (in the sense of the definition of finitely accessible category). But, if $\mathcal{G}$ is a set of finitely presented objects which is

---

[4] In category-theoretic terms it is the restriction of the contravariant functor F to the full subcategory on the objects with a morphism to U.

generating in the sense just defined, then $C \in \mathcal{C}$ will be finitely presented iff there is such a presentation where both H' and H are *finite* direct sums of copies of objects in $\mathcal{G}$ and it is the case that the collection of finitely presented objects of $\mathcal{C}$ will be $\varinjlim$-generating in $\mathcal{C}$, so $\mathcal{C}$ will be finitely accessible (indeed locally finitely presented).

The basic idea for setting up (finitary) model theory in any finitely accessible category (with products) $\mathcal{C}$ is that, since every object $C \in \mathcal{C}$ is determined by the morphisms to it from finitely presented objects, we take these morphisms to be the "elements" of C. But morphisms with different domains should be elements of different kinds - formally of different *sorts*. This means that the formal language we set up is naturally multisorted, with one sort, σ_A say, for each finitely presented object A of $\mathcal{C}$ and with the elements of $C \in \mathcal{C}$ of sort σ_A being the elements of the set, (A, C), of morphisms from A to C. We should use a set $\mathcal{G}$ (rather than a proper class) of sorts, so we restrict A to range over some *set* of finitely presented objects which contains at least one copy (to isomorphism) of each finitely presented object. This gives us the sorts of our language for $\mathcal{C}$. It does not depend, in any way that matters, on the actual choice of set of finitely presented objects that we use, as long as it has a copy of each (or indeed, "enough") of the finitely presented objects.

We also introduce function symbols, one for each morphism between objects in our chosen set $\mathcal{G}$ of finitely presented objects. If $f : A \to B$ is such a morphism, then the corresponding function symbol, for which we will use the same symbol f, has domain σ_B and codomain σ_A, reflecting the direction of the induced morphism $(f, -) : (B, -) \to (A, -)$, given by $g \mapsto gf$ (where $f \in (B, C)$ for any $C \in \mathcal{C}$), between representable functors.

This viewpoint on what are the "elements" of a structure might seem unfamiliar but it is exactly what is seen in the basic fact from module theory that a module M over a ring R is isomorphic to the module[5] of homomorphisms (R_R, M) from R_R, meaning R regarded as a right R-module, to M. That is because each homomorphism is determined by the image of 1 and every element of M is such an image.

For example, under this expanded viewpoint, if we take a direct product $R^{(n)}$ of copies of R_R, then the "elements" of a module M of sort $R^{(n)}$ are exactly the n-tuples of elements of M. More generally, if A is a finitely presented module, then the elements of a module M in sort A (that is, of $\sigma_A M$) could be regarded as the n-tuples (if A is n-generated) of elements of M which satisfy certain R-linear relations, namely those which generate all the R-linear relations on a chosen generating set of n elements for A.

This example of modules also shows that it is not necessary to represent every isomorphism class of finitely presented object when setting up the language - it is enough to have a set of sorts corresponding to a set of finitely presented objects which generate the category in the sense defined after Proposition 1 (that refers to the special case of a Grothendieck abelian category, but it is such categories with which we will be concerned). Any two such languages set up

---

[5] It is a right module *via* the left action of R on the module R_R.

using generating sets will be inter-interpretable (each formula in the one language can be translated to an equivalent formula in the other), so will give the same model theory for the structures in $\mathcal{C}$. In the case of categories, PreMod-R_X, of presheaves we will use the representable functors - the extensions by 0 - defined above. Let us return now to this case.

The presheaves $j_0 R_U$ as U ranges over subsets of X form a generating set of finitely presented objects of PreMod-R_X, so we will use these, hence the open subsets of X, to index the sorts of our language for PreMod-R_X. We should describe the functors $(j_0 R_U, -)$ and the morphisms between them in order to understand something of what can be expressed in this language.

Lemma 1. *Suppose that R_X is a ringed space. Let $j : U \to X$ be the inclusion of an open subset in X, let $F \in$ PreMod-R_U and let $G \in$ PreMod-R_X. Then there is a natural isomorphism of groups $(j_0 F, G) \simeq (F, G|_U)$.*

If U is an open subset of X, then the functor $(j_0 R_U, -)$ is, in the view of presheaves as contravariant functors on Op_X, the representable functor corresponding to $U \in \mathrm{Op}_X$ hence, by the Yoneda Lemma, $(j_0 R_U, G) \simeq GU$. In particular, if U, W are arbitrary open subsets of X, then $(j_0 R_U, j_0 R_W) \simeq$

$$j_0 R_W \cdot U = \begin{cases} R_X U & \text{if } U \subseteq W \\ 0 & \text{otherwise} \end{cases}.$$ We can understand this application of the

Yoneda Lemma more algebraically by noting that $j_0 R_U$ is generated by the section $1_U \in j_0 R_U \cdot U = R_X U$, in the sense that, for every open subset V of U, the image, $r_{UV}(1_U)$, of 1_U under the restriction map from U to V, is equal to $1_V \in R_X V$, which generates $R_X V = j_0 R_U \cdot V$ as an $R_X V$-module. Therefore any morphism $f : j_0 R_U \to G|_U$ will be determined by the image $f_U(1_U) \in GU$ where f_U is the component of f at U.

Thus, if we use the language for PreMod-R_X based on the generating set $(j_0 R_U)_{U \in \mathrm{Op}_X}$, then the function symbols of the language, beyond those used to express the abelian group structure of each sort $\sigma_U = (j_0 R_U, -)$ are as follows: given open subsets U, W of X then, if $U \subseteq W$ the function symbols from sort σ_W to sort σ_U are naturally indexed by the elements of $R_X U$, otherwise there is only the zero function symbol from σ_W to σ_U.

Using the Yoneda Lemma as above, we can explicitly describe the interpretations of these function symbols, as follows.

Given open sets $U \subseteq W$ and $t \in R_X U$, regarded (as above) as an element of $j_0 R_W \cdot U$, hence as a morphism from $j_0 R_U$ to $j_0 R_W$, and given any $G \in$ PreMod-R_X, we have the following diagram in PreMod-R_X showing the action of t:

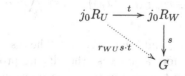

namely $t : (j_0 R_W, G) = GW = \sigma_W G \to (j_0 R_U, G) = GU = \sigma_U G$, takes a section $s \in GW$ to its restriction to U followed by multiplication by t - the result is a section in GU.

We can regard these actions, as multiplications by elements of a ring with many objects (see [13], also [12], for this point of view). Thus the R_X-presheaves become "modules over a ring with many objects" and, in fact (see [13] for an overview, [9] for details) the model theory of modules over the usual, 1-sorted, rings is applicable in its entirety. Let us give some examples (also see [14] and [17]) of what one can say with this language. After that, we will move on to the category of R_X-sheaves.

First, a notational point. If $V \subseteq U$ are open then, in any presheaf, F, the restriction map r_{UV}^F is the interpretation of a function symbol r (from sort σ_U to σ_V) in the language. The functional notation for the value of this map on a section $a \in FU$ is $r_{UV}(a)$, whereas the right module notation is ar. In practice, since perhaps the meaning is clearer, we shall use the former, functional notation, but bear in mind that it is naturally written in the module language as a right multiplication.

A presheaf F is said to be **separated** if, for every open set U, open cover $(U_\lambda)_\lambda$ of U and sections $s, t \in FU$, if, for every λ we have $r_{UU_\lambda}(s) = r_{UU_\lambda}(t)$ then $s = t$ (that is, if sections agree locally, then they are equal). Of course in the additive situation, it is enough to consider the case where $t = 0$. Given such an open set and open cover of it, note that there is a possibly infinitary sentence

$$\forall x, y \left(\left(\bigwedge_\lambda r_{UU_\lambda}(x) = r_{UU_\lambda}(y) \right) \to (x = y) \right)$$

- which expresses this condition (where the variables have sort σ_U). If the cover is, or may be taken to be, finite (so, in particular if U is compact) then this will be, or be equivalent to, a sentence of finitary model theory. As we let these sentences range over all open sets and open covers (finite if possible), then we see that the property of being a separated presheaf is expressible in an infinitary version of our language, finitary if every open set is compact (that is, if the space is **noetherian**). We deduce that over any noetherian space the category of separated presheaves is definable in the sense discussed in Sect. 2.

The other condition necessary for a presheaf F to be a sheaf is that, given any open set U, open cover $(U_\lambda)_\lambda$ of U and set $(s_\lambda \in FU_\lambda)_\lambda$ of compatible sections (meaning that, for every λ, μ, $r_{U_\lambda, U_\lambda \cap U_\mu}^F(s_\lambda) = r_{U_\mu, U_\lambda \cap U_\mu}^F(s_\mu)$), there is $s \in FU$ such that $r_{UU_\lambda}^F(s) = s_\lambda$ for every λ. We see that this can be expressed by the sentence, infinitary if the cover is infinite,

$$\forall (x_\lambda)_\lambda \left(\bigwedge_{\lambda\mu} r_{U_\lambda, U_\lambda \cap U_\mu}(x_\lambda) = r_{U_\mu, U_\lambda \cap U_\mu}(x_\mu) \right) \to \left(\exists x \bigwedge_\lambda r_{UU_\lambda}(x) = x_\lambda \right)$$

(where the variable x_λ has sort σ_λ and x has sort σ_U). These sentences, ranging over open sets and covers, therefore axiomatise the glueing property. Combining this with the observations on the separation property, we deduce that over any noetherian space the category of sheaves is a definable subcategory of the category of presheaves (one may also prove the converse, so this characterises noetherian spaces - see [15, 3.12]). But we will see in the next section that it is possible for the category of sheaves to be definable (that is a definable subcategory of *some* category of multisorted modules), without necessarily being a definable subcategory of the category of presheaves.

For another example of what can be expressed using this language, if U is any open set then the (closure under isomorphism) of the class of presheaves of the form $j_0 F$ for some $F \in \text{PreMod-}R_U$ is axiomatised by the set of sentences of the form $\forall x_V \, (x_V = 0)$ where V ranges over the open sets which are not contained in U (and the notation x_V indicates that x is a variable of sort σ_V).

As yet another type of example, the constant presheaves (those such that each restriction map is an isomorphism) are axiomatised by the set of sentences of the form $\big(\forall y \, \exists x \, (y = r_{UV}(x))\big) \wedge \big(\forall x \, (r_{UV}(x) = 0_V \to x = 0_U)\big)$ as V, U range over open subsets with $V \subseteq U$, where x has sort σ_U, y has sort σ_V and the subscripts on the constant symbols 0 indicate their sort.

5 Model Theory for Sheaves of Modules

There is a canonical functor, **sheafification**, which turns each presheaf F into the sheaf aF which best approximates it in the category of sheaves. More precisely, sheafification is left adjoint to the forgetful functor (that is, the inclusion) $u : \text{Mod-}R_X \to \text{PreMod-}R_X$. So for every $F \in \text{PreMod-}R_X$ and $M \in \text{Mod-}R_X$ there are natural isomorphisms $(F, uM) \simeq (aF, M)$. Roughly, aF is formed from F by first identifying every two sections which agree on some open cover, so as to obtain a separated presheaf, then adding, as new sections, the results of glueing together compatible families of sections. In the context of presheaves of modules, sheafification is localisation in the sense of Gabriel (see [20]). For, the presheaves F such that, for every $U \in \text{Op}_X$ and section $s \in FU$, there is an open cover of U such that each restriction of s is zero, are exactly those whose sheafification $aF = 0$, and these form a hereditary torsion class of presheaves, localisation with respect to which is the sheafification functor.

In the particular case where the space X is noetherian, this is a finite-type localisation [15, 3.8], which has the consequence that the sheaves form an axiomatisable, indeed, definable, subcategory of $\text{PreMod-}R_X$ [15, 3.12]. In that case, therefore, the language for presheaves described above also may be used for developing the model theory of $\text{Mod-}R_X$ since the objects in the latter category form an elementary class of presheaves. But it is not necessary that X be noetherian in order to have a good model theory for sheaves; a basis $\mathcal{B}$ of compact open sets closed under intersection is enough [16, 3.3] since, in that case, the category $\text{Mod-}R_X$ of sheaves is locally finitely presented [16, 3.5]. That is, rather than indexing the sorts of the language using *all* open sets, it is enough to use those from $\mathcal{B}$. Because we are using finitary model theory, we need such a basis consisting of compact open sets. That is because our "elements" of sorts of sheaves - that is, sections over open sets in the basis - should be "finitary elements", meaning that if such an element belongs to a directed sum or union, then it belongs to some member of that sum or union. Having just a *basis* of compact sets is enough because sections are locally determined. The requirement that $\mathcal{B}$ be closed under intersection enables us to write the compatibility-of-sections condition in the resulting formal language. We give some details, but quite briefly since they are very similar to those seen for presheaves.

Suppose then that $\mathcal{B}$ is a basis, closed under intersection, of open sets for the topology on X. Then it turns out that the sheaves $j_!R_U$ for $U \in \mathcal{B}$ form a generating set of finitely presented objects of Mod-R_X. Here $j_!R_U$ denotes the sheafification, $a(j_0R_U)$, of j_0R_U; it is the sheaf extension by 0 of R_U to X. In general, if $G \in$ Mod-R_U then its sheaf extension, $j_!G$, by 0 to X may be defined by

$$j_!G(V) = \{s \in G(V \cap U) : \mathrm{supp}(s) \text{ is closed in } V\}.$$

The functor $j_!$ from Mod-R_U to Mod-R_X is left adjoint to the restriction-to-U functor, so $(j_!G, F) \simeq (G, F|_U)$ for $G \in$ Mod-R_U and $F \in$ Mod-R_X, and $j_!$ is an equivalence between Mod-R_U and the subcategory of Mod-R_X consisting of the sheaves which have support contained in U (see [7, pp. 106/7]). Thus the $j_!R_U$ play a very similar role in Mod-R_X to that played in PreMod-R_X by the j_0R_U. But, in contrast to the presheaves j_0R_U, they are not necessarily finitely presented. However:

Proposition 2 *([16, 3.7]). If X has a basis, closed under intersection, of compact open subsets and if U is compact open, then $j_!R_U$ is a finitely presented sheaf.*

Furthermore, we don't need all the $j_!R_U$ in order to generate Mod-R_X:

Proposition 3 *([16, 3.2]). If $\mathcal{B}$ is a basis of open subsets for the topology on X then the $j_!R_U$ for $U \in \mathcal{B}$ together generate Mod-R_X.*

Corollary 1 *([16, 3.5]). If (X, R_X) is a ringed space and if $\mathcal{B}$ is a basis, closed under intersection, of compact open subsets of X, then the category Mod-R_X of R_X-modules is locally finitely presented, with the $j_!R_U$, for $U \in \mathcal{B}$ forming a generating set of finitely presented objects.*

In that case therefore, having fixed such a basis $\mathcal{B}$, what we have said about the multisorted model theory for a locally finitely presented category applies, to give us a language $\mathcal{L}_\mathcal{B}$ for R_X-modules. This language has a sort for each $U \in \mathcal{B}$ and a function symbol for each morphism in each $(j_!R_U, j_!R_V)$ with $U, V \in \mathcal{B}$. Given any $F \in$ Mod-R_X, the resulting $\mathcal{L}_\mathcal{B}$-structure has value $(j_!R_U, F) \simeq FU$ in sort σ_U and the description of the interpretations of the function symbols of $\mathcal{L}_\mathcal{B}$ as maps between these sorts is similar to that for presheaves. Namely ([14, p. 1189, 1.4]) the elements of $(j_!R_U, j_!R_V)$ may be identified with the sections r of $R_{U \cap V}$ which have support closed in U and the action of such r (that is, the interpretation of the corresponding function symbol on a sheaf F), regarded as a map from FV to FU, is restriction from FV to $F(U \cap V)$, followed by multiplication by r regarded as an element of $R_{U \cap V}$, followed by inclusion in FU.

We remark that in many examples, in particular those typically seen in algebraic geometry, it will be the case that the underlying space X has a basis of compact open sets which is closed under intersection, hence the category Mod-R_X will be locally finitely presented and we will have a good, finitary, model theory of sheaves. Indeed, in many examples, *every* open set will be compact, therefore sections over any open set can be referred to by variables in our language.

One can see a variety of examples of what can be expressed about sheaves using this language in [14], where strongly minimal sheaves are considered and there is some comparison of stalkwise, local and global properties. There are also many examples of definable subcategories (and definable = interpretation functors between them) in the recent paper [17], though that paper uses alternative, algebraic, characterisations of these concepts rather than explicitly introducing the formal language.

References

1. Adámek, J., Rosický, J.: Locally Presentable and Accessible Categories. London Mathematical Society Lecture Note Series, vol. 189. Cambridge University Press, Cambridge (1994)
2. Borceux, F., Van den Bossche, G.: Algebra in a Localic Topos with Applications to Ring Theory. LNM, vol. 1038. Springer, Heidelberg (1983). https://doi.org/10.1007/BFb0073030
3. Eklof, P., Sabbagh, G.: Model-completions and modules. Ann. Math. Logic **2**(3), 251–295 (1971)
4. Freyd, P.: Abelian Categories. Harper and Row, New York (1964)
5. Garavaglia, S.: Dimension and rank in the model theory of modules. Preprint, University of Michigan (1979). Revised 1980
6. Gruson, L., Jensen, C.U.: Modules algébriquement compact et foncteurs $\varprojlim^{(i)}$. C. R. Acad. Sci. Paris **276**, 1651–1653 (1973)
7. Iversen, B.: Cohomology of Sheaves. Springer, Heidelberg (1986). https://doi.org/10.1007/978-3-642-82783-9
8. Prest, M.: Model Theory and Modules. London Mathematical Society Lecture Note Series, vol. 130. Cambridge University Press, Cambridge (1988)
9. Prest, M.: Purity, Spectra and Localisation. Encyclopedia of Mathematics and its Applications, vol. 121. Cambridge University Press, Cambridge (2009)
10. Prest, M.: Definable Additive Categories: Purity and Model Theory. Memoirs of the American Mathematical Society, vol. 210/no. 987. American Mathematical Society, Providence (2011)
11. Prest, M.: Abelian categories and definable additive categories. arXiv:1202.0426
12. Prest, M.: Modules as exact functors. In: Proceedings of 2016 Auslander Distinguished Lectures and Conference, Contemporary Mathematics, vol. 716. American Mathematical Society. arXiv:1801.08015 (to appear)
13. Prest, M.: Multisorted modules and their model theory. In: Contemporary Mathematics. arXiv:1807.11889 (to appear)
14. Prest, M., Puninskaya, V., Ralph, A.: Some model theory of sheaves of modules. J. Symbolic Logic **69**(4), 1187–1199 (2004)
15. Prest, M., Ralph, A.: On sheafification of modules. Preprint, University of Manchester (2001). Revised 2004. https://personalpages.manchester.ac.uk/staff/mike.prest/publications.html

16. Prest, M., Ralph, A.: Locally finitely presented categories of sheaves of modules. Preprint, University of Manchester (2001). Revised 2004 and 2018. https://personalpages.manchester.ac.uk/staff/mike.prest/publications.html
17. Prest, M., Slávik, A.: Purity in categories of sheaves. Preprint. arXiv:1809.08981 (2018)
18. Rothmaler, Ph.: A trivial remark on purity. In: Proceedings of the 9th Easter Conference on Model Theory, Gosen 1991, Seminarber. 112, Humboldt-Univ. zu Berlin, p. 127 (1991)
19. Sabbagh, G.: Sous-modules purs, existentiellement clos et élementaires. C. R. Acad. Sci. Paris **272**, 1289–1292 (1971)
20. Stenström, B.: Rings of Quotients. Springer, Heidelberg (1975). https://doi.org/10.1007/978-3-642-66066-5
21. Tennison, B.R.: Sheaf Theory. London Mathematical Society Lecture Note Series, vol. 20. Cambridge University Press, Cambridge (1975)
22. Ziegler, M.: Model theory of modules. Ann. Pure Appl. Logic **26**(2), 149–213 (1984)

Transitivity and Equivalence in Decidable Fragments of First-Order Logic: A Survey

Ian Pratt-Hartmann[1,2(✉)]

[1] School of Computer Science, University of Manchester, Manchester, UK
[2] Wydział Matematyki, Informatyki i Mechaniki, Uniwersytet Warszawski,
Warsaw, Poland
ipratt@cs.man.ac.uk

Abstract. In this talk, I survey recent work on extensions of various well-known decidable fragments of first-order logic, in which certain distinguished predicates are required to denote transitive relations or equivalence relations. I explain the origins of this work in modal logic, and outline the current state-of-the-art.

Keywords: First-order logic · Transitivity · Equivalence · Complexity

1 Introduction

In their work *Grundzüge der theoretischen Logik*, Hilbert and Ackermann [8,9] formulated the so-called *Entscheidungsproblem* for (what we now call) first-order logic: is there an algorithm which, when given a first-order formula as input, will determine whether that formula is universally valid? This question was answered negatively by Church [2] and Turing [21]: no computer program can determine the universal validity—or, dually, the satisfiability—of a given first-order formula. Since it is impossible to write a first-order formula satisfied exactly when the domain of interpretation is finite, the *Entscheidungsproblem* has a natural finitary version: is there an algorithm which, when given a first-order formula as input, will determine whether that formula is satisfiable in some *finite* structure? This question too was answered negatively, by Trakhtenbrot [20].

There are two responses to this situation, both of which have led over the years to remarkable insights of both theoretical and practical significance. The first is to develop programs designed to test the satisfiability of arbitrary collections of first-order formulas, accepting that, however well they generally work in practice, there will always be instances that defeat them. The second is to restrict attention to a subset—or, as we say, *fragment*—of first-order logic, for which the satisfiability problem (or finite satisfiability problem) is decidable, exploiting the fact that, in many real-life situations, the formulas we encounter fit comfortably into such fragments. Of course, these two approaches are not antagonistic: many general first-order theorem provers borrow techniques developed in the context of research into restricted fragments of logic to guarantee termination on certain categories of problems; conversely, techniques developed for general first-order

© Springer-Verlag GmbH Germany, part of Springer Nature 2019
Md. A. Khan and A. Manuel (Eds.): ICLA 2019, LNCS 11600, pp. 103–107, 2019.
https://doi.org/10.1007/978-3-662-58771-3_10

theorem-proving have frequently proved useful in demonstrating the decidability of certain first-order fragments.

2   The Two-Variable Fragment and its Relatives

The two-variable-fragment, $\mathcal{FO}^2$, is the set of formulas of first-order logic (with equality) that do not contain function symbols and that use only two variables. For example, the formula $\forall x \exists y (r(x, y) \wedge \forall x (r(y, x) \rightarrow x = y))$, stating that every element is r-related to some element whose only r-successor is itself, is in $\mathcal{FO}^2$. Note in particular the 're-use' of the variable x by nested quantifiers in this example. The fragment $\mathcal{FO}^2$ is easily seen to include the (propositional) modal logic K under the standard translation, which maps modal formulas to $\mathcal{FO}^2$-formulas with a single free variable x. For example, the modal formula $\Box \Diamond \Box p$ corresponds to the $\mathcal{FO}^2$-formula $\forall y (r(x, y) \rightarrow \exists x (r(y, x) \wedge \forall y (r(x, y) \rightarrow p(y))))$. We remark that the equality predicate is not needed for translating modal formulas.

The two-variable fragment has an interesting history. Scott [18] showed that any formula of the two-variable fragment can be transformed into a formula in the so-called Gödel fragment: the set of first-order formulas which, when put in prenex form, have quantifier prefixes matching the pattern $\exists^* \forall\forall \exists^*$. Earlier, Gödel [3] showed that this fragment (*without equality*) has the finite model property and claimed *en passant* that this would still be the case with equality, a claim which was later shown to be false by Goldfarb [4]. Thus, Scott's reduction showed only that the satisfiability problem for the two-variable fragment without equality is decidable. That full $\mathcal{FO}^2$ has the finite model property was established by Mortimer [13], and an exponential size-bound on these models was eventually proved by Grädel, Kolaitis and Vardi [6], thus showing that the satisfiability problem for this fragment is NExpTime-complete. By contrast, the satisfiability problem for the modal logic K is only PSpace-complete, as shown by Ladner [12].

The *guarded fragment* of first-order logic, here denoted $\mathcal{G}$, was originally developed by Andréka, van Benthem and Németi [1] as a generalization of propositional modal logic K. The idea is that quantifiers must be *guarded* by atomic formulas featuring all the free variables in their scope. More precisely: a formula $\forall u.\varphi$ counts as guarded if φ is of the form $\alpha \rightarrow \psi$, where ψ is itself guarded, and α is an atomic formula featuring all free variables of ψ. (Existentially quantified formulas are treated dually.) The guarded fragment has a decidable satisfiability problem, with complexity 2-ExpTime-complete. However, if we consider the subfragment $\mathcal{G}^k$ obtained by restricting attention to formulas with k-variables, then, for any $k \geq 2$, the complexity of satisfiability falls to ExpTime-complete, as shown by Grädel [5]. It is easy to see that the standard translations of modal formulas into $\mathcal{FO}^2$ are all in fact guarded, and thus lie in the 2-variable guarded fragment $\mathcal{G}^2$. In terms of expressive power, $\mathcal{G}^2$ lies strictly in between K and $\mathcal{FO}^2$, and the complexity of its satisfiability problem is (on standard separation assumptions) correspondingly situated.

It is possible to extend all three fragments just mentioned—K, $\mathcal{G}^2$ and $\mathcal{FO}^2$—with *counting quantifiers* $\exists_{\leq C}$, $\exists_{\geq C}$ and $\exists_{=C}$, where C is a non-negative integer. We read $\exists_{\leq C}x.\varphi$ as "there exist at most C x such that φ", and similarly for $\exists_{\geq C}$ and $\exists_{=C}$; the formal semantics is as expected. Thus, for example "Every number has at most one successor" can be written using the formula $\forall x(\text{number}(x) \rightarrow \exists_{\leq 1}y.\text{succ}(x,y))$. The extension of the two-variable fragment, $\mathcal{FO}^2$, with counting quantifiers is standardly denoted $\mathcal{C}^2$. The definition of guarded quantification applies to counting quantifiers in a natural way. The extension of the guarded two-variable fragment, $\mathcal{G}^2$, with counting quantifiers will be denoted here $\mathcal{GC}^2$.

Counting quantification does not take us beyond first-order logic; however, it does strictly extend the fragments $\mathcal{FO}^2$ and $\mathcal{G}^2$. In particular, $\mathcal{C}^2$ and $\mathcal{GC}^2$ lack the finite model-property: that is, they contain axioms of infinity. For example, the $\mathcal{GC}^2$-formula $\forall x\exists y.s(x,y)\wedge\forall x\exists_{\leq 1}y.s(y,x)\wedge\exists x\forall y.\neg s(y,x)$ is easily seen to be satisfiable, but only in infinite structures. Thus, when it comes to these logics, we may ask separately about the decidability and computational complexity of the satisfiability *and* the finite satisfiability problems. For $\mathcal{C}^2$, both problems were shown to be decidable by Grädel, Otto and Rosen [7], and by Pacholski, Szwast and Tendera [14,15], and to be NExpTime-complete by Pratt-Hartmann [16]. Likewise, the satisfiability and finite satisfiability problems for $\mathcal{GC}^2$ are both ExpTime-complete (Kazakov [10], Pratt-Hartmann [17]).

When extending modal logic with counting quantification, one employs the *graded modalities* $\Diamond_{\leq C}\varphi$, $\Diamond_{\geq C}\varphi$ and $\Diamond_{=C}\varphi$ for $C \geq 0$. Specifically, the formula $\Diamond_{\leq C}\varphi$ may be glossed: "φ is true at no more than C accessible worlds," and similarly for $\Diamond_{\geq C}\varphi$ and $\Diamond_{=C}$. The semantics for graded modal logic generalize the relational semantics for ordinary modal logic in the expected way, and translate unproblematically into the logic $\mathcal{GC}^2$. We denote by $\mathcal{GM}$ the extension of basic modal logic K with graded modalities. Thus, for example, the $\mathcal{GM}$-formula $q_0\wedge\Diamond_{\geq 2}(\neg q_1\wedge\Diamond_{=1}q_2)$ translates to the $\mathcal{GC}^2$-formula $q_0(x)\wedge\exists_{\geq 2}y(r(x,y)\wedge\neg q_1(y)\wedge \exists_{=1}x(r(y,x) \wedge q_2(x)))$. The logic $\mathcal{GM}$ is not expressive enough to force infinite models. Its satisfiability problem is PSpace-complete, as shown by Tobies [19].

3 Transitivity and Equivalence

One significant expressive limitation of $\mathcal{FO}^2$ (and indeed of $\mathcal{C}^2$) is that it cannot express the property of *transitivity*. The same is true of many other familiar classes of relations, including, most saliently perhaps, *equivalence relations*. Modal logicians have always been at home in this situation. We cannot *state* in propositional modal logic that the underlying accessibility relation is transitive, or is an equivalence relation; however, we can *assume* such properties, and investigate the characteristics of the logic thus defined. Thus, for example, in the modal logic $K4$, we have (syntactically) the same formulas as K, but we assume that the modal accessibility relation is transitive. Similarly, the modal logic $S5$ is characterized by the assumption that the accessibility relation is an equivalence relation. The satisfiability problem for $K4$ remains PSpace-complete;

the satisfiability problem for $S5$ is NPTIME-complete. These results have counterparts for the corresponding graded systems. Thus, the logic $\mathcal{GM}1T$ (graded modal logic under the assumption that the accessibility relation is transitive) has NExpTime-complete satisfiability problem, while the corresponding problem for $\mathcal{GM}1E$ (accessibility is an equivalence relation) is still NPTime-complete (Kazakov and Pratt-Hartmann [11]).

The question then presents itself: what happens if the two-variable logics $\mathcal{FO}^2$, $\mathcal{G}^2$, $\mathcal{C}^2$ or $\mathcal{GC}^2$ are extended in a similar way? That is, what happens if, for any of these fragments, we add the stipulation that some number of distinguished binary predicates satisfy a property such as being transitive, or being an equivalence relation? What complexity-theoretic landscape presents itself? What tools and techniques are required to explore these questions? Denote by $\mathcal{FO}^2 k\mathrm{E}$ the logic $\mathcal{FO}^2$ in which k distinguished binary predicates are required to be interpreted as transitive relations, and by $\mathcal{FO}^2 k\mathrm{E}$ the logic $\mathcal{FO}^2$ in which k distinguished binary predicates are required to be interpreted as equivalence relations. Similarly for $\mathcal{G}^2$, $\mathcal{GC}^2$ and $\mathcal{C}^2$. In each case, we ask whether the satisfiability and finite satisfiability problems are decidable and, if so, what their computational complexity is. In my talk, I present a survey of these questions, outlining what is known, and which problems are still open. The results (due to many authors) are summarized in Table 1.

Table 1. Known complexity bounds for the satisfiability and finite satisfiability problems for logics with two-variables and a finite number of distinguished predicates required to be interpreted as transitive relations or as equivalence relations. All bounds are tight unless otherwise indicated.

		$1E$	$2E$	kE $(k \geq 3)$	$1T$	kT $(k \geq 2)$
K	PSPACE	NPTIME			PSPACE	
$\mathcal{G}^2$	EXPTIME	NEXPTIME	2-EXPTIME	Undec.	2-EXPTIME	Undec.
$\mathcal{FO}^2$	NEXPTIME	NEXPTIME	2-NEXPTIME	Undec.	$\leq$3-NEXPTIME[a]	Undec.
$\mathcal{GM}$	PSPACE	NPTIME			NEXPTIME	
$\mathcal{GC}^2$	EXPTIME	NEXPTIME	Undec.	Undec.	Undec.	Undec.
$\mathcal{C}^2$	NEXPTIME	NEXPTIME	Undec.	Undec.	Undec.	Undec.

[a]Upper bound only, and finite satisfiability only.

References

1. Andréka, H., van Benthem, J., Németi, I.: Modal languages and bounded fragments of predicate logic. J. Philos. Logic **27**, 217–274 (1998)
2. Church, A.: A note on the Entscheidungsproblem. J. Symb. Log. **1**(1), 40–41 (1936)
3. Gödel, K.: Zum Entscheidungsproblem des logischen Funktionenkalküls. Monatshefte für Mathematik und Physik **40**, 433–443 (1933)
4. Goldfarb, W.: The unsolvability of the Gödel class with identity. J. Symbolic Logic **49**, 1237–1252 (1984)

5. Grädel, E.: On the restraining power of guards. J. Symbolic Logic **64**, 1719–1742 (1999)
6. Grädel, E., Kolaitis, P., Vardi, M.: On the decision problem for two-variable first-order logic. Bull. Symbolic Logic **3**(1), 53–69 (1997)
7. Grädel, E., Otto, M., Rosen, E.: Two-variable logic with counting is decidable. In: Proceedings of the 12th IEEE Symposium on Logic in Computer Science, pp. 306–317. IEEE Online Publications (1997)
8. Hilbert, D., Ackermann, W.: Grundzüge der theoretischen Logik. Springer, Heidelberg (1928)
9. Hilbert, D., Ackermann, W.: Grundzüge der Theoretischen Logik. DGW, vol. 27. Springer, Heidelberg (1938). https://doi.org/10.1007/978-3-662-41928-1
10. Kazakov, Y.: A polynomial translation from the two-variable guarded fragment with number restrictions to the guarded fragment. In: Alferes, J.J., Leite, J. (eds.) JELIA 2004. LNCS (LNAI), vol. 3229, pp. 372–384. Springer, Heidelberg (2004). https://doi.org/10.1007/978-3-540-30227-8_32
11. Kazakov, Y., Pratt-Hartmann, I.: A note on the complexity of the satisfiability problem for graded modal logic. In: 24th IEEE Symposium on Logic in Computer Science, pp. 407–416. IEEE Online Publications (2009)
12. Ladner, R.: The computational complexity of provability in systems of modal propositional logic. SIAM J. Comput. **6**, 467–480 (1980)
13. Mortimer, M.: On languages with two variables. Zeitschrift für Mathematische Logik und Grundlagen der Mathematik **21**, 135–140 (1975)
14. Pacholski, L., Szwast, W., Tendera, L.: Complexity of two-variable logic with counting. In: Proceedings of the 12th IEEE Symposium on Logic in Computer Science, pp. 318–327. IEEE Online Publications (1997)
15. Pacholski, L., Szwast, W., Tendera, L.: Complexity results for first-order two-variable logic with counting. SIAM J. Comput. **29**(4), 1083–1117 (1999)
16. Pratt-Hartmann, I.: Complexity of the two-variable fragment with counting quantifiers. J. Logic Lang. Inform. **14**, 369–395 (2005)
17. Pratt-Hartmann, I.: Complexity of the guarded two-variable fragment with counting quantifiers. J. Logic Comput. **17**, 133–155 (2007)
18. Scott, D.: A decision method for validity of sentences in two variables. J. Symbolic Logic **27**, 477 (1962)
19. Tobies, S.: PSPACE reasoning for graded modal logics. J. Logic Comput. **11**(1), 85–106 (2001)
20. Trakhtenbrot, B.: The impossibility of an algorithm for the decision problem for finite models. Dokl. Akad. Nauk SSSR 70, 596–572 (1950). English translation in: AMS Trans. Ser. 2, vol. 23, 1–6 (1963)
21. Turing, A.M.: On computable numbers, with an application to the Entscheidungsproblem. Proc. London Math. Soc. **42**(2), 230–265 (1936)

The Undecidability of FO3 and the Calculus of Relations with Just One Binary Relation

Yoshiki Nakamura[✉]

Tokyo Institute of Technology, Tokyo, Japan
nakamura.y.ay@is.c.titech.ac.jp

Abstract. The validity problem for first-order logic is a well-known undecidable problem. The undecidability also holds even for FO3 and (equational formulas of) the calculus of relations. In this paper we tighten these undecidability results to the following: (1) FO3 with just one binary relation is undecidable even without equality; and (2) the calculus of relations with just one character and with only composition, union, and complement is undecidable. Additionally we prove that the finite validity problem is also undecidable for the above two classes.

Keywords: First-order logic · The calculus of relations · Undecidability

1 Introduction

The *validity problem* for first-order logic (a.k.a. the Entscheidungsproblem) is the problem to decide whether a given formula of first-order logic (FO) is valid or not. The problem is recursively enumerable by Gödel's completeness theorem [6], but is undecidable shown by Church [4] and Turing [23]. In connection with the undecidability result, many decidable and undecidable variants of FO are studied (we refer to the book [3]).

One restriction is to consider the validity problem over a restricted signature. In 1915, Löwenheim [12] proved that monadic FO with equality is decidable (more precisely, coNEXPTIME-complete [11, p. 318]). Monadic FO is FO with only unary relation symbols. In 1919, Skolem [19] extended the decidability result to monadic second-order logic (see also [24]). Subsequently Büchi [18] extended the decidability result to monadic second-order logic with the linear order (called S1S). In contrast to this, FO with just one binary relation symbol is undecidable even without equality. Löwenheim [12] proved the undecidability of FO with only binary relation symbols, but the number of binary relation symbols is countably infinite. Subsequently the number was reduced to three by Herbrand [8] and to just one by Kalmár [10], see e.g., [3, p. 6] or [2, Theorem 21.4

Supported by JSPS KAKENHI Grant Number 16J08119.

(The Church-Herbrand theorem)]. Over finite models, FO with just one binary relation symbol is also undecidable by Trakhtenbrot's Theorem [22] (moreover, the undecidability holds even over finite graphs of vertex-degree at most 3 [26]).

Another restriction is with respect to the number of variables occurring in formulas. FOk denotes the restriction of FO to formulas with at most k distinct variables. In these classes, it is known that FO3 with countably infinitely many binary relation symbols and without equality is undecidable (because the $\forall\exists\forall$ case is undecidable [9]), whereas FO2 (even with equality) is decidable [16] and coNEXPTIME-complete [7, Corollary 5.4]. FO2 has connection with modal logic in the sense of that some propositional modal logics can be embedded into FO2 [25]. With respect to the expressive power of unary relations, FO2 is equivalent to boolean modal logic with relational converse and the identity relation [13]. As for FO3, it has connection with *the calculus of (binary) relations* [20]. We denote the full signature of the calculus of relations by $\langle \cdot, \bullet^-, \cup, \bullet^\smile, 1 \rangle$, where $\cdot$ is relational composition, $\bullet^-$ is set-theoretic complement, $\cup$ is set-theoretic union, $\bullet^\smile$ is relational converse, and 1 is the identity relation. Actually, with respect to the expressive power of binary relations, FO3 with equality is equivalent to terms of the calculus of relations [21] (see also [5]).

In this paper we prove that the intersection of the above two restrictions is also undecidable, i.e., the validity problem for FO3 with just one binary relation symbol and without equality is undecidable. Moreover we prove the finite validity problem for the class is undecidable. The class is the minimal undecidable fragment with respect to the above two restrictions since both monadic FO and FO2 are decidable. As for the calculus of relations, we prove that the validity problem and the finite validity problem for equational formulas of the calculus of relations with just one character over the signature $\langle \cdot, \bullet^-, \cup \rangle$ are undecidable.

Outline

Figure 1 gives the outline of this paper. Every arrow denotes that there is a *conservative reduction* (a reduction preserving validity and finite validity) from the source to the target and A denotes a countably infinite set. In Sect. 2 we introduce first-order logic and the calculus of relations, and we show that FO3 with equality and the calculus of relations are equivalent in the sense of expressive power of binary relations. In this section we prove that relational converse $\bullet^\smile$ and the identity relation symbol 1 can be eliminated by using fresh variables preserving validity and finite validity. In Sect. 3 we give reductions for reducing the number of characters to one. Finally, in Sect. 4 we conclude this paper.

2   FO3 and the Calculus of Relations

Let V be a countably infinite set of *(first-order) variables*. In this paper we assume that every *signature* of first-order logic is a set of binary relation symbols. Let A be a set denoting a signature. The set of *formulas* of first-order logic over A, written $\mathsf{FO}_A^=$, is defined by the following grammar (we omit parentheses and we use them in ambiguous situations when it is not clear how to parse):

$$\varphi, \psi \in \mathsf{FO}_A^= ::= a(x,y) \mid x = y \mid \varphi \vee \psi \mid \exists x.\varphi \mid \neg\varphi \qquad (a \in A \text{ and } x, y \in V)$$

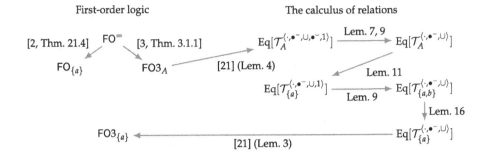

Fig. 1. An overview of reductions

We use the following abbreviations: (1) $\varphi \wedge \psi := \neg((\neg\varphi) \vee (\neg\psi))$; (2) $\varphi \rightarrow \psi :=$ $(\neg\varphi) \vee \psi$; (3) $\varphi \leftrightarrow \psi := (\varphi \rightarrow \psi) \wedge (\psi \rightarrow \varphi)$; (4) $\forall x.\varphi := \neg\exists x.\neg\varphi$. FO_A denotes the subset of $FO_A^=$ whose formulas do not contain the symbol $=$. $FOk_A^=$ (resp. FOk_A) denotes the subset of $FO_A^=$ (resp. FO_A) whose formulas contain at most k distinct variables[1]. $FO^=$ denotes the set of all formulas in $FO_A^=$ over a signature A.

A *model* over A is a tuple $M = \langle |M|, \mathcal{R}_M \rangle$, where $|M|$ is a nonempty set; and $\mathcal{R}_M : A \rightarrow \wp(|M|^2)$ is a function. Each $\mathcal{R}_M(a)$ denotes a binary relation on $|M|$. We may view every model as a graph, e.g., each element of $|M|$ is called a *vertex* and each element of $\mathcal{R}(a)$ is called an *edge* labelled with a. An *interpretation* I on M is a function from V to $|M|$. The *semantics* of every formula φ on $\langle M, I \rangle$ is a truth value. $\langle M, I \rangle \models \varphi$ (resp. $\langle M, I \rangle \not\models \varphi$) denotes that the value of φ on $\langle M, I \rangle$ is *true* (resp. *false*). $\langle M, I \rangle \models \varphi$ is defined as follows: (1) $\langle M, I \rangle \models a(x,y) :\Longleftrightarrow$ $\langle I(x), I(y) \rangle \in \mathcal{R}_M(a)$ for $a \in A$; (2) $\langle M, I \rangle \models x = y :\Longleftrightarrow I(x) = I(y)$; (3) $\langle M, I \rangle \models \varphi \vee \psi :\Longleftrightarrow \langle M, I \rangle \models \varphi$ or $\langle M, I \rangle \models \psi$; (4) $\langle M, I \rangle \models \exists x.\varphi :\Longleftrightarrow$ there is $v \in |M|$ s.t. $\langle M, [v/x]I \rangle \models \varphi$; (5) $\langle M, I \rangle \models \neg\varphi :\Longleftrightarrow \langle M, I \rangle \not\models \varphi$, where $[v/x]I$ denotes the function I in which $I(x)$ has been replaced by the element v. Then the *semantics* of φ on M is defined by $[\![\varphi]\!]_M := \{I : V \rightarrow |M| \mid \langle M, I \rangle \models \varphi\}$. $M \models \varphi$ denotes that, for any interpretation I, $\langle M, I \rangle \models \varphi$ holds. $[\![\varphi]\!]_M^{\vec{x}}$ denotes the *projection* of $[\![\varphi]\!]_M$ with respect to $\vec{x}$, i.e., $[\![\varphi]\!]_M^{\vec{x}} := \{\langle I(x_1), \ldots, I(x_n) \rangle \mid I \in [\![\varphi]\!]_M\}$, where $\vec{x} = \langle x_1, \ldots, x_n \rangle$.

A formula φ over the signature A is called *valid* (resp. *finitely valid*) if, for any model (resp. finite model) M over A, $M \models \varphi$ holds. The *validity problem* (resp. the *finite validity problem*) is the problem to decide whether a given formula is valid (resp. finitely valid). In this paper we rely on the undecidability of the $\forall\exists\forall$ case [9] and modify the result for FO3 and the calculus of relations. This class is a *conservative reduction* class [3, Definition 2.1.35] (i.e., there is a recursive function from every first-order logic formula over any signature to a formula of the class preserving validity and finite validity). From this, the following holds.

Lemma 1 (A corollary of [3, Theorem 3.1.1]). *Let A be any countably infinite set. Then*

---

[1] For example, $\exists x.\exists y.\exists z.\exists w.a(x,y)$ is not an $FO3_{\{a\}}$ formula (the formula is an $FO4_{\{a\}}$ formula), but $\exists x.\exists y.\exists z.\exists x.a(x,y)$ is an $FO3_{\{a\}}$ formula.

- *The validity problem for* FO3$_A$ *is undecidable; and*
- *The finite validity problem for* FO3$_A$ *is undecidable.*

We now define the calculus of relations. Let X be a set and let R and R' be binary relations. The *identity relation* $\triangle(Y)$ on a set Y is defined as $\{\langle v, v \rangle \mid v \in Y\}$. The *relational composition* $R \cdot R'$ is defined as $\{\langle v, v' \rangle \mid \exists v''.\langle v, v'' \rangle \in R \wedge \langle v'', v' \rangle \in R'\}$. The *relational converse* $R^\smile$ is defined as $\{\langle v', v \rangle \mid \langle v, v' \rangle \in R\}$. The *union* $R \cup R'$ and the *complement* R^- are defined as set-theoretic union and complement (in particular, $R^- := |X|^2 \setminus R$). Let A be a set of *characters*. The set of *terms* over A, written $\mathcal{T}_A$, is defined by the following grammar:

$$t, u \in \mathcal{T}_A ::= a \mid t \cdot u \mid t^- \mid t \cup u \mid t^\smile \mid 1 \qquad (a \in A)$$

We use the following abbreviations: (1) $t \cap u := (t^- \cup u^-)^-$; (2) $\top := a \cup a^-$; (3) $0 := a \cap a^-$; (4) $t^{n+1} := t^n \cdot t$ and $t^1 := t$, where a is some character in A and $n \geq 1$. $\mathcal{T}_A^\varsigma$ denotes the set of terms over A and over the signature ς, where ς is a subset of $\{\cdot, \bullet^-, \cup, \bullet^\smile, 1\}$. Note that the symbols, $\bullet^\smile$ and 1, are not used in these abbreviations.

Let M be a model over A. The semantics of every term t on M, written $[\![t]\!]_M$, is a binary relation on $|M|$, defined as follows: (1) $[\![a]\!]_M := \mathcal{R}_M(a)$; (2) $[\![t \cdot u]\!]_M := [\![t]\!]_M \cdot [\![u]\!]_M$; (3) $[\![t^-]\!]_M := |M|^2 \setminus [\![t]\!]_M$; (4) $[\![t \cup u]\!]_M := [\![t]\!]_M \cup [\![u]\!]_M$; (5) $[\![t^\smile]\!]_M := [\![t]\!]_M^\smile$; (6) $[\![1]\!]_M := \triangle(|M|)$. Note that $[\![t \cap u]\!]_M = [\![t]\!]_M \cap [\![u]\!]_M$, $[\![\top]\!]_M = |M|^2$, and $[\![0]\!]_M = \varnothing$ hold. We also define the set of *formulas of the calculus of the relations* over a set $\mathcal{T}$, written $\Phi[\mathcal{T}]$, by the following grammar:

$$\varphi, \psi \in \Phi[\mathcal{T}] ::= t = u \mid \varphi \vee \psi \mid \neg\varphi \qquad (t, u \in \mathcal{T})$$

where $\mathcal{T}$ is a set of terms. The semantics of every formula $\varphi \in \Phi[\mathcal{T}]$ on M is a truth value. $M \models \varphi$ (resp. $M \not\models \varphi$) denotes that the value of φ on M is *true* (resp. *false*) defined as follows: (1) $M \models t = u \iff [\![t]\!]_M = [\![u]\!]_M$; (2) $M \models \varphi \vee \psi \iff M \models \varphi$ or $M \models \psi$; (3) $M \models \neg\varphi \iff M \not\models \varphi$. A formula of the form $t = u$ is called an *equational formula*. We denote the set of *equational formulas* over a set $\mathcal{T}$ by $\text{Eq}[\mathcal{T}]$. Actually every formula is equivalent to an equational formula by the following lemma.

Lemma 2 (cf. [14, Theorem 1]). *For any formula* φ *in* $\Phi[\mathcal{T}_A]$, *there is an equational formula* ψ *of the form* $t = 0$ *in* $\text{Eq}[\mathcal{T}_A]$ *s.t. for any* M *over* A, $M \models \varphi \iff M \models \psi$.

Proof (Sketch). It is proved by using the following equivalences: (1) $M \models t = u \iff M \models (t \cap u^-) \cup (t^- \cap u) = 0$; (2) $M \models (t=0) \vee (u=0) \iff M \models t \cdot \top \cdot u = 0$; (3) $M \models \neg(t=0) \iff M \models (\top \cdot t \cdot \top)^- = 0$.

By Lemma 2, every formula of the calculus of relations can be translated to an equational formula preserving validity and finite validity. For simplicity, we may use formulas of the calculus of relations (but it is not essential). Note that the symbols, $\bullet^\smile$ and 1, are not used in the translation in Lemma 2.

FO3$^=$ and the calculus of relations are equivalent in the sense of expressive power of binary relations. We use $[n]$ to denote the set $\{1, \ldots, n\}$ for every natural number $n \geq 0$. The following lemma shows that every term of the calculus of relations has an equivalent FO3$^=$ formula.

Lemma 3 (e.g., [21, p. 28][5]). *Let x_1, x_2, and x_3 are three distinct variables. There is a recursive function G from $T_A \times [3]^2$ to* FO3$^=_{\overline{A}}$ *s.t., for any $\langle t, i, j \rangle$ and any model M over A, $[\![t]\!]_M = [\![G(t, i, j)]\!]_M^{x_i, x_j}$ holds.*

Proof (Sketch). We define $G(t, i, j)$ as follows: (1) $G(a, i, j) := a(x_i, x_j)$; (2) $G(t \cdot u, i, j) := \exists k. G(t, i, k) \wedge G(u, k, j)$; (3) $G(t^-, i, j) := \neg G(t, i, j)$; (4) $G(t \cup u, i, j) := G(t, i, j) \vee G(u, i, j)$; (5) $G(t^\smile, i, j) := G(t, j, i)$; (6) $G(1, i, j) := x_i = x_j$, where k is the minimum element in $[3] \setminus \{i, j\}$. Then $[\![t]\!]_M = [\![G(t, x_i, x_j)]\!]_M^{x_i, x_j}$ is proved by induction on the structure of t.

Actually the converse of Lemma 3 also holds, i.e., FO3$^=$ and the calculus of relations are equivalent in the sense of expressive power of binary relations.

Lemma 4 (e.g., [21, Sect. 3.9][5]). *There is a recursive function H from* FO3$^=_{\overline{A}} \times V^2$ *to T_A s.t., for any $\langle \varphi, v, w \rangle$ and any model M over A, $[\![\varphi]\!]_M^{v,w} = [\![H(\varphi, v, w)]\!]_M$.*

By Lemmas 3 and 4, every equational formula of the calculus of relations can be translated to an FO3 sentence, and vice versa, by the following:

$$M \models t = u \iff M \models \forall x_1. \forall x_2. G(t, 1, 2) \leftrightarrow G(u, 1, 2)$$
$$M \models \varphi \iff M \models H(\varphi, x, x) = \top$$

where x is an arbitrary variable. Therefore the following also holds.

Lemma 5. *Let A be any countably infinite set. Then both the validity problem and the finite validity problem for* Eq$[T_A^{\langle \cdot, \bullet^-, \cup, \bullet^\smile, 1 \rangle}]$ *are undecidable.*

2.1 Elimination of Relational Converse (Using the Identity Relation)

In this subsection, we show that relational converse $\bullet^\smile$ can be eliminated by using fresh variables preserving validity and finite validity (Lemma 7). Let $\breve{A}$ be a countably infinite set that is disjoint with A. We use $\breve{a}$ to denote the character in $\breve{A}$ denoting the converse of the character a in A. The following two axioms force that $\breve{a}$ is the converse of a: (Ca.1) $(a \cdot \breve{a}^-) \cap 1 = 0$; and (Ca.2) $(\breve{a} \cdot a^-) \cap 1 = 0$. (Ca.1-2) denotes the formula (Ca.1) $\wedge$ (Ca.2). In fact the following holds.

Proposition 6. *For any model M over the alphabet $A \cup \breve{A}$, the following hold: (1) $M \models$ (Ca.1-2) $\iff$ $M \models a^\smile = \breve{a}$; (2) $M \models (t \cdot u)^\smile = u^\smile \cdot t^\smile$; (3) $M \models (t^-)^\smile = (t^\smile)^-$; (4) $M \models (t \cup u)^\smile = t^\smile \cup u^\smile$; (5) $M \models (t^\smile)^\smile = t$; (6) $M \models (1)^\smile = 1$.*

Proof. (2)–(6) are easily proved by the definition of $[\![t]\!]_M$. We only prove (1). ($\Rightarrow$): We assume that there is a pair $\langle v, w \rangle$ such that $\langle v, w \rangle \in [\![a]\!]_M^{\smile} \setminus [\![ă]\!]_M$. Then, by $\langle w, v \rangle \in [\![a]\!]_M$ and $\langle v, w \rangle \in [\![ă^-]\!]_M$, $\langle w, w \rangle \in [\![(a \cdot ă^-) \cap 1]\!]_M$. However this contradicts to (Ca.1). Therefore $[\![a]\!]_M^{\smile} \subseteq [\![ă]\!]_M$. We assume that there is a pair $\langle v, w \rangle$ such that $\langle v, w \rangle \in [\![ă]\!]_M \setminus [\![a]\!]_M^{\smile}$. Then, by $\langle v, w \rangle \in [\![ă]\!]_M$ and $\langle w, v \rangle \in [\![a^-]\!]_M$, $\langle v, v \rangle \in [\![(ă \cdot a^-) \cap 1]\!]_M$. However this contradicts to (Ca.2). Therefore $[\![a]\!]_M^{\smile} \supseteq [\![ă]\!]_M$, and thus $M \models a^{\smile} = ă$. ($\Leftarrow$): We prove the contraposition. If $M \not\models$ (Ca.1), there are v and w such that $\langle v, w \rangle \in [\![a]\!]_M$ and $\langle w, v \rangle \in [\![ă^-]\!]_M$. Then by $\langle w, v \rangle \in [\![a^{\smile}]\!]_M$ and $\langle w, v \rangle \notin [\![ă]\!]_M$, $M \not\models a^{\smile} = ă$. If $M \not\models$ (Ca.2), there are v and w such that $\langle v, w \rangle \in [\![ă]\!]_M$ and $\langle w, v \rangle \in [\![a^-]\!]_M$. Then by $\langle v, w \rangle \notin [\![a^{\smile}]\!]_M$ and $\langle v, w \rangle \in [\![ă]\!]_M$, $M \not\models a^{\smile} = ă$. $\qquad\square$

Let $\mathrm{CF}(t)$ be the normal form of a term t w.r.t. the following rewriting rule: (1) $(a)^{\smile} \rightsquigarrow ă$; (2) $(t \cdot u)^{\smile} \rightsquigarrow u^{\smile} \cdot t^{\smile}$; (3) $(t^-)^{\smile} \rightsquigarrow (t^{\smile})^-$; (4) $(t \cup u)^{\smile} \rightsquigarrow t^{\smile} \cup u^{\smile}$; (5) $(t^{\smile})^{\smile} \rightsquigarrow t$; (6) $(1)^{\smile} \rightsquigarrow 1$. Note that, for any term t, $\mathrm{CF}(t)$ does not contain $\bullet^{\smile}$. We use A_φ to denote the finite set of all characters occurring in φ.

Lemma 7. *For any $t, u \in \mathcal{T}_A$, the following are equivalent: (1) $t = u$ is valid (resp. finitely valid). (2) $(\bigwedge_{a \in A_{t=u}} (\text{Ca.1-2})) \rightarrow \mathrm{CF}(t) = \mathrm{CF}(u)$ is valid (resp. finitely valid).*

Proof. (2) $\Rightarrow$ (1): Let M be a model over A such that $M \not\models t = u$. We define the model M' over $A \cup Ă$ by $M' = (|M|, \mathcal{R}')$, where $\mathcal{R}'(a) = \mathcal{R}_M(a)$ for $a \in A$; and $\mathcal{R}'(ă) = \mathcal{R}_M(a)^{\smile}$ for $ă \in Ă$. Then $M' \models \bigwedge_{a \in A_{t=u}} (\text{Ca.1-2})$ holds by Proposition 6 (1). Moreover, for any term $t \in \mathcal{T}_A$, $[\![t]\!]_M = [\![\mathrm{CF}(t)]\!]_{M'}$ holds by using Proposition 6. Therefore $M' \not\models \mathrm{CF}(t) = \mathrm{CF}(u)$. (1) $\Rightarrow$ (2): Let M be a model over A such that $M \models \bigwedge_{a \in A_{t=u}} (\text{Ca.1-2})$ holds, but $M \not\models \mathrm{CF}(t) = \mathrm{CF}(u)$. Then $[\![t]\!]_M = [\![\mathrm{CF}(t)]\!]_M$ holds for any term $t \in \mathcal{T}_A$ by using Proposition 6, and thus $M \not\models t = u$. $\qquad\square$

By Lemma 2, Each formula $(\bigwedge_{a \in A_{(t=u)}} (\text{Ca.1-2})) \rightarrow \mathrm{CF}(t) = \mathrm{CF}(u)$ can be translated to an equational formula without $\bullet^{\smile}$.

2.2 Elimination of the Identity Relation

In this subsection, we also eliminate the identity relation symbol 1 by using a fresh variable E. The key is to construct an equivalence relation denoting the identity relation not using 1 or $\bullet^{\smile}$. Let consider the following axioms: (E.1) $E \cdot \top = \top$; (E.2) $(E^- \cdot E) \cap E = 0$; (E.3) $(E \cdot E^-) \cap E = 0$; and (E.4) $(E \cdot E) \cap E^- = 0$. (E.1-4) denotes the formula (E.1) $\wedge \cdots \wedge$ (E.4).

Proposition 8. $M \models$ (E.1-4) $\iff [\![E]\!]_M$ *is an equivalence relation on* $|M|$.

Proof. ($\Leftarrow$): (E.1) is by the reflexivity. (E.4) is by the transitivity. (E.2) and (E.3) are by the symmetry and the transitivity. ($\Rightarrow$): The reflexivity is shown by (E.1) and (E.2). Let v be any vertex in $|M|$. Then let w be a vertex such that $\langle v, w \rangle \in [\![E]\!]_M$ holds (such w always exists by (E.1)). We assume that

$\langle v, v \rangle \notin [\![E]\!]_M$. Then, by $\langle v, v \rangle \in [\![E^-]\!]_M$, $\langle v, w \rangle \in [\![(E^- \cdot E) \cap E]\!]_M$ holds. However this contradicts to (E.2). Therefore $\langle v, v \rangle \in [\![E]\!]_M$. The symmetry is by the reflexivity and (E.3). Let $\langle v, w \rangle \in [\![E]\!]_M$. We assume $\langle w, v \rangle \notin [\![E]\!]_M$. Then $\langle v, v \rangle \in [\![(E \cdot E^-) \cap E]\!]_M$ by the reflexivity. However this contradicts to (E.3). Therefore $\langle w, v \rangle \in [\![E]\!]_M$. The transitivity is by (E.4).   □

For every term t, the term $\mathrm{IF}(t)$ is inductively defined as follows:(1) $\mathrm{IF}(a) := E \cdot a \cdot E$; (2) $\mathrm{IF}(t \cdot u) := \mathrm{IF}(t) \cdot \mathrm{IF}(u)$; (3) $\mathrm{IF}(t^-) := \mathrm{IF}(t)^-$; (4) $\mathrm{IF}(t \cup u) := \mathrm{IF}(t) \cup \mathrm{IF}(u)$; (5) $\mathrm{IF}(t^\smallsmile) := \mathrm{IF}(t)^\smallsmile$; (6) $\mathrm{IF}(1) := E$. Note that (i) $\mathrm{IF}(t)$ does not contain the symbol 1; and (ii) if t does not contain $\bullet^\smallsmile$, then $\mathrm{IF}(t)$ also does not contain $\bullet^\smallsmile$.

Lemma 9. *For any $t, u \in \mathcal{T}_A^{\langle \cdot, \bullet^-, \cup, 1, \bullet^\smallsmile \rangle}$, The following are equivalent: (1) $t = u$ is valid (resp. finitely valid). (2) (E.1-4) $\rightarrow \mathrm{IF}(t) = \mathrm{IF}(u)$ is valid (resp. finitely valid).*

Proof. (2) $\Rightarrow$ (1): Let M be a model over A such that $M \not\models t = u$. We define the model M' over $A \cup \{E\}$ by $M' := (|M|, \mathcal{R}')$, where $\mathcal{R}'(a) = \mathcal{R}_M(a)$ for $a \in A$ and $\mathcal{R}'(E) = \triangle(|M|)$. Then $M' \models$ (E.1-4) holds because $[\![E]\!]_{M'}$ is an equivalence relation. Also $M' \not\models \mathrm{IF}(t) = \mathrm{IF}(u)$ holds because $[\![t]\!]_M = [\![\mathrm{IF}(t)]\!]_{M'}$ holds by the definition of M'. (1) $\Rightarrow$ (2): Let M be a model over $A \cup \{E\}$ such that $M \models$ (E.1-4) holds, but $M \not\models \mathrm{IF}(t) = \mathrm{IF}(u)$. We define the model M' over A by $M' := (|M|/E, \mathcal{R}')$, where $|M|/E$ is the quotient set of $|M|$ by E; $[v]_E$ denotes the equivalence class of an element v with respect to E; and $\mathcal{R}'(a) = \{\langle [v]_E, [w]_E \rangle \mid \langle v, w \rangle \in \mathcal{R}_M(a)\}$ for $a \in A$. Then $\{\langle [v]_E, [w]_E \rangle \mid \langle v, w \rangle \in [\![\mathrm{IF}(t)]\!]_M\} = [\![t]\!]_{M'}$ holds for any term t. This is proved by induction on the structure of t. Therefore $M' \not\models t = u$.   □

We can eliminate both converse and the identity relation preserving validity and finite validity by Lemmas 7 and 9. Therefore the following holds.

Corollary 10. *Let A be a countably infinite set. Then both the validity problem and the finite validity problem for $\mathrm{Eq}[\mathcal{T}_A^{\langle \cdot, \bullet^-, \cup \rangle}]$ are undecidable.*

3   Reductions to the One Binary Relation Case

In this section we consider to reduce the number of characters. Let A be an ordered set $\{a_1, a_2, \dots\}$. Without loss of generality, we can assume that every term is in $\mathcal{T}_A^{\langle \cdot, \bullet^-, \cup \rangle}$ by Corollary 10. In this section we give two reductions to the one binary relation case. First we give a simple reduction, but this reduction uses the identity relation. The second reduction is a bit more complex than the first reduction, but the reduction does not use the identity relation.

3.1   A Conservative Reduction Using Identity

We first give a reduction T_1 from $\mathcal{T}_A^{\langle \cdot, \bullet^-, \cup, 1 \rangle}$ to $\mathcal{T}_{\{a\}}^{\langle \cdot, \bullet^-, \cup, 1 \rangle}$. The term $\mathrm{T}_1(t)$ is inductively defined as follows: (1) $\mathrm{T}_1(a_i) := (a \cap 1) \cdot a \cdot ((a^- \cap 1) \cdot a)^i \cdot (a \cap 1)$; (2) $\mathrm{T}_1(t \cdot u) := \mathrm{T}_1(t) \cdot \mathrm{T}_1(u)$; (3) $\mathrm{T}_1(t^-) := (a \cap 1) \cdot \mathrm{T}_1(t)^- \cdot (a \cap 1)$; (4) $\mathrm{T}_1(t \cup u) := \mathrm{T}_1(t) \cup \mathrm{T}_1(u)$; (5) $\mathrm{T}_1(1) := (a \cap 1)$.

Figure 2 is an example of transforming from models over A to models over $\{a\}$ in Lemma 11. Each blue (resp. red, gray) colored edge denotes an edge labeled with a_1 (resp. a_2, a).

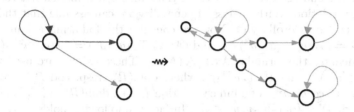

Fig. 2. A transformation to the one binary relation case using identity (Color figure online)

Lemma 11. *For any $t, u \in T_A^{\langle \cdot, \bullet^-, \cup, 1 \rangle}$, The following are equivalent: (1) $t = u$ is valid (resp. finitely valid). (2) $\mathrm{T}_1(t) = \mathrm{T}_1(u)$ is valid (resp. finitely valid).*

Proof. (2) $\Rightarrow$ (1): Let M be a model over $A_{t=u}$ such that $M \not\models t = u$. We define the model M' over $\{a\}$ by $M' := \langle |M'|, \mathcal{R}' \rangle$, where $|M'| := |M| \cup \{\langle v, w, i, j \rangle \mid \langle v, w \rangle \in \mathcal{R}_M(a_i), j \in [i]\}$ and $\mathcal{R}'(a) := \triangle(|M|) \cup \{\langle v, \langle v, w, i, 1 \rangle \rangle, \langle \langle v, w, i, i \rangle, w \rangle \mid \langle v, w \rangle \in \mathcal{R}_M(a_i)\} \cup \{\langle \langle v, w, i, j \rangle, \langle v, w, i, j+1 \rangle \rangle \mid \langle v, w \rangle \in \mathcal{R}_M(a_i), j \in [i-1]\}$. The right-hand side in Fig. 2 is an example of M'. Note that if M is finite, M' is also finite because $A_{t=u}$ is finite. Then $[\![t]\!]_M = [\![\mathrm{T}_1(t)]\!]_{M'}$ holds. This is easily proved by induction on the structure of t. Therefore $M' \not\models \mathrm{T}_1(t) = \mathrm{T}_1(u)$. (1) $\Rightarrow$ (2): Let M be a model over $\{a\}$ such that $M \not\models \mathrm{T}_1(t) = \mathrm{T}_1(u)$. Then we define the model M' over A by $M' := \langle |M'|, \mathcal{R}' \rangle$, where $|M'| := \{v \mid \langle v, v \rangle \in [\![a]\!]_M\}$ and $\mathcal{R}'(a_i) := |M'|^2 \cap [\![\mathrm{T}_1(a_i)]\!]_M$. Then $[\![t]\!]_{M'} = [\![\mathrm{T}_1(t)]\!]_M$ holds. This is easily proved by induction on the structure of t. Therefore $M' \not\models t = u$. □

By the above lemma, the following has been proved.

Theorem 12. *Let a be a character. Then both the validity problem and the finite validity problem for $\mathrm{Eq}[T_{\{a\}}^{\langle \cdot, \bullet^-, \cup, 1 \rangle}]$ are undecidable.*

Remark 13. The undecidability of the validity problem for the calculus of relations with just one character over the signature $\langle \cdot, \bullet^-, \cup, \bullet^\smile, 1 \rangle$ had been shown by Maddux [15, p. 399]. More strongly, Theorem 12 shows that the validity problem is undecidable even without $\bullet^\smile$ and shows that the finite validity problem is also undecidable. Furthermore, Theorem 12 will be strengthened to Theorem 17.

Combining Theorem 12 and Lemma 9, the following is also proved.

Corollary 14. *Let a_1, a_2 be two distinct characters. Then both the validity problem and the finite validity problem for $\mathrm{Eq}[T_{\{a_1,a_2\}}^{\langle \cdot, \bullet^-, \cup \rangle}]$ are undecidable.*

3.2   A Conservative Reduction Not Using Identity

In this subsection we give another reduction, which is not using the identity relation and relational converse. The key is how to distinguish the vertices of a given model and the other vertices on the transformed model, not relying on the identity relation. Without loss of generality we can assume that the size of character is 2 by Corollary 14. Now we consider the following axioms: (Ax.1) $(a \cap a^2) \cdot \top = (a \cap a^2) \cdot (a \cap a^3) \cdot \top$; and (Ax.2) $\top \cdot (a \cap a^3) = \top \cdot (a \cap a^2) \cdot (a \cap a^3)$. (Ax.1-2) denotes the formula (Ax.1) $\wedge$ (Ax.2). These axioms are used to force $\mathrm{cod}(\llbracket a \cap a^2 \rrbracket_M) = \mathrm{dom}(\llbracket a \cap a^3 \rrbracket_M)$, where $\mathrm{dom}(R)$ (resp. $\mathrm{cod}(R)$) denote the *domain* (resp. *codomain*) of a binary relation R, i.e., $\mathrm{dom}(R) := \{v \mid \exists w.\langle v, w \rangle \in R\}$ and $\mathrm{cod}(R) := \{w \mid \exists v.\langle v, w \rangle \in R\}$. In fact the following holds.

Proposition 15. $M \models$ (Ax.1-2) $\iff \mathrm{cod}(\llbracket a \cap a^2 \rrbracket_M) = \mathrm{dom}(\llbracket a \cap a^3 \rrbracket_M)$.

Proof. By $M \models$ (Ax.1) $\iff \mathrm{cod}(\llbracket a \cap a^2 \rrbracket_M) \subseteq \mathrm{dom}(\llbracket a \cap a^3 \rrbracket_M)$ and $M \models$ (Ax.2) $\iff \mathrm{cod}(\llbracket a \cap a^2 \rrbracket_M) \supseteq \mathrm{dom}(\llbracket a \cap a^3 \rrbracket_M)$. $\square$

We define a reduction T_2 from $\mathcal{T}_{\{a_1,a_2\}}^{\langle \cdot, \bullet^-, \cup \rangle}$ to $\mathcal{T}_{\{a\}}^{\langle \cdot, \bullet^-, \cup \rangle}$ as follows: (1) $\mathrm{T}_2(a_i) := (a \cap a^3) \cdot a^{i+1} \cdot (a \cap a^2)$; (2) $\mathrm{T}_2(t \cdot u) := \mathrm{T}_2(t) \cdot \mathrm{T}_2(u)$; (3) $\mathrm{T}_2(t^-) := ((a \cap a^3) \cdot \top) \cap (\top \cdot (a \cap a^2)) \cap \mathrm{T}_2(t)^-$; (4) $\mathrm{T}_2(t \cup u) := \mathrm{T}_2(t) \cup \mathrm{T}_2(u)$. Intuitively, both the codomain of $a \cap a^2$ and the domain of $a \cap a^3$ denote the set of vertices in the pre-transformed model in the above reduction.

Figure 3 is an example of transforming models over $\{a_1, a_2\}$ to models over $\{a\}$ in Lemma 16. Each blue (resp. red, gray) colored edge denotes an edge labeled with a_1 (resp. a_2, a).

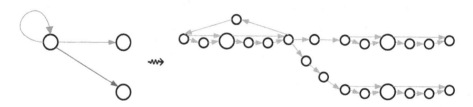

Fig. 3. A transformation to the one binary relation case not using identity (Color figure online)

Lemma 16. *For any $t, u \in \mathcal{T}_{\{a_1,a_2\}}^{\langle \cdot, \bullet^-, \cup \rangle}$, the following are equivalent: (1) $t = u$ is valid (resp. finitely valid). (2) (Ax.1-2) $\to \mathrm{T}_2(t) = \mathrm{T}_2(u)$ is valid (resp. finitely valid).*

Proof. (2) $\Rightarrow$ (1): Let M be a model over $\{a_1, a_2\}$ such that $M \not\models t = u$. We define the model M' over $\{a\}$ by $M' := \langle |M'|, \mathcal{R}' \rangle$, where $|M'| := (|M| \times [6]) \cup \{\langle v, w, i, j \rangle \mid \langle v, w \rangle \in \mathcal{R}_M(a_i), j \in [i]\}$ and $\mathcal{R}'(a)$ is the union of the following:

   (i) $\bigcup_{v \in |M|} (\{\langle \langle v, i \rangle, \langle v, i+1 \rangle \rangle \mid i \in [5]\} \cup \{\langle \langle v, 1 \rangle, \langle v, 3 \rangle \rangle, \langle \langle v, 3 \rangle, \langle v, 6 \rangle \rangle\})$;
  (ii) $\bigcup_{i \in \{1,2\}} \{\langle v, \langle v, w, i, 1 \rangle \rangle, \langle \langle v, w, i, i \rangle, w \rangle \mid (v, w) \in \mathcal{R}_M(a_i)\}$; and
 (iii) $\{\langle \langle v, w, 2, 1 \rangle, \langle v, w, 2, 2 \rangle \rangle \mid (v, w) \in \mathcal{R}_M(a_2)\}$.

The right-hand side in Fig. 3 is an example of M'.

First $M' \models$ (Ax.1-2) holds by $\mathrm{cod}(\llbracket a \cap a^2 \rrbracket_{M'}) = \{\langle v, 3 \rangle \mid v \in |M|\} = \mathrm{dom}(\llbracket a \cap a^3 \rrbracket_{M'})$. Moreover the following holds: $\{\langle \langle v, 3 \rangle, \langle w, 3 \rangle \rangle \mid (v, w) \in \llbracket t \rrbracket_M\} = \llbracket T_2(t) \rrbracket_{M'} \cdots (\heartsuit)$. $M' \not\models T_2(t) = T_2(u)$ is proved by $(\heartsuit)$. We prove $(\heartsuit)$ by induction on t.

If $t \equiv a_i$, let $\langle v, w \rangle$ be a pair such that $\langle v, w \rangle \in \llbracket a_i \rrbracket_M$. Then $\langle \langle v, 3 \rangle, \langle w, 3 \rangle \rangle \in \llbracket (a \cap a^3) \cdot a^{i+1} \cdot (a \cap a^2) \rrbracket_{M'}$ is easily checked. Conversely, if $\langle v, w \rangle \in \llbracket (a \cap a^3) \cdot a^{i+1} \cdot (a \cap a^2) \rrbracket_{M'}$, then then there are $v', w' \in |M|$ such that $\langle v, w \rangle = \langle \langle v', 3 \rangle, \langle w', 3 \rangle \rangle$ by $\mathrm{cod}(\llbracket a \cap a^2 \rrbracket_{M'}) = \mathrm{dom}(\llbracket a \cap a^3 \rrbracket_{M'}) = \{\langle v, 3 \rangle \mid v \in |M|\}$. Then $\{v \mid \langle \langle v', 3 \rangle, v \rangle \in \llbracket a \cap a^3 \rrbracket_{M'}\} = \{\langle v', 6 \rangle\}$ and $\{w \mid \langle w, \langle w', 3 \rangle \rangle \in \llbracket a \cap a^2 \rrbracket_{M'}\} = \{\langle w', 1 \rangle\}$ hold, and thus $\langle \langle v', 6 \rangle, \langle w', 1 \rangle \rangle \in \llbracket a^{i+1} \rrbracket_{M'}$. If we assume $\langle v', w' \rangle \notin \llbracket a_i \rrbracket_M$, then $\langle w', 1 \rangle$ is not reachable from $\langle v', 6 \rangle$ in $i+1$ steps because $i+1 \le 3$ and the length of paths from $\langle v', 6 \rangle$ to $\langle w', 1 \rangle$ via some $\langle w'', 1 \rangle$ is at least 6, where w'' is an element in $|M|$ not equal to w'. (More concretely, the following path is the shortest path from $\langle v', 6 \rangle$ to $\langle w', 1 \rangle$ via $\langle w'', 1 \rangle$: $\langle v', 6 \rangle \rightsquigarrow \langle v', w'', 1, 1 \rangle \rightsquigarrow \langle w'', 1 \rangle \rightsquigarrow \langle w'', 3 \rangle \rightsquigarrow \langle w'', 6 \rangle \rightsquigarrow \langle w'', w', 1, 1 \rangle \rightsquigarrow \langle w', 1 \rangle$.) However this contradicts to $\langle \langle v', 6 \rangle, \langle w', 1 \rangle \rangle \in \llbracket a^{i+1} \rrbracket_{M'}$. Therefore $\langle v', w' \rangle \in \llbracket a_i \rrbracket_M$.

If $t \equiv t \cdot u$:

$$\llbracket T_2(t \cdot u) \rrbracket_{M'} = \llbracket T_2(t) \cdot T_2(u) \rrbracket_{M'} = \llbracket T_2(t) \rrbracket_{M'} \cdot \llbracket T_2(u) \rrbracket_{M'}$$
$$= \{\langle \langle v, 3 \rangle, \langle v'', 3 \rangle \rangle \mid \langle v, v'' \rangle \in \llbracket t \rrbracket_M\} \cdot \{\langle \langle v'', 3 \rangle, \langle v', 3 \rangle \rangle \mid \langle v'', v' \rangle \in \llbracket u \rrbracket_M\} \quad \text{(I.H.)}$$
$$= \{\langle \langle v, 3 \rangle, \langle v', 3 \rangle \rangle \mid \langle v, v' \rangle \in \llbracket t \cdot u \rrbracket_M\}.$$

If $t \equiv t^-$:

$$\llbracket T_2(t^-) \rrbracket_{M'} = \llbracket ((a \cap a^3) \cdot \top) \cap (\top \cdot (a \cap a^2)) \cap T_2(t)^- \rrbracket_{M'}$$
$$= \{\langle \langle v, 3 \rangle, \langle w, 3 \rangle \rangle \mid \langle v, w \rangle \in |M|^2\} \cap \llbracket T_2(t)^- \rrbracket_{M'}$$
$$= \{\langle \langle v, 3 \rangle, \langle w, 3 \rangle \rangle \mid \langle v, w \rangle \in |M|^2\} \setminus \llbracket T_2(t) \rrbracket_{M'}$$
$$= \{\langle \langle v, 3 \rangle, \langle w, 3 \rangle \rangle \mid \langle v, w \rangle \in |M|^2\} \setminus \{\langle \langle v, 3 \rangle, \langle w, 3 \rangle \mid \langle v, w \rangle \in \llbracket t \rrbracket_M\} \quad \text{(I.H.)}$$
$$= \{\langle \langle v, 3 \rangle, \langle w, 3 \rangle \rangle \mid \langle v, w \rangle \in \llbracket t^- \rrbracket_M\}.$$

If $t \equiv t \cup u$:

$$\llbracket T_2(t \cup u) \rrbracket_{M'} = \llbracket T_2(t) \cup T_2(u) \rrbracket_{M'} = \llbracket T_2(t) \rrbracket_{M'} \cup \llbracket T_2(u) \rrbracket_{M'}$$
$$= \{\langle \langle v, 3 \rangle, \langle w, 3 \rangle \rangle \mid \langle v, w \rangle \in \llbracket t \rrbracket_M\} \cup \{\langle \langle v, 3 \rangle, \langle w, 3 \rangle \rangle \mid \langle v, w \rangle \in \llbracket u \rrbracket_M\} \quad \text{(I.H.)}$$
$$= \{\langle \langle v, 3 \rangle, \langle w, 3 \rangle \rangle \mid \langle v, w \rangle \in \llbracket t \cup u \rrbracket_M\}.$$

(1) $\Rightarrow$ (2): Let M be a model over $\{a\}$ such that $M \models$ (Ax.1-2), but $M \not\models T_2(t) = T_2(u)$. Then we define the model M over A by $M' := \langle |M'|, \mathcal{R}' \rangle$, where $|M'| := \mathrm{dom}(\llbracket a \cap a^3 \rrbracket_M)$ and $\mathcal{R}'(a_i) := \llbracket T_2(a_i) \rrbracket_M$. Note that $\mathcal{R}'(a_i) \subseteq |M'|^2$ holds because $\mathrm{dom}(\llbracket a \cap a^3 \rrbracket_M) = \mathrm{cod}(\llbracket a \cap a^2 \rrbracket_M)$ by Proposition 15. Then the following holds: $\llbracket t \rrbracket_{M'} = \llbracket T_2(t) \rrbracket_M \cdots (\heartsuit)$. $M' \not\models t = u$ is proved by $(\heartsuit)$. We now prove $(\heartsuit)$ by induction on the structure of t.

If $t \equiv a_i$, $[\![a_i]\!]_{M'} = [\![\mathrm{T}_2(a_i)]\!]_M$ is shown by the definition of M'.
If $t \equiv t \cdot u$,

$$[\![t \cdot u]\!]_{M'} = [\![t]\!]_{M'} \cdot [\![u]\!]_{M'} = [\![\mathrm{T}_2(t)]\!]_M \cdot [\![\mathrm{T}_2(u)]\!]_M \qquad \text{(I.H.)}$$
$$= [\![\mathrm{T}_2(t) \cdot \mathrm{T}_2(u)]\!]_M = [\![\mathrm{T}_2(t \cdot u)]\!]_M.$$

If $t \equiv t^-$,

$$[\![t^-]\!]_{M'} = |M'|^2 \setminus [\![t]\!]_{M'} = |M'|^2 \setminus [\![\mathrm{T}_2(t)]\!]_M \qquad \text{(I.H.)}$$
$$= [\![((a \cap a^3) \cdot \top) \cap (\top \cdot (a \cap a^2))]\!]_M \setminus [\![\mathrm{T}_2(t)]\!]_M$$
$$= [\![((a \cap a^3) \cdot \top) \cap (\top \cdot (a \cap a^2)) \cap \mathrm{T}_2(t)^-]\!]_M = [\![\mathrm{T}_2(t^-)]\!]_M.$$

If $t \equiv t \cup u$,

$$[\![t \cup u]\!]_{M'} = [\![t]\!]_{M'} \cup [\![u]\!]_{M'} = [\![\mathrm{T}_2(t)]\!]_M \cup [\![\mathrm{T}_2(u)]\!]_M \qquad \text{(I.H.)}$$
$$= [\![\mathrm{T}_2(t) \cup \mathrm{T}_2(u)]\!]_M = [\![\mathrm{T}_2(t \cup u)]\!]_M.$$

$\square$

By Lemma 16, the next theorem has been proved.

Theorem 17. *Let a be a character. Then both the validity problem and the finite validity problem for $\mathrm{Eq}[\mathcal{T}_{\{a\}}^{\langle \cdot, \bullet^-, \cup \rangle}]$ are undecidable.*

Combining Theorem 17 and Lemma 3, the following is also proved. Note that, if a term t does not contain 1, the formula $G(t, i, j)$ does not contain $=$.

Theorem 18. *Let a be a character. Then both the validity problem and the finite validity problem for $\mathrm{FO3}_{\{a\}}$ are undecidable.*

4   Conclusion

In this paper we showed that the validity problem and the finite validity problem are undecidable for the following classes: (1) FO3 with just one binary relation symbol and without equality; and (2) equational formulas of the calculus of relations with just one character over the signature $\langle \cdot, \bullet^-, \cup \rangle$. In connection with (2), the following decidable fragments are known.

Theorem 19 ([1, Theorem 5] **for (1)**). *The validity problem and the finite validity problem are decidable for the following classes: (1) the calculus of relations over the signature $\langle \cdot, \cup \rangle$ (even with $\bullet^\smile, 1, 0, \top, \cap$, and atomic negation); and (2) the calculus of relations over the signature $\langle \bullet^-, \cup \rangle$ (even with $\bullet^\smile, 1$).*

These decidability results are proved by the reduction to FO3 (Lemma 3) and using the decidability of the satisfiability problem and the finite model property of the $\exists^* \forall^*$ case with equality (proved by Ramsey [17] in 1930, see also [3, Sect. 6.2.2]). To the best of our knowledge, the (un)decidability of the validity problem and the finite validity problem for equational formulas of the calculus of relations over the signature $\langle \cdot, \bullet^- \rangle$ are open regardless of the number of characters.

References

1. Andréka, H., Bredikhin, D.A.: The equational theory of union-free algebras of relations. Algebr. Univers. **33**(4), 516–532 (1995). https://doi.org/10.1007/BF01225472
2. Boolos, G.S., Burgess, J.P., Jeffrey, R.C.: Computability and Logic. 5th edn., p. 350. Cambridge University Press, Cambridge (2007). https://doi.org/10.1017/CBO9780511804076
3. Börger, E., Grädel, E., Gurevich, Y.: The Classical Decision Problem, p. 482. Springer, Heidelberg (1997)
4. Church, A.: A note on the Entscheidungsproblem. J. Symb. Log. **1**(1), 40–41 (1936). https://doi.org/10.2307/2269326
5. Givant, S.: The calculus of relations as a foundation for mathematics. J. Autom. Reason. **37**(4), 277–322 (2007). https://doi.org/10.1007/s10817-006-9062-x
6. Gödel, K.: Die Vollständigkeit der Axiome des logischen Funktionenkalküls. Monatshefte für Mathematik und Physik **37**(1), 349–360 (1930). https://doi.org/10.1007/BF01696781
7. Grädel, E., Kolaitis, P.G., Vardi, M.Y.: On the decision problem for two-variable first-order logic. Bull. Symb. Log. **3**(01), 53–69 (1997). https://doi.org/10.2307/421196
8. Herbrand, J.: Sur le problème fondamental de la logique mathématique. Sprawozdania z posiedezen Towarzysta Naukowego Warszawskiego, Wydzial III **24**, 12–56 (1931)
9. Kahr, A.S., Moore, E.F., Wang, H.: Entscheidungsproblem reduced to the AEA case. Proc. Natl. Acad. Sci. **48**(3), 365–377 (1962). https://doi.org/10.1073/pnas.48.3.365
10. Kalmár, L.: Zuriickftihrung des Entscheidungsproblems auf den Fall von Formeln mit einer einzigen, bindren, Funktionsvariablen. In: Compositiomathematica, p. 4 (1936)
11. Lewis, H.R.: Complexity results for classes of quantificational formulas. J. Comput. Syst. Sci. **21**(3), 317–353 (1980). https://doi.org/10.1016/0022-0000(80)90027-6
12. Löwenheim, L.: Über Möglichkeiten im Relativkalköl. Mathematische Annalen **76**(4), 447–470 (1915)
13. Lutz, C., Sattler, U., Wolter, F.: Modal logic and the two-variable fragment. In: Fribourg, L. (ed.) CSL 2001. LNCS, vol. 2142, pp. 247–261. Springer, Heidelberg (2001). https://doi.org/10.1007/3-540-44802-0_18
14. Maddux, R.D.: Relation Algebras. Studies in Logic and the Foundations of Mathematics, vol. 150, p. 758. Elsevier (2006). https://doi.org/10.1016/S0049-237X(06)80023-6
15. Maddux, R.D.: Undecidable semiassociative relation algebras. J. Symb. Log. **59**(02), 398–418 (1994). https://doi.org/10.2307/2275397
16. Mortimer, M.: On languages with two variables. Math. Log. Q. **21**(1), 135–140 (1975). https://doi.org/10.1002/malq.19750210118
17. Ramsey, F.P.: On a problem of formal logic. In: Classic Papers in Combinatorics, pp. 1–24. Birkhäuser-Verlag (2009). https://doi.org/10.1007/978-0-8176-4842-8_1
18. Büchi, J.R.: On a decision method in restricted second-order arithmetic. In: International Congress on Logic, Method, and Philosophy of Science, pp. 1–11. Stanford University Press (1962)

19. Skolem, T.: Untersuchungen über die Axiome des Klassenkalküls und ÜberProduktations- und Summationsprobleme, welche gewisse Klassen von Aussagen betreffen. In: Videnskapsselskapets skrifter, I. Matematisk-naturvidenskabelig, vol. 3, p. 37 (1919)

20. Tarski, A.: On the calculus of relations. J. Symb. Log. **6**(3), 73–89 (1941). https://doi.org/10.2307/2268577

21. Tarski, A., Givant, S.: A Formalization of Set Theory Without Variables, vol. 41, p. 318. Colloquium Publications. American Mathematical Society (1987)

22. Trakhtenbrot, B.A.: The Impossibility of an algorithm for the decision problem in finite classes. In: Detlovs, V.K. (ed.) Nine Papers on Logic and Quantum Electro-dynamics. American Mathematical Society Translations Series 2, vol. 23, pp. 1–5. American Mathematical Society (1963)

23. Turing, A.M.: On computable numbers, with an application to the Entschei-dungsproblem. Proc. Lond. Math. Soc. **s2–42**(1), 230–265 (1937). https://doi.org/10.1112/plms/s2-42.1.230

24. Vardi, M.Y.: From philosophical to industrial logics. In: Ramanujam, R., Sarukkai, S. (eds.) ICLA 2009. LNCS (LNAI), vol. 5378, pp. 89–115. Springer, Heidelberg (2008). https://doi.org/10.1007/978-3-540-92701-3_7

25. Vardi, M.Y.: Why is modal logic so robustly decidable? Technical report, p. 24 (1997)

26. Willard, R.: Hereditary undecidability of some theories of finite structures. J. Symb. Log. **59**(04), 1254–1262 (1994). https://doi.org/10.2307/2275703

Satisfaction Classes via Cut Elimination

Cezary Cieśliński[(✉)]

Institute of Philosophy, University of Warsaw, Warsaw, Poland
c.cieslinski@uw.edu.pl

Abstract. We present a construction of a satisfaction class in an arbitrary countable recursively saturated models of first-order arithmetic. Our construction is fully classical, namely, it employs nothing more than the classical techniques of formal proof theory.

1 Introduction

The goal of this paper is to sketch a fully classical construction of a satisfaction class for the language of first-order arithmetic. The initial introductory remarks describe the motivation for this endeavor.

It is well-known that non-standard models of arithmetic contain nonstandard arithmetical formulas. Indeed, for an arbitrary nonstandard model M there will be an $a \in M$ such that $M \models$ 'a is an arithmetical formula' even though in the real world a is not a formula at all. A natural question arises whether it is possible to develop semantics for such objects (we will call them 'formulas in the sense of the model'). First attempts in this direction were made by Robinson [8] and Krajewski [7]. To this aim, the notion of a satisfaction class has been introduced. Roughly, a satisfaction class is a subset of the model which can be treated as a reasonable interpretation of the satisfaction predicate obeying the usual Tarski-style compositional clauses.

Further work on satisfaction classes brought remarkable results. In particular, it transpired that a non-inductive satisfaction class can be constructed in an arbitrary countable recursively saturated model of arithmetic. This is the famous theorem of Kotlarski, Krajewski and Lachlan (KKL in short), which demonstrates the conservativity of non-inductive satisfaction axioms over first-order arithmetic.[1]

However, the original proof of the theorem uses techniques which many readers found exotic. From the author's experience, the machinery of (so-called) 'approximations', developed by KKL in their paper, remains one of the main stumbling blocks in the wider dissemination of this important result. Accordingly, the question has been asked whether the result can be proved by purely classical methods. One successful attempt in this direction has been recently made by Enayat and Visser [2]. In their paper, they showed how to construct

---

[1] See [5]. For an overview, see also [4,6].

© Springer-Verlag GmbH Germany, part of Springer Nature 2019
Md. A. Khan and A. Manuel (Eds.): ICLA 2019, LNCS 11600, pp. 121–131, 2019.
https://doi.org/10.1007/978-3-662-58771-3_12

a satisfaction class using classical techniques of formal semantics (namely, compactness and the union of elementary chain theorem).[2]

In the present paper I propose to prove the theorem by the classical techniques of formal proof theory, namely, by cut elimination. Coupled with Enayat and Visser construction, this makes the fascinating field of satisfaction classes accessible to the students and the logicians, whose primary interest is either model theory or proof theory.

2 Preliminaries

The language of first-order Peano arithmetic will be denoted here as L_{PA}. Primitive non-logical symbols of L_{PA} are '$+$', '$\times$', '0', 'S'. The expressions Var, Tm, Tm^c, $Fm_{L_{PA}}$ and $Sent_{L_{PA}}$ will be used as referring (respectively) to the sets of variables, terms, constant terms, formulas and sentences of L_{PA}. We will also use these expressions as shorthands for arithmetical predicates representing the relevant sets in PA. Given a model M, we write $Sent_{L_{PA}}(M)$ for the set of all objects a such that $a \in M$ and $M \models Sent_{L_{PA}}(a)$.

The perspective adopted in this paper is that of truth, not satisfaction. Accordingly, we will consider the language L_T obtained from L_{PA} by adding the unary truth predicate '$T(x)$' (instead of a binary satisfaction predicate). $Sent_{L_T}$ is the set of sentences of L_T.

We introduce now the basic theory of truth, denoted as CT^-. The acronym 'CT' stands for 'compositional truth'; the superscript indicates that in this theory we have induction only for the arithmetical language (not with the truth predicate).

Definition 1. CT^- *is the theory in the language L_T axiomatized by all the usual axioms of Peano arithmetic together with the following truth axioms:*

- $\forall s, t \in Tm^c\big(T(s = t) \equiv val(s) = val(t)\big)$
- $\forall \varphi\big(Sent_{L_{PA}}(\varphi) \to (T\neg\varphi \equiv \neg T\varphi)\big)$
- $\forall \varphi \forall \psi\big(Sent_{L_{PA}}(\varphi \vee \psi) \to (T(\varphi \vee \psi) \equiv (T\varphi \vee T\psi))\big)$
- $\forall v \forall \varphi(x)\big(Sent_{L_{PA}}(\forall v\varphi(v)) \to (T(\forall v\varphi(v)) \equiv \forall x T(\varphi(\dot{x})))\big)$

In effect, the truth axioms of CT^- follow the familiar pattern of Tarski's inductive truth definition. Let us only emphasize that the quantifier axiom given here employs numerals. A numeral is an arithmetical term of the form '$S \ldots S(0)$'; in other words, numerals are expressions obtained by preceding the symbol '0' with arbitrarily many successor symbols. Accordingly, the intended meaning of the quantifier axiom is that '$\forall v\varphi(v)$' is true iff the result of substituting an arbitrary numeral for v in $\varphi(v)$ is true.

---

[2] In their proof it is assumed that arithmetic is formulated in the relational language. See [1] for extending the result to the language with function symbols.

A 'truth class' in a model M is a subset T of M which makes all the axioms of CT^- true. Now, the conservativity theorem states that the truth axioms of CT^- are conservative not just over full Peano arithmetic, but also over the fragments of arithmetic sufficient for reconstructing basic theory of syntax. In particular, adding the truth axioms to $I\Sigma_1$ (the theory just like Peano arithmetic but with induction restricted to Σ_1 arithmetical formulas) produces a conservative extension of $I\Sigma_1$. This is a direct corollary of the KKL theorem, which can be formulated as an expandability result concerning recursively saturated models of $I\Sigma_1$.

Two definitions below introduce the notion of a recursive type and the concept of a recursively saturated model.

Definition 2. *Let Z be a set of formulas with one free variable x and with parameters $a_1...a_n$ from a model M. We say that:*

(a) Z is realized in M iff there is an $s \in M$ such that every formula in Z is satisfied in M under a valuation assigning s to x.

(b) Z is a type of M iff every finite subset of Z is realized in M.

(c) Z is a recursive type of M iff apart from being a type of M, Z is also recursive.

Definition 3. *M is recursively saturated iff every recursive type of M is realized in M.*

The KKL theorem can now be formulated as follows.

Theorem 4. *For every $M \models I\Sigma_1$, if M is countable and recursively saturated, then there is a set $T \subset M$ such that $(M, T) \models CT^-$.*

3   From Consistent M-Logic to a Truth Class

From now on, we will work with a fixed countable and recursively saturated model M of $I\Sigma_1$. As in the original KKL's argument, our first step is the development of a proof system called 'M-logic' (ML in short). Intuitively, ML is a system which permits us to process arbitrary sentences in the sense of M, including the nonstandard ones. The system is described externally (not in the model) in the form of a sequent calculus.[3] We will use '$\Rightarrow$' for the sequent arrow, with expressions of the form '$\Gamma \Rightarrow \Delta$' referring to sequents. We shall always assume that both Γ and Δ are externally finite sequences of M-sentences. Note that we do not admit formulas with free variables in the sequents.

The definition of M-logic is framed after Gentzen's original system LK (see [3]). All the initial sequents have the form $\varphi \Rightarrow \varphi$, for an arbitrary

---

[3] This is the first difference between our proof and the original KKL's construction.

$\varphi \in Sent_{LPA}(M)$. The following rules of ML are copied directly from Gentzen's system:

- Weakening, left and right (W-left and W-right):

$$\frac{\Gamma \Rightarrow \Delta}{\Gamma \Rightarrow \Delta, \varphi} \qquad \frac{\Gamma \Rightarrow \Delta}{\varphi, \Gamma \Rightarrow \Delta}$$

- Exchange, left and right (E-left and E-right):

$$\frac{\Gamma, \psi, \varphi, \Gamma' \Rightarrow \Delta}{\Gamma, \varphi, \psi, \Gamma' \Rightarrow \Delta} \qquad \frac{\Gamma \Rightarrow \Delta, \psi, \varphi, \Delta'}{\Gamma \Rightarrow \Delta, \varphi, \psi, \Delta'}$$

- Contraction, left and right (C-left and C-right):

$$\frac{\varphi, \varphi, \Gamma \Rightarrow \Delta}{\varphi, \Gamma \Rightarrow \Delta} \qquad \frac{\Gamma \Rightarrow \Delta, \varphi, \varphi}{\Gamma \Rightarrow \Delta, \varphi}$$

- Cut:

$$\frac{\Gamma \Rightarrow \Delta, \varphi \qquad \varphi, \Sigma \Rightarrow \Lambda}{\Gamma, \Sigma \Rightarrow \Delta, \Lambda}$$

- $\neg$-left and $\neg$-right:

$$\frac{\Gamma \Rightarrow \Delta, \varphi}{\neg\varphi, \Gamma \Rightarrow \Delta} \qquad \frac{\varphi, \Gamma \Rightarrow \Delta}{\Gamma \Rightarrow \Delta, \neg\varphi}$$

- $\wedge$-left and $\wedge$-right (for arbitrary sentences A and B such that one of them is φ):

$$\frac{\varphi, \Gamma \Rightarrow \Delta}{A \wedge B, \Gamma \Rightarrow \Delta} \qquad \frac{\Gamma \Rightarrow \Delta, \varphi \qquad \Gamma \Rightarrow \Delta, \psi}{\Gamma \Rightarrow \Delta, \varphi \wedge \psi}$$

- $\vee$-left and $\vee$-right (for arbitrary sentences A and B such that one of them is φ):

$$\frac{\varphi, \Gamma \Rightarrow \Delta \qquad \psi, \Gamma \Rightarrow \Delta,}{\varphi \vee \psi, \Gamma \Rightarrow \Delta} \qquad \frac{\Gamma \Rightarrow \Delta, \varphi}{\Gamma \Rightarrow \Delta, A \vee B}$$

- $\rightarrow$-left and $\rightarrow$-right:

$$\frac{\Gamma \Rightarrow \Delta, \varphi \qquad \psi, \Sigma \Rightarrow \Lambda}{\varphi \rightarrow \psi, \Gamma, \Sigma \Rightarrow \Delta, \Lambda} \qquad \frac{\varphi, \Gamma \Rightarrow \Delta, \psi}{\Gamma \Rightarrow \Delta, \varphi \rightarrow \psi}$$

In addition, M-logic has the following rules of inference:

- The truth rule for literals (Tr-lit). Let φ be of the form $t = s$ with $M \models t = s$ or of the form $t \neq s$ with $M \models t \neq s$:

$$\frac{\varphi, \Gamma \Rightarrow \Delta}{\Gamma \Rightarrow \Delta}$$

- The M-rule, left and right (M-left, M-right):

$$\frac{\{\varphi(a), \Gamma \Rightarrow \Delta : a \in M\}}{\exists x \varphi(x), \Gamma \Rightarrow \Delta} \qquad \frac{\{\Gamma \Rightarrow \Delta, \varphi(a) : a \in M\}}{\Gamma \Rightarrow \Delta, \forall x \varphi(x)}$$

- $\exists$-right and $\forall$-left:

$$\frac{\Gamma \Rightarrow \Delta, \varphi(a)}{\Gamma \Rightarrow \Delta, \exists x \varphi(x)} \qquad \frac{\varphi(a), \Gamma \Rightarrow \Delta}{\forall x \varphi(x), \Gamma \Rightarrow \Delta}$$

Proofs in ML are (possibly infinite) trees of finite height, where the height of a proof is defined (as usual) as the length of the maximal path. By definition, trees with no maximal path do not qualify as proofs in ML. Observe that in ML, the infinitary rules M-left and M-right replace the original rules $\exists$-left and $\forall$-right of Gentzen.[4] It should be also emphasized that in *all* the quantifier rules of ML we employ numerals. Thus, for example, in order to apply $\exists$-right, we need a sentence $\varphi(a)$ with a numeral for a. In contrast, in Gentzen's original system the rule $\exists$-right would permit us to derive $\Gamma \Rightarrow \Delta, \exists x \varphi(x)$ from $\Gamma \Rightarrow \Delta, \varphi(t)$ for an arbitrary term t, not necessarily a numeral. The effect of this modification of Gentzen's system is that the truth class which we construct can contain term pathologies. Thus, in a model (M, T) of CT^- which we eventually obtain there can exist a nonstandard formula $\varphi(x)$ such that for some term t, $\varphi(t)$ belongs to T (so that, loosely speaking, the model thinks that $\varphi(t)$ is true), while the sentence $\neg \exists x \varphi(x)$ also belongs to T. In this way we obtain a disconcerting effect: the model thinks that $\neg \exists x \varphi(x)$ is true even though it considers as true some term instantiation of $\varphi(x)$.[5] However, this accords with our formulation of CT^-, where all the quantifier axioms employ numerals.

Lemma 5. *If M-logic is consistent, then M can be expanded to a model of CT^-.*

For the proof of the lemma, we introduce first the family of unary arithmetical predicates '$Pr_n(S)$' with the intuitive reading 'sequent S has a proof in M-logic of height at most n' (in short, S is n-provable). Observe that for each rule R of M-logic, the relation 'S can be obtained by R from n-provable sequents' can always be expressed by an arithmetical formula, provided that n-provability is arithmetically expressible. In view of this, we introduce the following definition.

Definition 6

- $Pr_0(S) := S$ is an initial sequent,
- $Pr_{n+1}(S) := Pr_n(S) \vee \bigvee_{R \in ML} (S$ can be obtained by R from n-provable sequents).

---

[4] Proof systems with similar infinitary rules have already been studied in the literature in the context of cut elimination. See, for example, [9].

[5] This will happen if all the *numerical* instantiations of $\varphi(x)$ are seen as false by the model, that is, if for all numerals a, the sentence $\neg \varphi(a)$ belongs to T.

By external induction on natural numbers it can be demonstrated that:

Observation 7. $\forall k \in \omega \; \forall S [ML \vdash_k S \equiv M \models Pr_k(S)].$

We can now turn to the proof of Lemma 5.

Proof. Let $\varphi_0, \varphi_1, \ldots$ be an enumeration of the set of M-sentences (this is the only place where the countability assumption is used).

We define:
$T_0 = \emptyset$

$$T_{n+1} = \begin{cases} T_n \cup \{\varphi_n\} & \text{if } ML \nvdash (T_n \to \neg\varphi_n) \text{ and } \varphi_n \text{ is not existential,} \\ T_n \cup \{\exists x \psi(x)\} \cup \{\psi(a)\} & \text{if } \varphi_n = \exists x \psi(x) \text{ and } ML \nvdash (T_n \to \neg\varphi_n), \\ & \quad \text{for an } a \in M \text{ such that } ML \nvdash (T_n \to \neg\psi(a)), \\ T_n \cup \{\neg\varphi_n\} & \text{otherwise.} \end{cases}$$

The expression 'T_n' on the right side of the definition (as in '$ML \nvdash (T_n \to \neg\varphi_n)$') stands for the conjunction of all the sentences φ_i or their negations, whichever of them were added on previous levels. We need to verify that whenever $ML \nvdash (T_n \to \neg\exists x \psi(x))$, there will exist an $a \in M$ such that $ML \nvdash (T_n \to \neg\psi(a))$. This is the only place where recursive saturation is employed.

Thus, assume that $ML \nvdash (T_n \to \neg\exists x \psi(x))$. Define:

$$p(x) = \{\neg Pr_k(T_n \to \neg\psi(x)) : k \in \omega\}.$$

We observe that $p(x)$ is a type. Otherwise there is a natural number k such that $M \models \forall a Pr_k(T_n \to \neg\psi(a))$. Hence for all a, $ML \vdash_k T_n \to \neg\psi(a)$. But then by the M-rule and cut, $ML \vdash T_n \to \neg\exists x \psi(x)$, which is a contradiction.

Since $p(x)$ is a type, by recursive saturation there is an $a \in M$ which realizes it and we have: $\forall k M \models \neg Pr_k(T_n \to \neg\psi(a))$, hence the sentence $T_n \to \neg\psi(a)$ is not provable in M-logic, as required.

Checking that $(M, T) \models CT^-$ provided that M-logic is consistent is now routine and we leave it to the reader. □

4 Consistency of M-Logic

At this stage all that is missing is the proof of consistency of M-logic. In KKL [5] the consistency of M-logic is proved by the technique of approximations. Here we propose cut elimination as the proof method. Let us start by the following simple observation.

Observation 8. *If every sequent provable in M-logic has a cut-free proof, then M-logic is consistent.*

Proof. If M-logic is inconsistent, then it proves that $0 = 1$. By cut elimination, take a cut-free proof P of $0 = 1$. It is easy to observe that every sentence in

P has to be either atomic or negated atomic.[6] For a sequent S belonging to P, let the height of S in P be defined as the length of maximal path generated by S in P.[7] Let $Tr_0(x)$ be the arithmetical truth predicate for atomic sentences and their negations. By external induction on the height of sequents in P, it can be demonstrated that for every sequent S in P, if all sentences in the antecedent of S are Tr_0, then some sentence in the succedent of S is Tr_0. It immediately follows that $M \models Tr_0(0 = 1)$, which is impossible. □

Lemma 9. *For every sequent S, if S is provable in ML, then S has a cut-free proof in ML.*

The aim of the remaining part of the paper is to lead the proof of Lemma 9 to the point at which it can be completed simply by repeating Gentzen's original argument for cut elimination. It should be emphasized that we are not there yet. Our setting is that of possibly nonstandard sentences (sentences in the sense of M) and this generates an obstacle which first has to be removed.

In order to see the obstacle, let us recap the classical argument. The aim is to show that the system with the following mix rule (which is a generalized version of cut) admits mix elimination:

$$\frac{\Gamma \Rightarrow \Delta \qquad \Sigma \Rightarrow \Lambda}{\Gamma, \Sigma^* \Rightarrow \Delta^*, \Lambda} \ (\varphi)$$

where Σ and Δ contain φ (the mix formula); Σ^* and Δ^* differ from Σ and Δ only in that they do not contain any occurrence of φ. Since mix and cut produce equivalent proof systems, mix elimination gives us the desired result.

In the next stage it is demonstrated that mix can be eliminated from any proof which contains only a single application of the mix rule in the last step. This is done by double induction on the *degree* of proofs (main induction) and on the *rank* of proofs (subinduction). For proofs with mix used only in the last step, we define:

- The *left rank* of the proof is the largest number of consecutive sequents in a path starting with the left-hand upper sequent of the mix and such that every sequent in the path contains the mix formula in the succedent.
- The *right rank* of the proof is the largest number of consecutive sequents in a path starting with the right-hand upper sequent of the mix and such that every sequent in the path contains the mix formula in the antecedent.
- The *rank* of the proof is the left rank of the proof + the right rank of the proof.
- The *degree* of the proof is the syntactic complexity of the mix formula.

There is no problem in our setting with induction on the rank of proofs, since both the left and the right rank of the proof in ML will always be a (standard)

---

[6] Without cut, (Tr-Lit) is the only rule that permits us to eliminate sentences in the proof and (Tr-Lit) can eliminate literals only.

[7] Thus, sequents which are initial in P have height 0 and the maximal height of a sequent in P is not larger than the height of P.

natural number, restricted by the height of the proof. However, the induction on the degree of proofs is quite problematic. Since the mix formula might be non-standard, its syntactic complexity might be a non-standard element of M. Arguing externally by induction on non-standard numbers is clearly an invalid move and this is the main obstacle complicating the situation.

Our remedy is to replace the general notion of a degree with a notion relativized to a proof. Assume that we are given a proof P with mix applied only in the last step, that eliminates the (possibly non-standard) mix formula φ. The guiding intuition formalized below is that in the cut elimination proof the syntactic shape of φ matters only comparatively. For example, φ might have the form $\neg\psi$. The intuition is that this will matter only provided that ψ itself (without negation) appears somewhere in P; otherwise φ might just as well be treated as a formula of complexity 0, even if it is non-standard.

Our objective is to make these ideas precise. In what follows the word 'sequence' should always be interpreted externally; in other words, sequences are finite or infinite objects in the real world, not necessarily elements of M. The length of a finite sequence $a = (a_0 \ldots a_k)$ is the number of its elements, that is, $lh(a) = k + 1$. For an infinite sequence a we define $lh(a)$ as ω.

Definition 10

- $x \lhd y$ ('x is a direct subsentence of y') is an abbreviation of the following arithmetical formula:

  $Sent_{LPA}(x) \wedge Sent_{LPA}(y) \wedge$
  $\left(\exists \psi \in Sent_{LPA}(y = \ulcorner \neg\psi \urcorner \wedge x = \psi) \right.$
  $\vee \, \exists \varphi, \psi \in Sent_{LPA}(y = \ulcorner \varphi \circ \psi \urcorner \wedge x = \varphi \vee x = \psi)$
  $\left. \vee \, \exists \theta(x) \in Fm_{LPA} \exists a \exists v \in Var(y = \ulcorner Qv\theta(v) \urcorner \wedge x = \ulcorner \theta(a) \urcorner) \right).$

- Let $\varphi \in Sent_{LPA}(M)$. We say that s is a $\lhd$-sequence for φ iff $s_0 = \varphi$ and for every $k < lh(s) - 1$ $s_{k+1} \lhd s_k$.

The notion of a degree can now be defined in the following way.

Definition 11. Let P be an arbitrary proof in ML with mix used only in the last step. Let φ be the mix formula in P. We define:

- $d(\varphi, P)$ (the degree of φ in P) $= \sup\{lh(s) : s$ is a $\lhd$-sequence for φ such that for every $k < lh(s)$ $s_k \in P\}$.
- $d(P)$ (the degree of P) is defined as $d(\varphi, P)$.

Lemma 12. Let P be an arbitrary proof in ML with mix used only in the last step. Then $d(P)$ is a natural number (in other words, it is never ω).

In order to prove the lemma, we introduce first the function $str(x)$ ('the structure of a formula x'). Let the letter p be a new symbol (it will be treated as a propositional variable). The function is defined as follows (Q is either $\exists$ or $\forall$ and $\circ$ is an arbitrary binary connective).

Definition 13

- $str(\ulcorner t = s \urcorner) = \ulcorner p \urcorner$
- $str(\ulcorner \neg \psi \urcorner) = \neg str(\ulcorner \psi \urcorner)$
- $str(\ulcorner \varphi \circ \psi \urcorner) = str(\ulcorner \varphi \urcorner) \circ str(\ulcorner \psi \urcorner)$
- $str(\ulcorner Qx\varphi \urcorner) = Qx \, str(\ulcorner \varphi \urcorner)$

Intuitively, given a formula φ, the function produces a formula which is exactly like φ, except that the letter 'p' is substituted for all occurences of atomic formulas in φ. Abbreviate $str(\varphi) = str(\psi)$ as $\varphi \sim \psi$. The key property of the equivalence relation $\sim$ is encapsulated in the following observation.

Observation 14. *Let $Z \subseteq Sent_{LPA}(M)$. For every s, if s is a $\lhd$-sequence with elements from Z, then $lh(s) \leq card(\{[\varphi]_\sim : \varphi \in Z\})$, where $[\varphi]_\sim$ is a class of sentences ψ from Z such that $\varphi \sim \psi$.*

Let $compl(\varphi)$ be the number of connectives and quantifiers in φ. Observation 14 follows immediately from the following fact (we use $\lhd^*$ for the transitive closure of $\lhd$).

Fact 15

(a) $\forall \varphi, \psi(\varphi \lhd^* \psi \to compl(\varphi) < compl(\psi))$.
(b) $\forall \varphi, \psi(\varphi \sim \psi \to compl(\varphi) = compl(\psi))$.

The proof of Fact 15 is done by easy induction and does not contain any surprises. For part (a), proceed with induction on the length of the $\lhd$-sequence s leading from ψ to φ. Part (b) can be done by induction on the complexity of ψ.

Proof of Lemma 12 (idea). Fix a proof P in ML which contains mix only in the last step. Let Z be the set of all sentences which appear in P. We demonstrate that $\{[\varphi]_\sim : \varphi \in Z\}$ is finite, which by Observation 14 guarantees the conclusion of Lemma 12.

For an arbitrary sequent S in P, let $l(S)$ (the level of S) be the length of the path leading from S to the end sequent of P. We denote by S_i the set of all sequents in P whose level is not greater than i. Let $Sent_i$ be defined as the set of all sentences which appear in some element of S_i. Let k be the height of P. The task is to show that:

$$\forall i \leq k \{[\varphi]_\sim : \varphi \in Sent_i\} \text{ is finite.}$$

This will end the proof, since $Sent_k = Z$.

We proceed by induction. Observe that for $i = 0$ the conclusion is trivial, as $Sent_0$ itself is finite ($Sent_0$ is the set of sentences which appear in the end sequent of P). The proof is concluded by demonstrating that $\{[\varphi]_\sim : \varphi \in Sent_{i+1}\}$ is finite, under the assumption that $\{[\varphi]_\sim : \varphi \in Sent_i\}$ is finite.[8] $\square$

---

[8] Here the argument proceeds by cases, corresponding to various ways in which elements of $Sent_i$ can be obtained from elements of $Sent_{i+1}$.

In effect, Definition 11 and Lemma 12 give us a notion of a degree of the proof which can be used in a Gentzen-style proof even in a non-standard setting. The way to proving cut elimination, and thus the consistency of ML, is now open.

I will not present the whole cut elimination proof, since it is mostly a repetition of Gentzen's reasoning. Instead, I will restrict myself to discussing one example of a new rule (the one not present in the original Gentzen's system).

Our task is to demonstrate that mix can be eliminated from any proof which contains only a single application of the mix rule in the last step. Let us assume (main induction) that cut can be eliminated in every proof of a degree $< n$. Let us also assume (subinduction) that cut can be eliminated in every proof of a degree n but with rank $< k$. Our task is to show that cut can be eliminated in proofs of degree n and rank k.

The proof starts with the case of $k = 2$ (the lowest possible rank) and proceeds by analysing subcases. Here we analyse only one subcase corresponding to a rule of ML absent in LK. Namely, let us assume that the mix formula of the form $\forall x\varphi(x)^9$ is obtained by a logical rule in both the succedent of the left-hand upper sequent of the mix and in the antecedent of the right-hand upper sequent of the mix. Then the last stage of the proof runs as follows:

$$M\text{-right} \frac{\{\Gamma \Rightarrow \Delta, \varphi(a) : a \in M\}}{\Gamma \Rightarrow \Delta, \forall x\varphi(x)} \quad \frac{\varphi(c), \Sigma \Rightarrow \Lambda}{\forall x\varphi(x), \Sigma \Rightarrow \Lambda} \forall\text{-left}}{\Gamma, \Sigma \Rightarrow \Delta, \Lambda} \text{mix}$$

We can then eliminate mix in the following way:

$$\frac{\dfrac{\Gamma \Rightarrow \Delta, \varphi(c) \quad \varphi(c), \Sigma \Rightarrow \Lambda}{\Gamma, \Sigma^* \Rightarrow \Delta^*, \Lambda} \text{mix}}{\Gamma, \Sigma \Rightarrow \Delta, \Lambda} \text{possibly, some weakenings and exchanges}$$

We use the inductive assumption here, namely, we show that the same end sequent can be obtained by applying mix to the formula $\varphi(c)$, which has the degree $n - 1$ in P (the sentence $\forall x\varphi(x)$ has the degree n). Observe that in the modified proof $\varphi(c)$ will preserve the same degree $n - 1$; observe also that the modification did not involve adding to the proof any new formula (in general: in the present setting new proofs without mix are produced from sentences belonging to the initial proof P).

Acknowledgements. The author was supported by a grant from the National Science Centre in Cracow (NCN), project number 2017/27/B/HS1/01830.

---

[9] The case of the existential mix formula is closely analogous and I do not discuss it separately.

References

1. Cieśliński, C.: The Epistemic Lightness of Truth: Deflationism and its Logic. Cambridge University Press, Cambridge (2017)
2. Enayat, A., Visser, A.: New constructions of satisfaction classes. In: Achourioti, T., Galinon, H., Martínez Fernández, J., Fujimoto, K. (eds.) Unifying the Philosophy of Truth. LEUS, vol. 36, pp. 321–335. Springer, Dordrecht (2015). https://doi.org/10.1007/978-94-017-9673-6_16
3. Gentzen, G.: Investigations into logical deduction. Am. Philos. Q. **1**(4), 288–306 (1964)
4. Kotlarski, H.: Full satisfaction classes: a survey. Notre Dame J. Formal Logic **32**(4), 573–579 (1991)
5. Kotlarski, H., Krajewski, S., Lachlan, A.: Construction of satisfaction classes for nonstandard models. Can. Math. Bull. **24**(3), 283–293 (1981)
6. Kotlarski, H., Ratajczyk, Z.: Inductive full satisfaction classes. Ann. Pure Appl. Logic **47**(3), 199–223 (1990)
7. Krajewski, S.: Non-standard satisfaction classes. In: Marek, W., Srebrny, M., Zarach, A. (eds.) Set Theory and Hierarchy Theory: A Memorial Tribute to Andrzej Mostowski. LNM, vol. 537, pp. 121–144. Springer, Heidelberg (1976). https://doi.org/10.1007/BFb0096898
8. Robinson, A.: On languages which are based on non-standard arithmetic. Nagoya Math. J. **22**, 83–117 (1963)
9. Yasugi, M.: Cut elimination theorem for second order arithmetic with the Π_1^1-comprehension axiom and the ω-rule. J. Math. Soc. Jpn. **22**(3), 308–324 (1970)

Sequent Calculi for Normal Update Logics

Katsuhiko Sano[1(⊠)] and Minghui Ma[2]

[1] Graduate School of Letters, Hokkaido University, Sapporo, Japan
`v-sano@let.hokudai.ac.jp`
[2] Institute of Logic and Cognition, Sun Yat-Sen University, Guangzhou, China
`mamh6@mail.sysu.edu.cn`

Abstract. Normal update logic is the temporalization of normal conditional logic. Sequent calculi for the least normal update logic **UCK** by Andreas Herzig (1998) and some of its extensions are developed. The subformula property of these sequent calculi is shown by Takano's semantic method. Consequently we prove the finite model property and decidability of these sequent calculi.

1 Introduction

The *normal update logic* in Herzig [9] is an extension of Chellas' (normal) conditional logic [3,4] by update operators. Let $[\varphi]\psi$ be the conditional with antecedent φ and consequent ψ. In Herzig's normal update logic, the formula $[\varphi]\psi$ (usually called *conditional operator*) is read as a hypothetical update, i.e., if the belief base (a finite set of beliefs) is updated by φ, then ψ follows. Herzig introduced the update operator as the left adjoint of the conditional operator. We express the formula constructed by the update operator as $\langle\varphi^{-}\rangle\psi$ in the present paper which can be read "ψ has been updated by φ". These operators are widely used in database theory, and they are linked to belief change in formal epistemology. Herzig [9] provided the formal account of the update operator in terms of models for conditionals. The fundamental feature of Herzig's account is that the following adjointness law holds in the normal update logic:

$$\text{(Adjointness)} \ \langle\varphi^{-}\rangle\psi \vdash \chi \text{ if and only if } \psi \vdash [\varphi]\chi.$$

It follows that the normal update logic can be viewed as a temporal extension of the normal conditional logic since the adjointness law is fundamental for the definition of basic tense logic (cf. [2]). But there is still a difference: the conditional and update operators are binary while tense operators in basic tense logic are unary. In a normal update logic, the antecedent of the conditional operator

K. Sano—Partially supported by JSPS KAKENHI Grant-in-Aid for Young Scientists (B) Grant Number 15K21025 and Grant-in-Aid for Scientific Research (B) Grant Number 17H02258, and JSPS Core-to-Core Program (A. Advanced Research Networks).
M. Ma—Supported by the key project of National Social Science Found of China (Grant no. 18ZDA033).

can be replaced with an equivalent formula. This point makes the normal update logic different from basic tense logic.

There are several proof-theoretic studies [1,5,6,8,11,14,19,24] on the normal conditional logic and its extensions in terms of sequent calculi. As far as the authors know, however, there is no proof-theoretic study on normal update logics. The aim of the present paper is to investigate logical properties of normal update logic, including the finite model property and decidability, in the setting of proof theory. The most fundamental result in proof theory is cut-elimination (cf. [7,15,23]). A form of cut rule in a sequent calculus is formulated as follows:

$$\frac{\Gamma \Rightarrow \Delta, \varphi \quad \varphi, \Pi \Rightarrow \Theta}{\Gamma, \Pi \Rightarrow \Delta, \Theta} \ (Cut)$$

where Γ, Δ, Π, Θ are finite multisets of formulas, and the formula φ is said to be the cut formula. A sequent $\Gamma \Rightarrow \Delta$ can be read as "if all formulas in Γ hold, then some formulas in Δ holds." Clearly the cut formula φ may not be a subformula of a formula in the lower sequent of the cut rule. If the cut rule is eliminable from a given system, then a derivation possibly with the cut rule can be transformed into a derivation without any application of the cut rule. As a result, a provable sequent in the calculus under consideration has a derivation such that every formula occurred in the derivation is a subformula of a formula in the sequent (subformula property), provided that all other rules in the system satisfy the condition that each formula in the upper sequent is a subformula of a formula in the lower sequent.

It is well known that a sequent calculus for the modal logic **S5** based on the ordinary notion of sequent does not enjoy the cut-elimination (see [17, p. 116]). By a syntactic argument, however, Takano [20] showed that, for a provable sequent, we have a derivation such that each formula in it consists of subformulas of the provable sequent, where it is noted that the cut formula in an application of (Cut) can be restricted to a subformula of the provable sequent. That is, the modal logic **S5** still enjoys the subformula property. A semantic proof of Takano's result on **S5** can be found in [18]. Takano [21,22] also gives the semantic proof of the subformula property for all fifteeen modal logics determined by the modal axioms **T**, **B**, **4**, **5** and **D**. Furthermore, in a sequent calculus for the basic tense logic $\mathbf{K}_t$, which was first proposed by [16], the cut-elimination theorem also fails, and Takano [20] commented that the analytic cut property holds for the sequent calculus of $\mathbf{K}_t$. A semantic proof of the analytic cut property for the tense extension of the modal logic **K4** can be found in [12,13]. A semantic proof of the subformula property of bi-intuitionistic logic was given by Kowalski and Ono [10].

We develop sequent calculi for four normal update logics from [9]. Our sequent calcului can be regarded as combinations of one-sided cut-free sequent calcului for normal conditional logics in [19] and two-sided sequent calculus for basic tense logic $\mathbf{K}_t$ in [16]. We shall apply a Takano-style semantic proof to show the subformula property for these sequent calculi. By the semantic proof, we obtain the finite model property and hence the decidability of these logics.

This paper is structured as follows. Section 2 presents some preliminaries on normal update logics. Section 3 gives sequent calculi for normal update logics.

Section 4 proves the subformula property, finite model property and decidability of some sequent calculi.

2   Preliminaries

In this section, we recall some basic concepts and results in normal update logic which can be found in Herzig [9].

Definition 1. Given a countably infinite set Prop of propositional variables, the set of all formulas $\mathcal{L}_{\mathrm{UC}}$ is defined inductively as follows:

$$\mathcal{L}_{\mathrm{UC}} \ni \varphi ::= p \mid \neg\varphi \mid (\varphi \wedge \varphi) \mid (\varphi \vee \varphi) \mid (\varphi \to \varphi) \mid [\varphi]\varphi \mid \langle \varphi^- \rangle\varphi,$$

where $p \in$ Prop. Define $\varphi \leftrightarrow \psi := (\varphi \to \psi) \wedge (\psi \to \varphi)$. The dual of $[\varphi]\psi$ is defined as $\langle \varphi \rangle\psi := \neg[\varphi]\neg\psi$. The dual of $\langle \varphi^- \rangle\psi$ is defined as $[\varphi^-]\psi := \neg\langle \varphi^- \rangle\neg\psi$. The logical constants $\bot$ and $\top$ are defined as usual.

For any formula $\varphi \in \mathcal{L}_{\mathrm{UC}}$, the set of all subformulas of φ, denoted by $\mathrm{Sub}(\varphi)$, is defined recursively as usual. For any set of formulas Γ, we define the set of all subformulas occurred in Γ as the set $\mathrm{Sub}(\Gamma) = \bigcup\{\mathrm{Sub}(\varphi) \mid \varphi \in \Gamma\}$. A set of formulas Γ is called *subformula closed* if $\Gamma = \mathrm{Sub}(\Gamma)$.

Definition 2. A *conditional frame* is a pair $\mathfrak{F} = (W, f)$ where W is a non-empty set of states, and $f : W \times \mathcal{P}(W) \to \mathcal{P}(W)$ is a function from the product $W \times \mathcal{P}(W)$ to the powerset $\mathcal{P}(W)$. The function f in a conditional frame is called a *selection* function. A *conditional model* is a triple $\mathfrak{M} = (W, f, V)$ where (W, f) is a conditional frame and $V : \mathsf{Prop} \to \mathcal{P}(W)$ is a valuation.

Definition 3. Given a formula $\varphi \in \mathcal{L}_{\mathrm{UC}}$, a conditional model $\mathfrak{M} = (W, f, V)$, and a state $w \in W$, we define recursively the notion of a formula φ being true (or satisfied) at w in $\mathfrak{M}$, notation $\mathfrak{M}, w \models \varphi$, as follows:

$\mathfrak{M}, w \models p$ iff $w \in V(p)$,
$\mathfrak{M}, w \models \neg\varphi$ iff $\mathfrak{M}, w \not\models \varphi$,
$\mathfrak{M}, w \models \varphi \wedge \psi$ iff $\mathfrak{M}, w \models \varphi$ and $\mathfrak{M}, w \models \psi$,
$\mathfrak{M}, w \models \varphi \vee \psi$ iff $\mathfrak{M}, w \models \varphi$ or $\mathfrak{M}, w \models \psi$,
$\mathfrak{M}, w \models \varphi \to \psi$ iff $\mathfrak{M}, w \models \varphi$ implies $\mathfrak{M}, w \models \psi$,
$\mathfrak{M}, w \models [\varphi]\psi$ iff for any $u \in W$, if $u \in f(w, [\![\varphi]\!]_{\mathfrak{M}})$, then $\mathfrak{M}, u \models \psi$,
$\mathfrak{M}, w \models \langle \varphi^- \rangle\psi$ iff there exists $u \in W$ s.t. $w \in f(u, [\![\varphi]\!]_{\mathfrak{M}})$ and $\mathfrak{M}, u \models \psi$,

where the set $[\![\varphi]\!]_{\mathfrak{M}} := \{u \in W \mid \mathfrak{M}, u \models \varphi\}$ (the subscript $\mathfrak{M}$ is dropped from $[\![\varphi]\!]_{\mathfrak{M}}$ if it is clear from the context). A formula φ is *valid* in a conditional model $\mathfrak{M}$, notation $\mathfrak{M} \models \varphi$, if $[\![\varphi]\!]_{\mathfrak{M}} = W$. A formula φ is *valid* in a conditional frame $\mathfrak{F} = (W, f)$, notation $\mathfrak{F} \models \varphi$, if $[\![\varphi]\!]_{(\mathfrak{F}, V)} = W$ for any valuation V on $\mathfrak{F}$.

A formula φ is *valid* in a class of conditional frames $\mathbb{C}$, notation $\mathbb{C} \models \varphi$, if $\mathfrak{F} \models \varphi$ for all $\mathfrak{F} \in \mathbb{C}$. For any class of conditional frames $\mathbb{C}$, the *update logic* of $\mathbb{C}$ is defined as the set $\mathsf{Log}(\mathbb{C}) = \{\varphi \in \mathcal{L}_{\mathrm{UC}} \mid \mathbb{C} \models \varphi\}$.

A set of formulas Γ *defines* or *corresponds to* a class of conditional frames $\mathbb{C}$ if, for any conditional frame $\mathfrak{F}$, all formulas in Γ are valid in $\mathfrak{F}$ if and only if $\mathfrak{F} \in \mathbb{C}$. If Γ is a singleton set $\{\varphi\}$, we say that φ defines a class $\mathbb{C}$.

Definition 4. The class of all conditional frames is denoted by $\mathbb{CF}$. We define $\mathbb{CID}$ as the class of all conditional frames $\mathfrak{F} = (W, f)$ satisfying the following condition for all $w \in W$ and $X \subseteq W$:

$$f(w, X) \subseteq X. \qquad (F_{\mathrm{CID}})$$

We define $\mathbb{CMP}$ as the class of all conditional frames $\mathfrak{F} = (W, f)$ satisfying the following condition for all $w \in W$ and $X \subseteq W$:

$$w \in X \text{ implies } w \in f(w, X). \qquad (F_{\mathrm{CMP}})$$

We define $\mathbb{IDMP} = \mathbb{CID} \cap \mathbb{CMP}$. Let $\mathsf{UCK} = \mathrm{Log}(\mathbb{CF})$, $\mathsf{UID} = \mathrm{Log}(\mathbb{CID})$, $\mathsf{UMP} = \mathrm{Log}(\mathbb{CMP})$ and $\mathsf{UIDMP} = \mathrm{Log}(\mathbb{IDMP})$ be update logics.

Proposition 1. *The following definability results hold:*

(1) $[p]p$ *defines* $\mathbb{CID}$.
(2) $(p \wedge [p]q) \to q$ *defines* $\mathbb{CMP}$.
(3) $[p]p \wedge ((q \wedge [q]r) \to r)$ *defines* $\mathbb{IDMP}$.

Herzig [9] proves that the update logics are finitely axiomatized by the following Hilbert-style systems.

Definition 5. The Hilbert-style system HUCK consists of the following axiom schema and inference rules:

(1) (**Taut**) All instances of classical propositional tautologies.
(2) Inference rules:

$$\frac{\varphi_1 \leftrightarrow \varphi_2}{\langle \varphi_1^- \rangle \psi \leftrightarrow \langle \varphi_2^- \rangle \psi}(\mathsf{UEA}) \quad \frac{\varphi \quad \varphi \to \psi}{\psi}(\mathsf{MP}) \quad \frac{\langle \varphi^- \rangle \psi \to \chi}{\psi \to [\varphi]\chi}(\mathsf{Ad1}) \quad \frac{\psi \to [\varphi]\chi}{\langle \varphi^- \rangle \psi \to \chi}(\mathsf{Ad2})$$

Given a set $\{\varphi_1, \ldots, \varphi_n\}$ of formulas, let $\mathsf{HUCK} \oplus \varphi_1 \oplus \ldots \oplus \varphi_n$ be the axiomatic extension of HUCK by adding $\varphi_1, \ldots, \varphi_n$ as new axiom schemata. Consider the following two additional axiom schemata:

$$(\mathbf{CID}) \ [\varphi]\varphi, \qquad (\mathbf{CMP}) \ (\varphi \wedge [\varphi]\psi) \to \psi.$$

We define the axiomatic extensions $\mathsf{HUID} = \mathsf{HUCK} \oplus (\mathbf{CID})$, $\mathsf{HUMP} = \mathsf{HUCK} \oplus (\mathbf{CMP})$, and $\mathsf{HUIDMP} = \mathsf{HUCK} \oplus (\mathbf{CID}) \oplus (\mathbf{CMP})$.

For any axiomatic system H, the notation $\mathsf{H} \vdash \varphi$ stands for that φ is *provable* (or a *theorem*) in the system H.

Proposition 2. *The following formulas and rules are provable in* HUCK:

$$(\mathbf{Cnv_1}) \ \varphi \to [\psi]\langle \psi^- \rangle \varphi, \ (\mathbf{Cnv_2}) \ \langle \psi^- \rangle [\psi]\varphi \to \varphi,$$
$$(\mathbf{CN}) \ [\varphi]\top \leftrightarrow \top, \qquad (\mathbf{CR}) \ [\varphi](\psi \wedge \chi) \leftrightarrow ([\varphi]\psi \wedge [\varphi]\chi),$$
$$(\mathbf{UR}) \ \langle \varphi^- \rangle \bot \leftrightarrow \bot, \qquad (\mathbf{UN}) \ \langle \varphi^- \rangle (\psi \vee \chi) \leftrightarrow (\langle \varphi^- \rangle \psi \vee \langle \varphi^- \rangle \chi),$$

$$\frac{\varphi_1 \leftrightarrow \varphi_2}{[\varphi_1]\psi \leftrightarrow [\varphi_2]\psi}(\mathbf{CEA}), \quad \frac{\varphi_1 \to \varphi_2}{[\psi]\varphi_1 \to [\psi]\varphi_2}(\mathbf{CRM}), \quad \frac{\varphi_1 \to \varphi_2}{\langle \psi^- \rangle \varphi_1 \to \langle \psi^- \rangle \varphi_2}(\mathbf{URM}).$$

Moreover, $\mathsf{HUID} \vdash \langle \varphi^- \rangle \psi \to \varphi$ *and* $\mathsf{HUMP} \vdash (\varphi \wedge \psi) \to \langle \varphi^- \rangle \psi.$

Herzig [9] introduced the canonical model for a Hilbert-style system of normal update logic and the following soundness and completeness results are obtained (see [9, Theorem 3.14]):

Fact 1. *For any formula $\varphi \in \mathcal{L}_{UC}$, the following hold:*

(1) $\mathsf{HUCK} \vdash \varphi$ *if and only if* $\mathbb{CF} \models \varphi$,
(2) $\mathsf{HUID} \vdash \varphi$ *if and only if* $\mathbb{CID} \models \varphi$,
(3) $\mathsf{HUMP} \vdash \varphi$ *if and only if* $\mathbb{CMP} \models \varphi$,
(4) $\mathsf{HUIDMP} \vdash \varphi$ *if and only if* $\mathbb{IDMP} \models \varphi$.

3   Sequent Calculi for Normal Update Logics

Let Γ, Δ, Σ etc. with or without subscripts denote finite (possibly empty) multisets of formulas. A *sequent* is an expression of the form $\Gamma \Rightarrow \Delta$ where Γ and Δ are finite multisets of formulas. When we write Γ, Δ, the comma "," is understood as the multiset union. A *sequent rule* is an inference rule of the form

$$\frac{\Gamma_1 \Rightarrow \Delta_1 \quad \cdots \quad \Gamma_n \Rightarrow \Delta_n}{\Gamma_0 \Rightarrow \Delta_0}(\mathbf{r})$$

where $\Gamma_i \Rightarrow \Delta_i$ for $1 \leqslant i \leqslant n$ are called *premisses* of $(\mathbf{r})$, and $\Gamma_0 \Rightarrow \Delta_0$ is called the *conclusion* of $(\mathbf{r})$. For any finite multiset Γ of formulas, $[\varphi]\Gamma$ and $\langle \varphi^- \rangle \Gamma$ are defined as the multisets where $[\varphi]$ and $\langle \varphi^- \rangle$ are prefixed to all elements in Γ, respectively. In particular, if Γ is empty, then it is noted that both $[\varphi]\Gamma$ and $\langle \varphi^- \rangle \Gamma$ are empty.

Definition 6. For any $\varphi_0, \ldots, \varphi_n \in \mathcal{L}_{UC}$, we write $\varphi_0 \Leftrightarrow \cdots \Leftrightarrow \varphi_n$ to mean all the set of all sequents of the form $\varphi_0 \Rightarrow \varphi_i$ and $\varphi_i \Rightarrow \varphi_0$ for all $1 \leqslant i \leqslant n$. The sequent calculi GUCK, GUID, GUMP and GUIDMP are defined in Table 1.

We use G to denote any sequent calculus in Table 1. Note that the sequent calculus LK for classical propositional logic is taken as the basis, and additional rules for conditional and update operators are supplied. In these additional rules, note that Θ_i and Σ_i are possibly empty in $([\cdot])$, $(\langle \cdot^- \rangle)$ and $([\cdot]_{\mathtt{CID}})$. If each Θ_i is empty in the rules $([\cdot])$ and $([\cdot]_{\mathtt{CID}})$, we obtain the following two rules:

$$\frac{\varphi_0 \Leftrightarrow \cdots \Leftrightarrow \varphi_n \quad \Sigma_1, \ldots, \Sigma_n \Rightarrow \psi}{[\varphi_1]\Sigma_1, \ldots, [\varphi_n]\Sigma_n \Rightarrow [\varphi_0]\psi} \ ([\cdot]),$$

$$\frac{\varphi_0 \Leftrightarrow \cdots \Leftrightarrow \varphi_n \quad \varphi_0, \Sigma_1, \ldots, \Sigma_n \Rightarrow \psi}{[\varphi_1]\Sigma_1, \ldots, [\varphi_n]\Sigma_n \Rightarrow [\varphi_0]\psi} \ ([\cdot]_{\mathtt{CID}}).$$

They can be regarded as the two-sided reformulation of the inference rules (CK_g) and (CKID_g) in [19, p. 14, Fig. 3].

A *derivation* in G is a finite tree $\mathcal{D}$ in which each node is a sequent obtained by an axiom or a rule in G. The *height* of a derivation $\mathcal{D}$ in G is the maximal length of a branch from the root node. We say that a sequent $\Gamma \Rightarrow \Delta$ is *provable* in G, notation $\mathsf{G} \vdash \Gamma \Rightarrow \Delta$, if there is a derivation $\mathcal{D}$ in G such that the root node of $\mathcal{D}$ is the sequent $\Gamma \Rightarrow \Delta$.

Table 1. Sequent calculi for normal update logics

- Axioms:

$$\varphi \Rightarrow \varphi \ (\mathbf{id})$$

- Structural Rules:

$$\frac{\Gamma \Rightarrow \Delta, \varphi}{\Gamma \Rightarrow \Delta} \ (\Rightarrow w) \qquad \frac{\varphi, \Gamma \Rightarrow \Delta}{\Gamma \Rightarrow \Delta} \ (w \Rightarrow)$$

$$\frac{\Gamma \Rightarrow \Delta, \varphi, \varphi}{\Gamma \Rightarrow \Delta, \varphi} \ (\Rightarrow c) \qquad \frac{\varphi, \varphi, \Gamma \Rightarrow \Delta}{\varphi, \Gamma \Rightarrow \Delta} \ (c \Rightarrow)$$

- Propositional Rules:

$$\frac{\Gamma \Rightarrow \Delta, \varphi_1 \quad \Gamma \Rightarrow \Delta, \varphi_2}{\Gamma \Rightarrow \Delta, \varphi_1 \wedge \varphi_2} \ (\Rightarrow \wedge) \qquad \frac{\varphi_i, \Gamma \Rightarrow \Delta}{\varphi_1 \wedge \varphi_2, \Gamma \Rightarrow \Delta} \ (\wedge \Rightarrow)$$

$$\frac{\Gamma \Rightarrow \Delta, \varphi_i}{\Gamma \Rightarrow \Delta, \varphi_1 \vee \varphi_2} \ (\Rightarrow \vee) \qquad \frac{\varphi_1, \Gamma \Rightarrow \Delta \quad \varphi_2, \Gamma \Rightarrow \Delta}{\varphi_1 \vee \varphi_2, \Gamma \Rightarrow \Delta} \ (\vee \Rightarrow)$$

$$\frac{\varphi, \Gamma \Rightarrow \Delta}{\Gamma \Rightarrow \Delta, \neg\varphi} \ (\Rightarrow \neg) \qquad \frac{\Gamma \Rightarrow \Delta, \varphi}{\neg\varphi, \Gamma \Rightarrow \Delta} \ (\neg \Rightarrow)$$

$$\frac{\varphi, \Gamma \Rightarrow \Delta, \psi}{\Gamma \Rightarrow \Delta, \varphi \rightarrow \psi} \ (\Rightarrow\rightarrow) \qquad \frac{\Gamma \Rightarrow \Delta, \varphi \quad \psi, \Pi \Rightarrow \Sigma}{\varphi \rightarrow \psi, \Gamma, \Pi \Rightarrow \Delta, \Sigma} \ (\rightarrow\Rightarrow)$$

- Cut rule:

$$\frac{\Gamma \Rightarrow \Delta, \varphi \quad \varphi, \Pi \Rightarrow \Theta}{\Gamma, \Pi \Rightarrow \Delta, \Theta} \ (Cut)$$

- Additional Rules for GUCK: ([·]) and ($\langle \cdot^- \rangle$), where

$$\frac{\varphi_0 \Leftrightarrow \cdots \Leftrightarrow \varphi_n \quad (\langle \varphi_i^- \rangle \Theta_i, \Sigma_i)_{1 \leqslant i \leqslant n} \Rightarrow \psi}{(\Theta_i, [\varphi_i]\Sigma_i)_{1 \leqslant i \leqslant n} \Rightarrow [\varphi_0]\psi} \ ([\cdot])$$

$$\frac{\varphi_0 \Leftrightarrow \cdots \Leftrightarrow \varphi_n \quad \psi \Rightarrow (\Sigma_i, [\varphi_i]\Theta_i)_{1 \leqslant i \leqslant n}}{\langle \varphi_0^- \rangle \psi \Rightarrow (\langle \varphi_i^- \rangle \Sigma_i, \Theta_i)_{1 \leqslant i \leqslant n}} \ (\langle \cdot^- \rangle)$$

- Additional Rules for GUID: ($\langle \cdot^- \rangle$), ([·]$_{\text{CID}}$) and (CID$_{\langle \cdot^- \rangle}$), where

$$\frac{\varphi_0 \Leftrightarrow \cdots \Leftrightarrow \varphi_n \quad \varphi, (\langle \varphi_i^- \rangle \Theta_i, \Sigma_i)_{1 \leqslant i \leqslant n} \Rightarrow \psi}{(\Theta_i, [\varphi_i]\Sigma_i)_{1 \leqslant i \leqslant n} \Rightarrow [\varphi_0]\psi} \ ([\cdot]_{\text{CID}}) \qquad \frac{\varphi_0, \Gamma \Rightarrow \Delta}{\langle \varphi_0^- \rangle \psi, \Gamma \Rightarrow \Delta} \ (\text{CID}_{\langle \cdot^- \rangle})$$

- Additional Rules for GUMP: ([·]), ($\langle \cdot^- \rangle$), (CMP$_{[\cdot]}$) and (CMP$_{\langle \cdot^- \rangle}$) where

$$\frac{\Gamma \Rightarrow \Delta, \varphi \quad \psi, \Gamma \Rightarrow \Delta}{[\varphi]\psi, \Gamma \Rightarrow \Delta} \ (\text{CMP}_{[\cdot]}) \qquad \frac{\Gamma \Rightarrow \Delta, \varphi \quad \Gamma \Rightarrow \Delta, \psi}{\Gamma \Rightarrow \Delta, \langle \varphi^- \rangle \psi} \ (\text{CMP}_{\langle \cdot^- \rangle})$$

- Additional Rules for GUIDMP:

$$(\langle \cdot^- \rangle), \ ([\cdot]_{\text{CID}}), \ (\text{CID}_{\langle \cdot^- \rangle}), \ (\text{CMP}_{[\cdot]}), \text{ and } (\text{CMP}_{\langle \cdot^- \rangle}).$$

Definition 7. For any conditional model $\mathfrak{M} = (W, f, V)$, we say that a sequent $\Gamma \Rightarrow \Delta$ is *true* in $\mathfrak{M}$, notation $\mathfrak{M} \models \Gamma \Rightarrow \Delta$, if for any state $w \in W$, $\mathfrak{M}, w \models \bigwedge \Gamma$ implies $\mathfrak{M}, w \models \bigvee \Delta$. A sequent rule

$$\frac{\Gamma_1 \Rightarrow \Delta_1 \dots \Gamma_n \Rightarrow \Delta_n}{\Gamma_0 \Rightarrow \Delta_0}(\mathbf{r})$$

preserves truth in a model $\mathfrak{M}$, if $\mathfrak{M} \models \Gamma_0 \Rightarrow \Delta_0$ whenever $\mathfrak{M} \models \Gamma_i \Rightarrow \Delta_i$ for all $1 \leqslant i \leqslant n$. A sequent $\Gamma \Rightarrow \Delta$ is *valid* in a conditional frame $\mathfrak{F}$, notation $\mathfrak{F} \models \Gamma \Rightarrow \Delta$, if $\mathfrak{F}, V \models \Gamma \Rightarrow \Delta$ for all valuations V in $\mathfrak{F}$. Given a class $\mathbb{C}$ of conditional frames, by $\mathbb{C} \models \Gamma \Rightarrow \Delta$ we mean that $\mathfrak{F} \models \Gamma \Rightarrow \Delta$ for all $\mathfrak{F} \in \mathbb{C}$.

Proposition 3 (Soundness). *For any sequent $\Gamma \Rightarrow \Delta$, the following hold:* (1) *if* GUCK $\vdash \Gamma \Rightarrow \Delta$, *then* $\mathbb{CF} \models \Gamma \Rightarrow \Delta$; (2) *if* GUID $\vdash \Gamma \Rightarrow \Delta$, *then* $\mathbb{CID} \models \Gamma \Rightarrow \Delta$; (3) *if* GUMP $\vdash \Gamma \Rightarrow \Delta$, *then* $\mathbb{CMP} \models \Gamma \Rightarrow \Delta$; (4) *if* GUIDMP $\vdash \Gamma \Rightarrow \Delta$, *then* $\mathbb{IDMP} \models \Gamma \Rightarrow \Delta$.

Given any finite multiset Γ of formulas, $\bigwedge \Gamma$ and $\bigvee \Gamma$ are defined to be the conjunction and the disjunction of all elements of Γ, respectively, where $\bigwedge \emptyset$ is understood as $\top$ and $\bigvee \emptyset$ as $\bot$.

Theorem 1. *Let* H *be the Hilbert-style system corresponding to a sequent calculus* G. *Then the following hold:*

(1) *If* G $\vdash \Gamma \Rightarrow \Delta$ *then* H $\vdash \bigwedge \Gamma \to \bigvee \Delta$.
(2) *If* H $\vdash \varphi$ *then* G $\vdash \Rightarrow \varphi$.
(3) H $\vdash \varphi$ *if and only if* G $\vdash \Rightarrow \varphi$.
(4) G $\vdash \Gamma \Rightarrow \Delta$ *if and only if* H $\vdash \bigwedge \Gamma \to \bigvee \Delta$.

Note that the formulas $\varphi \to [\psi]\langle\psi^-\rangle\varphi$ and $\langle\psi^-\rangle[\psi]\varphi \to \varphi$ which are obtained from the adjointness in Hilbert-style systems are provable in GUCK as follows:

$$\frac{\dfrac{\psi \Leftrightarrow \psi \quad \langle\psi^-\rangle\varphi \Rightarrow \langle\psi^-\rangle\varphi}{\varphi \Rightarrow [\psi]\langle\psi^-\rangle\varphi}([\cdot])}{\Rightarrow \varphi \to [\psi]\langle\psi^-\rangle\varphi}(\Rightarrow\to) \quad , \quad \frac{\dfrac{\psi \Leftrightarrow \psi \quad [\psi]\varphi \Rightarrow [\psi]\varphi}{\langle\psi^-\rangle[\psi]\varphi \Rightarrow \varphi}(\langle\cdot^-\rangle)}{\langle\psi^-\rangle[\psi]\varphi \to \varphi}(\Rightarrow\to) \quad .$$

Theorem 2 (Completeness). *For any sequent $\Gamma \Rightarrow \Delta$, the following hold:* (1) *if* $\mathbb{CF} \models \Gamma \Rightarrow \Delta$, *then* GUCK $\vdash \Gamma \Rightarrow \Delta$; (2) *if* $\mathbb{CID} \models \Gamma \Rightarrow \Delta$, *then* GUID $\vdash \Gamma \Rightarrow \Delta$; (3) *if* $\mathbb{CMP} \models \Gamma \Rightarrow \Delta$, *then* GUMP $\vdash \Gamma \Rightarrow \Delta$; (4) *if* $\mathbb{IDMP} \models \Gamma \Rightarrow \Delta$, *then* GUIDMP $\vdash \Gamma \Rightarrow \Delta$.

Proof. Let G be a sequent calculus, H be the corresponding Hilbert-style system, and $\mathbb{C}$ be corresponding class of frames. Assume $\mathbb{C} \models \Gamma \Rightarrow \Delta$. Then $\mathbb{C} \models \bigwedge \Gamma \to \bigvee \Delta$. By Fact 1, H $\vdash \bigwedge \Gamma \to \bigvee \Delta$. By Theorem 1, G $\vdash \Gamma \Rightarrow \Delta$. □

Now we shall prove that the sequent calculi in Table 1 does not enjoy the cut-elimination. For any formula φ, let φ^n denote the multiset containing exactly n occurrences of φ. In particular, if $n = 0$, φ^n is understood as the empty multiset.

Proposition 4. *Let* G *be any sequent calculus in Table 1 and* G$^-$ *be the sequent calculus obtained from* G *by deleting the cut rule. The sequent* $p, \langle q^-\rangle[q]\neg p \Rightarrow$ *is not derivable in* G$^-$. *Therefore (Cut) is not eliminable from* G.

The following is a derivation of the sequent $\Rightarrow p \to \neg\langle q^-\rangle[q]\neg p$ in any sequent calculus defined in Table 1, where (Cut) is not eliminable by Proposition 4:

$$\frac{\dfrac{\dfrac{[q]\neg p \Rightarrow [q]\neg p}{\langle q^-\rangle[q]\neg p \Rightarrow \neg p}\;((\cdot^-))\qquad \dfrac{p \Rightarrow p}{\neg p, p \Rightarrow}\;(\neg\Rightarrow)}{p, \langle q^-\rangle[q]\neg p \Rightarrow}\;(Cut)}{\dfrac{p \Rightarrow \neg\langle q^-\rangle[q]\neg p}{\Rightarrow p \to \neg\langle q^-\rangle[q]\neg p}\;(\Rightarrow\neg)}\;(\Rightarrow\to)\;.$$

All sequent calculi G in Table 1 lacks the cut-elimination. We can apply Takano's semantic method to show the subformula property of G.

4   Subformula Property and Decidability

For any sequent calculus, we say that an inference rule $(\mathbf{r})$ has the *subformula property* if every formula in the upper sequent(s) of the rule is a subformula of a formula in the lower sequent. Clearly the rules (Cut), $([\cdot])$ and $((\cdot^-))$ in GUCK have no subformula property. As a result, some derivations of a sequent, say, in GUCK, may contain a formula which is not a subformula of a formula in the sequent. Our derivation of the sequent $\Rightarrow p \to \neg\langle q^-\rangle[q]\neg p$ in the previous section, however, suggest that we may add restrictions on the rules $([\cdot])$, $((\cdot^-))$, $([\cdot]_{\mathrm{CID}})$ and (Cut) to obtain the subformula property. We say that a sequent system has the *subformula property* if, whenever a sequent $\Gamma \Rightarrow \Delta$ is derivable in the system, there is a derivation $\mathcal{D}$ such that all formulas occurring in $\mathcal{D}$ are subformulas of formulas in the sequent $\Gamma \Rightarrow \Delta$. Obviously, if all the rules of a system have the subformula property, then the system has the subformula property. The goal of this section is to establish that any sequent calculus G in Table 1 has the subformula property. For this purpose, we are going to show that, if a sequent $\Gamma \Rightarrow \Delta$ is derivable in G, then it has a derivation in G where all formulas in the derivation are subformulas of the original sequent $\Gamma \Rightarrow \Delta$.

In what follows in this section, we use the lower roman letters a, b, c, etc. to denote *sets* of formulas. Basically we follow a stragety employed in [10] for bi-intuitionistic logic to establish the subformula property of normal update logics.

Definition 8. Let Ξ be a subformula closed finite set of formulas. We say that $\Gamma \Rightarrow \Delta$ is Ξ-*provable in* G (simply, Ξ-provable, if no confusion arises, and it is written as $\mathsf{G} \vdash_\Xi \Gamma \Rightarrow \Delta$) if it is provable by a derivation such that all formulas in the derivation belong to Ξ. A pair (a, b) of two subsets of Ξ is Ξ-*disjoint in* G if $a \Rightarrow b$ is not Ξ-provable in G, i.e., $\mathsf{G} \nvdash_\Xi a \Rightarrow b$. A pair $(a, b) \in \Xi^2$ is Ξ-*saturated in* G if it is maximally Ξ-disjoint, i.e., (a, b) is Ξ-disjoint, and the following two conditions hold:

(i) For all $\gamma \in \Xi \setminus a$, $\gamma, a \Rightarrow b$ is Ξ-provable,
(ii) For all $\gamma \in \Xi \setminus b$, $a \Rightarrow b, \gamma$ is Ξ-provable.

We say that $a \subseteq \Xi$ is Ξ-*saturated* in G if the pair $(a, \Xi \setminus a)$ is Ξ-saturated in G. A pair (a, b) is Ξ-*complete*, if $\varphi \in a$ or $\varphi \in b$ for all $\varphi \in \Xi$.

In what follows in this section, we always use Ξ as a subformula closed finite set of formulas. We note that $a \cap b = \emptyset$ for a Ξ-disjoint pair (a, b). Given any $a \subseteq \Xi$, we use $\bar{a}$ to mean $\Xi \setminus a$.

Lemma 1. *Let $(a, b) \in \Xi^2$ be Ξ-disjoint in G. Then the pair (a, b) is Ξ-saturated in G if and only if (a, b) is Ξ-complete.*

Lemma 2. *If (a, b) is Ξ-disjoint in G, then there exists a Ξ-saturated pair (a^+, b^+) in G such that $a \subseteq a^+$ and $b \subseteq b^+$.*

Definition 9. Let Ξ be subformula closed finite set of formulas and G be a sequent calculus. We define the set W_{G}^{Ξ} as follows:

$$W_{\mathsf{G}}^{\Xi} = \{\, a \mid a \text{ is } \Xi\text{-saturated in } \mathsf{G} \,\}.$$

For any formula $\varphi \in \Xi$, we define $|\varphi| = \{\, a \in W^{\Xi} \mid \varphi \in a \,\}$.

Let $f : W_{\mathsf{G}}^{\Xi} \times \mathcal{P}(W_{\mathsf{G}}^{\Xi}) \to \mathcal{P}(W_{\mathsf{G}}^{\Xi})$ be a function. We say that f is $[.]$-*saturated,* if for any $a \in W_{\mathsf{G}}^{\Xi}$, the following holds:

$$[\varphi]\psi \in a, \text{ if and only if, for all } b \in W_{\mathsf{G}}^{\Xi}, b \in f_0^{\Xi}(a, |\varphi|) \text{ implies } b \in \psi.$$

We also say that f is $\langle.^- \rangle$-*saturated,* if for any $a \in W_{\mathsf{G}}^{\Xi}$, the following holds:

$$\langle \varphi^- \rangle \psi \in b, \text{ if and only if, for some } a \in W_{\mathsf{G}}^{\Xi}, b \in f_0^{\Xi}(a, |\varphi|) \text{ and } a \in \psi.$$

We define that $\mathfrak{M}_{\mathsf{G}}^{\Xi} = (W_{\mathsf{G}}^{\Xi}, f, V^{\Xi})$ is a *conditional Ξ-model for* G if f is both $[.]$-saturated and $\langle.^- \rangle$-saturated, and $V^{\Xi}(p) = |p|$ for all variables $p \in \Xi$.

Lemma 3. *Let $a, b \in W_{\mathsf{G}}^{\Xi}$ and $\varphi, \psi \in \Xi$. Whenever the formula on the left belongs to Ξ, the appropriate equivalence below holds.*

(1) $\varphi \wedge \psi \in a$, *if and only if,* $\varphi \in a$ *and* $\psi \in a$.
(2) $\varphi \vee \psi \in a$, *if and only if,* $\varphi \in a$ *or* $\psi \in a$.
(3) $\varphi \to \psi \in a$, *if and only if,* $\varphi \in a$ *implies* $\psi \in a$.
(4) $\neg\varphi \in a$, *if and only if,* $\varphi \in \bar{a}$.

Proof. The items from (1) to (4) are easily shown by definition of Ξ-saturation and the corresponding propositional rules of G. □

Lemma 4. *Let $\mathfrak{M}_{\mathsf{G}}^{\Xi}$ be the conditional Ξ-model for G. For all formulas $\varphi \in \Xi$ and $a \in W^{\Xi}$, $\varphi \in a$ if and only if $\mathfrak{M}_{\mathsf{G}}^{\Xi}, a \models \varphi$.*

Proof. By induction on construction of formulas with the help of Lemma 3 and $[.]$-saturation and $\langle.^- \rangle$-saturation. □

Lemma 5. *Let $\mathsf{G} \in \{\, \mathsf{GUCK}, \mathsf{GUMP} \,\}$. For any $a \in W_{\mathsf{G}}^{\Xi}$ and $X \subseteq W_{\mathsf{G}}^{\Xi}$, we define the function $f_0^{\Xi}(a, X)$ as follows:*

– Let $X = |\gamma|$ for some $\gamma \in \Xi$. We define $b \in f_0^\Xi(a, X)$ if and only if:

$$\begin{cases} \text{(i)} & [\alpha]\beta \in a, \text{ then } \beta \in b; \text{ and} \\ \text{(ii)} & \langle \alpha^- \rangle \beta \in \bar{b}, \text{ then } \beta \in \bar{a}. \end{cases}$$

for all formulas $\alpha, \beta \in \Xi$ with $|\alpha| = X$.
– When $X \neq |\gamma|$ for any $\gamma \in \Xi$, then $f_0^\Xi(a, X) := X$.

Then the following hold: (1) f_0^Ξ is $[.]$-saturated; (2) f_0^Ξ is $\langle .^- \rangle$-saturated; (3) if $\mathsf{G} = \mathsf{GUMP}$, then f_0^Ξ satisfies the condition (F_{UMP}) in Definition 4.

Proof. We only check the item (1). We establish $[.]$-saturation. The 'only if' direction is immediately obtained by the condition (i) of the definition of $f_0^\Xi(a, |\varphi|)$. So we focus on the 'if' direction below. Suppose $[\varphi]\psi \in \bar{a}$. Let $\varphi_1, \ldots, \varphi_n$ be all the formulas in Ξ such that $|\varphi| = |\varphi_i|$ for all $1 \leqslant i \leqslant n$. Now we show that, for all $1 \leqslant i \leqslant n$, $\mathsf{G} \vdash_\Xi \varphi \Rightarrow \varphi_i$ and $\mathsf{G} \vdash_\Xi \varphi_i \Rightarrow \varphi$. Without loss of generality, it suffices to show $\mathsf{G} \vdash_\Xi \varphi \Rightarrow \varphi_i$. Assume that $\mathsf{G} \nvdash_\Xi \varphi \Rightarrow \varphi_i$. By Lemma 2, there is a Ξ-saturated pair $(b, \bar{b})$ such that $\varphi \in b$ and $\varphi_i \in \bar{b}$. Since $|\varphi| = |\varphi_i|$, we have $\varphi_i \in b$. Then $\mathsf{G} \vdash_\Xi b \Rightarrow \bar{b}$, a contradiction with Ξ-disjointness of $(b, \bar{b})$. Next, for all $1 \leqslant i \leqslant n$, we define

$$\Pi_i = \{ \gamma \mid [\varphi_i]\gamma \in a \} \text{ and } \Theta_i = \{ \delta \in a \mid \langle \varphi_i^- \rangle \delta \in \Xi \}.$$

Now we show that $\mathsf{G} \nvdash_\Xi (\langle \varphi_i^- \rangle \Theta_i, \Pi_i)_{1 \leqslant i \leqslant n} \Rightarrow \psi$. Suppose not. Then we have

$$\frac{\varphi \Leftrightarrow \cdots \Leftrightarrow \varphi_n \quad (\langle \varphi_i^- \rangle \Theta_i, \Pi_i)_{1 \leqslant i \leqslant n} \Rightarrow \psi}{(\Theta_i, [\varphi_i]\Pi_i)_{1 \leqslant i \leqslant n} \Rightarrow [\varphi]\psi} \ ([\cdot]^\circ)$$

where we note that $\langle \varphi_i^- \rangle \Theta_i \subseteq \Xi$. Since $(\Theta_i, [\varphi_i]\Pi_i)_{1 \leqslant i \leqslant n} \subseteq a$ and $[\varphi]\psi \in \bar{a}$, we have $\mathsf{G} \vdash_\Xi a \Rightarrow \bar{a}$, a contradiction with Ξ-disjointness.

Since $\mathsf{G} \nvdash_\Xi (\langle \varphi_i^- \rangle \Theta_i, \Pi_i)_{1 \leqslant i \leqslant n} \Rightarrow \psi$, by Lemma 2, there exists $b \in W_{\mathsf{G}}^\Xi$ such that $(\langle \varphi_i^- \rangle \Theta_i, \Pi_i)_{1 \leqslant i \leqslant n} \subseteq b$ and $\psi \in \bar{b}$. Now we show that $b \in f_0^\Xi(a, |\varphi|)$. We need to check two conditions (i) and (ii). We fix any $\alpha \in \Xi$ such that $|\alpha| = |\varphi|$. Let $\alpha = \varphi_i$ (recall the definition of φ_is in the beginning of the proof). The condition (i) follows directly from $\Pi_i \subseteq b$ as follows. Suppose that $[\varphi_i]\delta \in a$. By the definition of Π_i, $\delta \in \Pi_i$. It follows from $\Pi \subseteq b$ that $\delta \in b$. This finishes to establish (i). For the condition (ii), suppose $\langle \varphi_i^- \rangle \delta \in \bar{b}$. Our goal is to show $\delta \in \bar{a}$. Suppose for contradiction that $\delta \in a$. Since $\langle \varphi_i^- \rangle \delta \in \bar{b} \subseteq \Xi$, we derive from $\delta \in a$ that $\langle \varphi_i^- \rangle \delta \in \Theta_i \subseteq b$. This is a contradiction with the Ξ-disjointness of $b \Rightarrow \bar{b}$. So we can conclude that $\delta \in \bar{a}$. $\square$

Lemma 6. *Let $\mathsf{G} \in \{ \mathsf{GUID}, \mathsf{GUIDMP} \}$. For any $a \in W^\Xi$ and $X \subseteq W^\Xi$, the function $f_{ID}^\Xi(a, X)$ is defined as follows.*

– Let $X = |\gamma|$ for some $\gamma \in \Xi$.

$$b \in f_{ID}^\Xi(a, X) \text{ iff } \begin{cases} \text{(i)} & [\alpha]\beta \in a \text{ implies } \beta \in b, \\ \text{(ii)} & \langle \alpha^- \rangle \beta \in \bar{b} \text{ implies } \beta \in \bar{a}, \\ \text{(iii)} & b \in |\alpha|, \end{cases}$$

for all formulas $\alpha, \beta \in \Xi$ such that $|\alpha| = X$.

– *Let $X \neq |\alpha|$ for any $\alpha \in \Xi$. Then $f_{ID}^{\Xi}(a, X) := X$.*

Then the following hold.

(1) f_{ID}^{Ξ} *is* $[.]$-*saturated.*
(2) f_{ID}^{Ξ} *is* $\langle .^{-} \rangle$-*saturated.*
(3) *If* $\mathsf{G} = \mathsf{GUIDMP}$, *then* f_{ID}^{Ξ} *satisfies the condition* (F_{UMP}) *of Definition 4.*
(4) f_{ID}^{Ξ} *satisfies the condition* (F_{CID}) *of Definition 4.*

Lemma 7. *Let* G *be any sequent calculus given in Table 1. Let* $\Gamma \Rightarrow \Delta$ *be a sequent and* $\Xi := \mathrm{Sub}(\Gamma, \Delta)$. *If* $\Gamma \Rightarrow \Delta$ *is not* Ξ-*provable in* G, *then the sequent is refuted at some state in a finite conditional model* $\mathfrak{M}_{\mathsf{G}}^{\Xi}$ *whose frame is belonging to the corresponding class of conditonal frames to* G.

Proof. Suppose that $\Gamma \Rightarrow \Delta$ is not Ξ-provable in G. By Lemma 2, there is $a \in W_{\mathsf{G}}^{\Xi}$ such that $\Gamma \subseteq a$ and $\Delta \subseteq \bar{a}$. By Lemma 4, $\mathfrak{M}_{\mathsf{G}}^{\Xi}, a \not\models \Gamma \Rightarrow \Delta$. Let $\mathfrak{F}^{\Xi}$ be the frame part of $\mathfrak{M}^{\Xi}$. When $\mathsf{G} \in \{\mathsf{GUID}, \mathsf{GUIDMP}\}$, $\mathfrak{F}^{\Xi} \in \mathbb{CID}$ holds by definition of f_{ID}^{Ξ}. When $\mathsf{G} \in \{\mathsf{GUMP}, \mathsf{GUIDMP}\}$, $\mathfrak{F}^{\Xi} \in \mathbb{CMP}$ is obtained by Lemmas 5 (3) and 6 (3). □

Theorem 3 (Subformula Property). *Let* G *be any sequent calculus given in Table 1. If* $\Gamma \Rightarrow \Delta$ *is provable in* G, *then it is* $\mathrm{Sub}(\Gamma, \Delta)$-*provable in* G.

Proof. Put $\Xi := \mathrm{Sub}(\Gamma, \Delta)$. We prove the contrapositive implication of it. Suppose that $\Gamma \Rightarrow \Delta$ is not Ξ-provable in G. By Lemma 7, the sequent is refuted in a finite conditional frame which belongs to the corresponding frame class to G. By soundness of G for the corresponding conditional frames (Proposition 3), we obtain that $\Gamma \Rightarrow \Delta$ is not provable in G, as desired. □

Given any $\mathsf{X} \in \{\mathsf{UCK}, \mathsf{UID}, \mathsf{UMP}, \mathsf{UIDMP}\}$, we say that Hilbert system HX has the *finite model property* if there is a class $\mathbb{C}$ of conditonal frames such that $\mathbb{C} \models \varphi$ for all theorems φ of H and every non-theorem ψ of H is refuted in a *finite* conditional frame in $\mathbb{C}$.

Corollary 1. *Let* $\mathsf{X} \in \{\mathsf{UCK}, \mathsf{UID}, \mathsf{UMP}, \mathsf{UIDMP}\}$. *Then,* HX *has the finite model property and is decidable.*

Proof. By Proposition 3, Theorem 1 and Lemma 7. We note that the non-theoremhood of φ in H implies the unprovability of $\Rightarrow \varphi$ in G because provability of a sequent in G implies provability of the same sequent in G. □

References

1. Alenda, R., Olivetti, N., Pozzato, G.L.: Nested sequent calculi for normal conditional logics. J. Logic Comput. **26**(1), 7–50 (2013)
2. Burgess, J.P.: Basic tense logic. In: Gabbay, D., Guenthner, F. (eds.) Handbook of Philosophical Logic, Reidel, Dordrecht, vol. II, pp. 89–133 (1984)
3. Chellas, B.F.: Basic conditional logic. J. Philosophcal Logic **4**(2), 133–153 (1975)
4. Chellas, B.F.: Modal Logic. Cambridge University Press, Cambridge (1980)

5. de Swart, H.C.: A Gentzen-or Beth-type system, a practical decision procedure and a constructive completeness proof for the counterfactual logics VC and VCS. J. Symbolic Logic **48**(1), 1–20 (1983)
6. Gent, P.: A sequent- or tableau-style system for Lewis's counterfactual logic. Notre Dame J. Formal Logic **33**(3), 369–382 (1992)
7. Gentzen, G.: Untersuchungen über das logische Schließen, Mathematische Zeitschrift, **39**(1), pp. 176–210 (1935)
8. Girlando, M., Lellmann, B., Olivetti, N., Pozzato, G.L.: Standard sequent calculi for Lewis' logics of counterfactuals. In: Michael, L., Kakas, A. (eds.) JELIA 2016. LNCS (LNAI), vol. 10021, pp. 272–287. Springer, Cham (2016). https://doi.org/10.1007/978-3-319-48758-8_18
9. Herzig, A.: Logics for belief base updating. In: Dubois, D., Przde, H. (eds.) Handbook of Defeasible Reasoning and Uncertainty Management System, vol. 3, pp. 189–231. Springer, Dordrecht (1998). https://doi.org/10.1007/978-94-011-5054-5_5
10. Kowalski, T., Ono, H.: Analytic cut and interpolation for bi-intuisionistic logic. The Review of Symbolic Logic, **10**(2), pp. 259–283 (2017)
11. Lellmann, B., Pattinson, D.: Sequent systems for Lewis' conditional logics. In: del Cerro, L.F., Herzig, A., Mengin, J. (eds.) JELIA 2012. LNCS (LNAI), vol. 7519, pp. 320–332. Springer, Heidelberg (2012). https://doi.org/10.1007/978-3-642-33353-8_25
12. Maruyama, A.: Towards combined system of modal logics - a syntactic and semantic study, Ph.D. thesis, School of Information Science, Japan Advanced Institute of Science and Technology (2003)
13. Maruyama, A., Tojo, S., Ono, H.: Decidability of temporal epistemic logics for multi-agent models. In: Proceedings of the ICLP'01 Workshop on Computational Logic in Multi-Agent Systems (CLIMA-01), pp. 31–40 (2001)
14. Negri, S., Sabrdolini, G.: Proof analysis for Lewis counterfactuals. Rev. Symbolic Logic **9**(1), 44–75 (2016)
15. Negri, S., Von Plato, J.: Structural Proof Theory. Cambridge University Press, Cambridge (2001)
16. Nishimura, H.: A study of some tense logics by Gentzen's sequential method. Publ. Res. Inst. Math. Sci. **16**, 343–353 (1980)
17. Ohnishi, M., Matsumoto, K.: Gentzen method in modal calculi II. Osaka J. Math. **11**(2), 115–120 (1959)
18. Ono, H.: Semantical approach to cut elimination and subformula property in modal logic. In: Yang, S.C.-M., Deng, D.-M., Lin, H. (eds.) Structural Analysis of Non-Classical Logics. LASLL, pp. 1–15. Springer, Heidelberg (2016). https://doi.org/10.1007/978-3-662-48357-2_1
19. Pattinson, D., Schröder, L.: Generic modal cut elimination applied to conditional logic. Logical Methods Comput. Sci. **7**(1:4), 1–28 (2011)
20. Takano, M.: Subformula property as a substitute for cut-elimination in modal propositional logics. Math. Jpn. **37**(6), 1129–1145 (1992)
21. Takano, M.: A modified subformula property for the modal logics K5 and K5D. Bull. Sect. Logic **30**(2), 115–122 (2001)
22. Takano, M.: A semantical analysis of cut-free calculi for modal logics. Rep. Math. Logic **53**, 43–65 (2018)
23. Troelstra, A.S., Schwichtenberg, H.: Basic Proof Theory, 2nd edn. Cambridge University Press, Cambridge (2000)
24. Zach, R.: Non-analytic tableaux for Chellas's conditional logic CK and Lewis's logic of counterfactuals VC. Australas. J. Logic **15**(3), 609–628 (2018)

Logics for Rough Concept Analysis

Giuseppe Greco[1] (iD), Peter Jipsen[2], Krishna Manoorkar[3], Alessandra Palmigiano[4,5] (iD), and Apostolos Tzimoulis[2(✉)]

[1] Utrecht University, Utrecht, Netherlands
[2] Chapman University, Orange, USA
apostolos@tzimoulis.eu
[3] Indian Institute of Technology, Kanpur, India
[4] Delft University of Technology, Delft, Netherlands
[5] Department of Pure and Applied Mathematics, University of Johannesburg, Johannesburg, South Africa

Abstract. Taking an algebraic perspective on the basic structures of Rough Concept Analysis as the starting point, in this paper we introduce some varieties of lattices expanded with normal modal operators which can be regarded as the natural rough algebra counterparts of certain subclasses of rough formal contexts, and introduce proper display calculi for the logics associated with these varieties which are sound, complete, conservative and with uniform cut elimination and subformula property. These calculi modularly extend the multi-type calculi for rough algebras to a 'nondistributive' (i.e. general lattice-based) setting.

Keywords: Rough set theory · Formal Concept Analysis · Modal logic · Lattice-based logics · Algebras for rough sets · Proper display calculi

1 Introduction

This paper continues a line of investigation started in [10] and aimed at introducing sequent calculi for the logics of varieties of 'rough algebras', introduced and discussed in [1,20]. The 'rough algebras' considered in the present paper are *nondistributive* (i.e. general lattice-based) generalizations of those of [20,21]; specifically, they are varieties of lattices expanded with normal modal operators, natural examples of which arise in connection with (certain subclasses of) *rough formal contexts*, introduced by Kent in [15] as the basic notion of *Rough Concept Analysis* (RCA), a synthesis of Rough Set Theory [19] and Formal Concept Analysis [8]. The core idea of Kent's approach is to use a given indiscernibility relation E on the objects of a formal context (A, X, I) to generate E-definable approximations R and S of the relation I such that $S \subseteq I \subseteq R$. The starting point of our approach is that R and S can be used to generate tuples of adjoint normal modal operators $\langle S \rangle \dashv [S]$ and $\langle R \rangle \dashv [R]$. We identify conditions under

The research of the fourth author is supported by the NWO Vidi grant 016.138.314, the NWO Aspasia grant 015.008.054, and a Delft Technology Fellowship awarded in 2013.

Md. A. Khan and A. Manuel (Eds.): ICLA 2019, LNCS 11600, pp. 144–159, 2019.
https://doi.org/10.1007/978-3-662-58771-3_14

which $[S]$ and $\langle R \rangle$ are interior operators and $[R]$ and $\langle S \rangle$ are closure operators. This provides the basic algebraic framework, which we axiomatically extend so as to define 'nondistributive' counterparts of the varieties introduced in [21], whenever possible.

From an algebraic perspective, it is interesting to observe that, unlike $\langle S \rangle$ and $[S]$, the modal operators $\langle R \rangle$ and $[R]$ play the reverse roles they usually have in rough set theory: namely, $[R]$, being an *inflationary* map, plays naturally the role of the *closure* operator providing the upper lax approximation of a given formal concept, and similarly $\langle R \rangle$, being a *deflationary* map, plays the role of the *interior* operator, providing the lower lax approximation of a given formal concept.

From a proof-theoretic perspective, these properties make it possible to extend the *multi-type approach* (thanks to which, a modular family of analytic calculi was introduced in [10] for the logics of 'rough algebras') to varieties of 'rough algebras' on a 'nondistributive' propositional base. In particular, the calculi defined in Sect. 6 are all *proper display calculi* (cf. [24]), the cut elimination and subformula property of which can be straightforwardly verified by appealing to the meta-theorem of [5].[1] An interesting departure from the calculi of [10] concerns the counterparts of the IA3 condition, which in the present paper comes in two variants: the lower (strict), and the upper (lax). The inequality corresponding to the lower variant of IA3, which was analytic in the presence of distributivity, is not analytic inductive in the absence of distributivity (cf. [7, Definition 55]). However, the inequality corresponding to the upper variant of IA3 is analytic inductive, and hence can be captured by an analytic structural rule.

This paper contains the first algebraic and proof-theoretic contribution to a line of research aimed at integrating Rough Set Theory and Formal Concept Analysis, and at building the necessary logical machinery to support formal reasoning about categorization decisions under the assumption that categories and concepts can be *vague*. Future directions concern enriching this basic framework so as to formally account for the fact that the *dynamics* of categories also affect their becoming vaguer or sharper.

2 Preliminaries

The purpose of this section, which is based on [3, Appendix] and [2] and [18, Sects. 2.3 and 2.4], is to briefly recall the basic notions of the theory of *enriched formal contexts* (cf. Definition 2) while introducing the notation which will be used throughout the paper. For any relation $T \subseteq U \times V$, and any $U' \subseteq U$ and $V' \subseteq V$, let

$$T^{(0)}[V'] := \{u \mid \forall v(v \in V' \Rightarrow uTv)\} \qquad T^{(1)}[U'] := \{v \mid \forall u(u \in U' \Rightarrow uTv)\}.$$

It can be easily verified that $U' \subseteq T^{(0)}[V']$ iff $V' \subseteq T^{(1)}[U']$, that $V_1 \subseteq V_2 \subseteq V$ (resp. $U_1 \subseteq U_2 \subseteq U$) implies that $T^{(0)}[V_2] \subseteq T^{(0)}[V_1]$ (resp. $T^{(1)}[U_2] \subseteq T^{(1)}[U_1]$), and $S \subseteq T \subseteq U \times V$ implies that $S^{(0)}[V'] \subseteq T^{(0)}[V']$ and $S^{(1)}[U'] \subseteq T^{(1)}[U']$ for all $V' \subseteq V$ and $U' \subseteq U$.

Formal contexts, or *polarities*, are structures $\mathbb{P} = (A, X, I)$ such that A and X are sets, and $I \subseteq A \times X$ is a binary relation. Intuitively, formal contexts can be understood as abstract representations of databases [8], so that A represents a collection of *objects*, X

---

[1] In [22], sequent calculi for non-distributive versions of the logics associated with varieties of 'rough algebras' are introduced, which are sound and complete but without cut elimination.

as a collection of *features*, and for any object a and feature x, the tuple (a, x) belongs to I exactly when object a has feature x. In what follows, we use a, b (resp. x, y) for elements of A (resp. X), and B (resp. Y) for subsets of A (resp. of X).

As is well known, for every formal context $\mathbb{P} = (A, X, I)$, the pair of maps

$$(\cdot)^{\uparrow} : \mathcal{P}(A) \to \mathcal{P}(X) \quad \text{and} \quad (\cdot)^{\downarrow} : \mathcal{P}(X) \to \mathcal{P}(A),$$

respectively defined by the assignments $B^{\uparrow} := I^{(1)}[B]$ and $Y^{\downarrow} := I^{(0)}[Y]$, form a Galois connection and hence induce the closure operators $(\cdot)^{\uparrow\downarrow}$ and $(\cdot)^{\downarrow\uparrow}$ on $\mathcal{P}(A)$ and on $\mathcal{P}(X)$ respectively.[2] Moreover, the fixed points of these closure operators form complete sub-$\cap$-semilattices of $\mathcal{P}(A)$ and $\mathcal{P}(X)$ respectively, and hence are complete lattices which are dually isomorphic to each other via the restrictions of the maps $(\cdot)^{\uparrow}$ and $(\cdot)^{\downarrow}$. This motivates the following.

Definition 1. *For every formal context $\mathbb{P} = (A, X, I)$, a* formal concept *of $\mathbb{P}$ is a pair $c = (B, Y)$ such that $B \subseteq A$, $Y \subseteq X$, and $B^{\uparrow} = Y$ and $Y^{\downarrow} = B$. The set B is the* extension *of c, which we will sometimes denote $[\![c]\!]$, and Y is the* intension *of c, sometimes denoted $([c])$. Let $L(\mathbb{P})$ denote the set of the formal concepts of $\mathbb{P}$. Then the* concept lattice *of $\mathbb{P}$ is the complete lattice*

$$\mathbb{P}^{+} := (L(\mathbb{P}), \wedge, \vee),$$

where for every $X \subseteq L(\mathbb{P})$,

$$\wedge X := (\cap_{c \in X} [\![c]\!], (\cap_{c \in X} [\![c]\!])^{\uparrow}) \quad \text{and} \quad \vee X := ((\cap_{c \in X} ([c]))^{\downarrow}, \cap_{c \in X} ([c])).$$

Then clearly, $\top^{\mathbb{P}^{+}} := \wedge \emptyset = (A, A^{\uparrow})$ and $\bot^{\mathbb{P}^{+}} := \vee \emptyset = (X^{\downarrow}, X)$, and the partial order underlying this lattice structure is defined as follows: for any $c, d \in L(\mathbb{P})$,

$$c \leq d \quad \text{iff} \quad [\![c]\!] \subseteq [\![d]\!] \quad \text{iff} \quad ([d]) \subseteq ([c]).$$

Theorem 1 *(Birkhoff's theorem, main theorem of FCA). Any complete lattice $\mathbb{L}$ is isomorphic to the concept lattice $\mathbb{P}^{+}$ of some formal context $\mathbb{P}$.*

Definition 2. *An* enriched formal context *is a tuple $\mathbb{F} = (\mathbb{P}, R_{\Box}, R_{\Diamond})$ such that $\mathbb{P} = (A, X, I)$ is a formal context, and $R_{\Box} \subseteq A \times X$ and $R_{\Diamond} \subseteq X \times A$ are I-compatible* relations, *that is, $R_{\Box}^{(0)}[x]$ (resp. $R_{\Diamond}^{(0)}[a]$) and $R_{\Box}^{(1)}[a]$ (resp. $R_{\Diamond}^{(1)}[x]$) are Galois-stable for all $x \in X$ and $a \in A$. The* complex algebra *of $\mathbb{F}$ is*

$$\mathbb{F}^{+} = (\mathbb{P}^{+}, [R_{\Box}], \langle R_{\Diamond} \rangle),$$

where $\mathbb{P}^{+}$ is the concept lattice of $\mathbb{P}$, and $[R_{\Box}]$ and $\langle R_{\Diamond} \rangle$ are unary operations on $\mathbb{P}^{+}$ defined as follows: for every $c \in \mathbb{P}^{+}$,

$$[R_{\Box}]c := (R_{\Box}^{(0)}[([c])], (R_{\Box}^{(0)}[([c])])^{\uparrow}) \quad \text{and} \quad \langle R_{\Diamond} \rangle c := ((R_{\Diamond}^{(0)}[[\![c]\!]])^{\downarrow}, R_{\Diamond}^{(0)}[[\![c]\!]]).$$

Since $R_{\Box}$ and $R_{\Diamond}$ are I-compatible, $[R_{\Box}], \langle R_{\Diamond} \rangle, [R_{\Box}^{-1}], \langle R_{\Diamond}^{-1} \rangle : \mathbb{P}^{+} \to \mathbb{P}^{+}$ are well-defined.

---

[2] When $B = \{a\}$ (resp. $Y = \{x\}$) we write $a^{\uparrow\downarrow}$ for $\{a\}^{\uparrow\downarrow}$ (resp. $x^{\downarrow\uparrow}$ for $\{x\}^{\downarrow\uparrow}$).

Lemma 1 *(cf. [18, Lemma 3]). For any enriched formal context* $\mathbb{F} = (\mathbb{P}, R_\square, R_\lozenge)$, *the algebra* $\mathbb{F}^+ = (\mathbb{P}^+, [R_\square], \langle R_\lozenge \rangle)$ *is a complete lattice expanded with normal modal operators such that* $[R_\square]$ *is completely meet-preserving and* $\langle R_\lozenge \rangle$ *is completely join-preserving.*

Definition 3. *For any formal context* $\mathbb{P} = (A, X, I)$ *and any* I-*compatible relations* $R, T \subseteq A \times X$, *the* composition $R ; T \subseteq A \times X$ *is defined as follows: for any* $a \in A$ *and* $x \in X$,

$$(R;T)^{(1)}[a] = R^{(1)}[I^{(0)}[T^{(1)}[a]]] \text{ or equivalently } (R;T)^{(0)}[x] = R^{(0)}[I^{(1)}[T^{(0)}[x]]].$$

3 Motivation: Kent's Rough Concept Analysis

Below, we report on the basic definitions and constructions in Rough Concept Analysis [15], cast in the notational conventions of Sect. 2.

Rough formal contexts (abbreviated as *Rfc*) are tuples $\mathbb{G} = (\mathbb{P}, E)$ such that $\mathbb{P} = (A, X, I)$ is a polarity (cf. Sect. 2), and $E \subseteq A \times A$ is an equivalence relation (the *indiscernibility* relation between objects). For every $a \in A$ we let $(a)_E := \{b \in A \mid aEb\}$. The relation E induces two relations $R, S \subseteq A \times I$ approximating I, defined as follows: for every $a \in A$ and $x \in X$,

$$aRx \text{ iff } bIx \text{ for some } b \in (a)_E; \qquad aS x \text{ iff } bIx \text{ for all } b \in (a)_E. \tag{1}$$

By definition, R, S are E-definable (i.e. $R^{(0)}[x] = \cup_{aRx}(a)_E$ and $S^{(0)}[x] = \cup_{aSx}(a)_E$ for any $x \in X$), and E being reflexive immediately implies that

Lemma 2. *For any Rfc* $\mathbb{G} = (\mathbb{P}, E)$, *if* R *and* S *are defined as in* (1), *then*

$$S \subseteq I \quad and \quad I \subseteq R. \tag{2}$$

Intuitively, we can think of R as the *lax* version of I determined by E, and S as its *strict* version determined by E. Following the methodology introduced in [4] and applied in [2,3] to introduce a polarity-based semantics for the modal logics of formal concepts, under the assumption that R and S are I-compatible (cf. Definition 2), the relations R and S can be used to define normal modal operators $[R], \langle R \rangle, [S], \langle S \rangle$ on $\mathbb{P}^+$ defined as follows: for any $c \in \mathbb{P}^+$,

$$[\![[R]c]\!] := R^{(0)}[([\![c]\!])] = \{a \in A \mid \forall x (x \in ([\![c]\!]) \Rightarrow aRx)\} \tag{3}$$

$$[\![[S]c]\!] := S^{(0)}[([\![c]\!])] = \{a \in A \mid \forall x (x \in ([\![c]\!]) \Rightarrow aS x)\}. \tag{4}$$

That is, the members of $[R]c$ are exactly those objects that satisfy (possibly by proxy of some object equivalent to them) all features in the description of c, while the members of $[S]c$ are exactly those objects that not only satisfy all features in the description of c, but that 'force' all their equivalents to also satisfy them. The assumption that $S \subseteq I$ implies that $[\![[S]c]\!] = S^{(0)}[([\![c]\!])] \subseteq I^{(0)}[([\![c]\!])] = [\![c]\!]$, hence $[S]c$ is a sub-concept of c. The assumption that $I \subseteq R$ implies that $[\![c]\!] = I^{(0)}[([\![c]\!])] \subseteq R^{(0)}[([\![c]\!])] = [\![[R]c]\!]$, hence $[R]c$ is a super-concept of c. Moreover, for any $c \in \mathbb{P}^+$,

$$([\![\langle R \rangle c]\!]) := R^{(1)}[[\![c]\!]] = \{x \in X \mid \forall a (a \in [\![c]\!] \Rightarrow aRx)\} \tag{5}$$

$$([\langle S \rangle c]) := S^{(1)}[[[c]]] = \{x \in X \mid \forall a(a \in [[c]] \Rightarrow aS\,x)\}. \tag{6}$$

That is, $\langle R \rangle c$ is the concept described by those features shared not only by each member of c but also by their equivalents, while $\langle S \rangle c$ is the concept described by the common features of those members of c which 'force' each of their equivalents to share them. The assumption that $I \subseteq R$ implies that $([c]) = I^{(1)}[[[c]]] \subseteq R^{(1)}[[[c]]] = ([\langle R \rangle c])$, and hence $\langle R \rangle c$ is a sub-concept of c. The assumption that $S \subseteq I$ implies that $([\langle S \rangle c]) = S^{(1)}[[[c]]] \subseteq I^{(1)}[[[c]]] = ([c])$, and hence $\langle S \rangle c$ is a super-concept of c. Summing up the discussion above, we have verified that the conditions $I \subseteq R$ and $S \subseteq I$ imply that the following sequents of the modal logic of formal concepts are valid on Kent's basic structures:

$$\Box_s \phi \vdash \phi \quad \phi \vdash \Box_\ell \phi \quad \phi \vdash \Diamond_s \phi \quad \Diamond_\ell \phi \vdash \phi, \tag{7}$$

where $\Box_s$ is interpreted as $[S]$, $\Box_\ell$ as $[R]$, $\Diamond_s$ as $\langle S \rangle$ and $\Diamond_\ell$ as $\langle R \rangle$. Translated algebraically, these conditions say that $\Box_s$ and $\Diamond_\ell$ are *deflationary*, as *interior* operators are, $\Diamond_s$ and $\Box_\ell$ are *inflationary*, as *closure* operators are. Hence, it is natural to ask under which conditions they (i.e. their semantic interpretations) are indeed closure/interior operators. The next definition and lemma provide answers to this question.

Definition 4. *An Rfc* $\mathbb{G} = (\mathbb{P}, E)$ *is* amenable *if E, R and S (defined as in (1)) are I-compatible.*[3]

Lemma 3. *For any amenable Rfc* $\mathbb{G} = (\mathbb{P}, E)$, *if and R and S are defined as in (1), then*

$$R;R \subseteq R \quad \text{and} \quad S \subseteq S;S. \tag{8}$$

Proof. Let $x \in X$. To show that $R^{(0)}[I^{(1)}[R^{(0)}[x]]] \subseteq R^{(0)}[x]$, let $a \in R^{(0)}[I^{(1)}[R^{(0)}[x]]]$. By adjunction, this is equivalent to $I^{(1)}[R^{(0)}[x]] \subseteq R^{(1)}[a]$, which implies that $I^{(0)}[R^{(1)}[a]] \subseteq I^{(0)}[I^{(1)}[R^{(0)}[x]]] = R^{(0)}[x]$, the last equality holding since R is I-compatible by assumption. Moreover, $I \subseteq R$ (cf. Lemma 2) implies that $I^{(1)}[a] \subseteq R^{(1)}[a]$, which implies that $I^{(0)}[R^{(1)}[a]] \subseteq I^{(0)}[I^{(1)}[a]] \subseteq (a)_E$, the last inclusion holding since E is I-compatible by assumption. Hence, $I^{(0)}[R^{(1)}[a]] \subseteq R^{(0)}[x] \cap (a)_E$. Suppose for contradiction that $a \notin R^{(0)}[x]$. By the E-definability of R, this is equivalent to $R^{(0)}[x] \cap (a)_E = \varnothing$. Hence $I^{(0)}[R^{(1)}[a]] = \varnothing$, from which it follows that $R^{(1)}[a] = I^{(1)}[I^{(0)}[R^{(1)}[a]]] = I^{(1)}[\varnothing] = X$. Hence, $x \in R^{(1)}[a]$, i.e. $a \in R^{(0)}[x]$, against the assumption that $a \notin R^{(0)}[x]$.

Let $x \in X$. To show that $S^{(0)}[x] \subseteq S^{(0)}[I^{(1)}[S^{(0)}[x]]]$, assume that $a \in S^{(0)}[x]$. Since S is E-definable by construction, this is equivalent to $(a)_E \subseteq S^{(0)}[x]$. To show that $a \in S^{(0)}[I^{(1)}[S^{(0)}[x]]]$, we need to show that bIy for any $b \in (a)_E$ and any $y \in I^{(1)}[S^{(0)}[x]]$. Let $y \in I^{(1)}[S^{(0)}[x]]$. Hence, by definition, $b'Iy$ for every $b' \in S^{(0)}[x]$. Since $(a)_E \subseteq S^{(0)}[x]$, this implies that bIy for any $b \in (a)_E$, as required.

By the general theory developed in [4] and applied to enriched formal contexts in [18, Proposition 5], properties (8) guarantee that the following sequents of the modal logic of formal concepts are also valid on amenable Rfc's:

$$\Box_s \phi \vdash \Box_s \Box_s \phi \quad \Box_\ell \Box_\ell \phi \vdash \Box_\ell \phi \quad \Diamond_s \Diamond_s \phi \vdash \Diamond_s \phi \quad \Diamond_\ell \phi \vdash \Diamond_\ell \Diamond_\ell \phi. \tag{9}$$

---

[3] The assumption that E is I-compatible does not follow from R and S being I-compatible. Let $\mathbb{G} = (\mathbb{P}, Id_A)$ for any polarity $\mathbb{P}$ such that not all singleton sets of objects are Galois-stable. Hence $E = Id_A$ is not I-compatible. However, if $E = Id_A$, then $R = S = I$ are I-compatible.

Finally, again by [18, Proposition 5], the fact that by construction $\Box_s$ and $\Diamond_s$ (resp. $\Box_\ell$ and $\Diamond_\ell$) are interpreted by operations defined in terms of the same relation guarantees the validity of the following sequents on amenable Rfc's:

$$\phi \vdash \Box_s \Diamond_s \phi \qquad \Diamond_s \Box_s \phi \vdash \phi \qquad \phi \vdash \Box_\ell \Diamond_\ell \phi \qquad \Diamond_\ell \Box_\ell \phi \vdash \phi. \tag{10}$$

Axioms (7), (9) and (10) constitute the starting point and motivation for the proof-theoretic investigation of the logics associated to varieties of algebraic structures which can be understood as abstractions of amenable Rfc's. We define these varieties in the next section.

4 Kent Algebras

In the present section, we introduce *basic Kent algebras* (and the variety of *abstract Kent algebras* (aKa) to which they naturally belong), as algebraic generalizations of amenable Rfc's, and then introduce some subvarieties of aKas in the style of [20,21].

Definition 5. *A basic Kent algebra is a structure* $\mathbb{A} = (\mathbb{L}, \Box_s, \Diamond_s, \Box_\ell, \Diamond_\ell)$ *such that* $\mathbb{L}$ *is a complete lattice, and* $\Box_s, \Diamond_s, \Box_\ell, \Diamond_\ell$ *are unary operations on* $\mathbb{L}$ *such that for all* $a, b \in \mathbb{L}$,

$$\Diamond_s a \leq b \text{ iff } a \leq \Box_s b \quad \text{and} \quad \Diamond_\ell a \leq b \text{ iff } a \leq \Box_\ell b, \tag{11}$$

and for any $a \in \mathbb{L}$,

$$\Box_s a \leq a \qquad a \leq \Diamond_s a \qquad a \leq \Box_\ell a \qquad \Diamond_\ell a \leq a \tag{12}$$

$$\Box_s a \leq \Box_s \Box_s a \qquad \Diamond_s \Diamond_s a \leq \Diamond_s a \qquad \Box_\ell \Box_\ell a \leq \Box_\ell a \qquad \Diamond_\ell a \leq \Diamond_\ell \Diamond_\ell a \tag{13}$$

We let **KA**$^+$ *denote the class of basic Kent algebras.*

From (11) it follows that, in basic Kent algebras, $\Box_s$ and $\Box_\ell$ are completely meet-preserving, $\Diamond_s$ and $\Diamond_\ell$ are completely join-preserving. For any amenable Rfc $\mathbb{G} = (\mathbb{P}, E)$, if R and S are defined as in (1), then

$$\mathbb{G}^+ := (\mathbb{P}^+, [S], \langle S \rangle, [R], \langle R \rangle)$$

where $\mathbb{P}^+$ is the concept lattice of the formal context $\mathbb{P}$ and $[S], \langle S \rangle, [R], \langle R \rangle$ are defined as in (3)–(6). The following proposition is an immediate consequence of [18, Proposition 5], using Lemmas 2 and 3, and the fact that $[R]$ and $\langle R \rangle$ (resp. $[S]$ and $\langle S \rangle$) are defined using the same relation.

Proposition 1. *If* $\mathbb{G} = (\mathbb{P}, E)$ *is an amenable Rfc, then* $\mathbb{G}^+$ *is a basic Kent algebra.*

The natural variety containing basic Kent algebras is defined as follows.

Definition 6. *An* abstract Kent algebra *(aKa) is a structure* $\mathbb{A} = (\mathbb{L}, \Box_s, \Diamond_s, \Box_\ell, \Diamond_\ell)$ *such that* $\mathbb{L}$ *is a lattice, and* $\Box_s, \Diamond_s, \Box_\ell, \Diamond_\ell$ *are unary operations on* $\mathbb{L}$ *validating* (11), (12) *and* (13). *We let* **KA** *denote the class of abstract Kent algebras.*

From (11) it follows that, in aKas, $\Box_s$ and $\Box_\ell$ are finitely meet-preserving, $\Diamond_s$ and $\Diamond_\ell$ are finitely join-preserving.

Lemma 4. *For any aKa* $\mathbb{A} = (\mathbb{L}, \Box_s, \Diamond_s, \Box_\ell, \Diamond_\ell)$ *and every* $a \in \mathbb{L}$,

$$\Box_s a \vee \Diamond_\ell a \leq \Box_\ell a \wedge \Diamond_s a. \tag{14}$$

$$a \leq \Box_s \Diamond_s a \qquad \Diamond_s \Box_s a \leq a \qquad a \leq \Box_\ell \Diamond_\ell a \qquad \Diamond_\ell \Box_\ell a \leq a \tag{15}$$

$$\Box_s a \leq \Box_s \Diamond_s a \qquad \Diamond_s \Box_s a \leq \Diamond_s a \qquad \Diamond_\ell a \leq \Box_\ell \Diamond_\ell a \qquad \Diamond_\ell \Box_\ell a \leq \Box_\ell a. \tag{16}$$

$$\Diamond_s \Box_s a \leq \Box_s a \qquad \Diamond_s a \leq \Box_s \Diamond_s a \qquad \Diamond_\ell \Box_\ell a \leq \Box_\ell a \qquad \Diamond_\ell a \leq \Box_\ell \Diamond_\ell a. \tag{17}$$

Proof. The inequalities in (15) are straightforward consequences of (11). The inequalities in (14) and (16) follow from (12) and (15), using the transitivity of the order. The inequalities in (17) follow from those in (13) using (11).

Conditions (17) define the 'Kent algebra' counterparts of topological quasi Boolean algebras 5 (tqBa5) [21]. In the next definition, we introduce 'Kent algebra' counterparts of some other varieties considered in [21] (omitting those the axiomatization of which involves negation and those that cannot be captured by multi-type analytic rules in the present setting), and also varieties characterized by interaction axioms between lax and strict connectives which follow the pattern of the 5-axioms in rough algebras.

Definition 7. *An aKa* $\mathbb{A}$ *as above is an* aKa5' *if for any* $a \in \mathbb{L}$,

$$\Diamond_\ell a \leq \Box_s \Diamond_\ell a \qquad \Diamond_s \Box_\ell a \leq \Box_\ell a \qquad \Box_s a \leq \Diamond_\ell \Box_s a \qquad \Box_\ell \Diamond_s a \leq \Diamond_s a; \tag{18}$$

is a K-IA3$_s$ *if for any* $a, b \in \mathbb{L}$,

$$\Box_s a \leq \Box_s b \text{ and } \Diamond_s a \leq \Diamond_s b \text{ imply } a \leq b, \tag{19}$$

and is a K-IA3$_\ell$ *if for any* $a, b \in \mathbb{L}$,

$$\Box_\ell a \leq \Box_\ell b \text{ and } \Diamond_\ell a \leq \Diamond_\ell b \text{ imply } a \leq b. \tag{20}$$

Notice that the axioms above do not need to be analytic inductive in order for the resulting logic to be (multi-type) properly displayable: interestingly, the third and fourth inequality in (18) are not analytic inductive (cf. [7, Definition 55]), but are equivalent to analytic inductive inequalities in the multi-type language of the heterogeneous algebras discussed in the next section. This is an illustration of the technical advantage of moving to the multi-type setting (see also [10, Introduction, Sect. 4], where it is discussed how the multi-type approach was key in overcoming the difficulties encountered by the authors of [17] in introducing an analytic calculus for IA3).

5   Multi-type Presentation of Kent Algebras

Similarly to what holds for rough algebras (cf. [10, Sect. 3]), since the modal operations of any aKa $\mathbb{A} = (\mathbb{L}, \Box_s, \Diamond_s, \Box_\ell, \Diamond_\ell)$ are either interior operators or closure operators, each of them factorizes into a pair of adjoint normal modal operators which are retractions or co-retractions, as illustrated in the following table:

$\Box_s = \circ_I \cdot \blacksquare_I$	$\blacksquare_I \cdot \circ_I = id_{S_I}$	$\Diamond_s = \circ_C \cdot \blacklozenge_C$	$\blacklozenge_C \cdot \circ_C = id_{S_C}$
$\circ_I : S_I \hookrightarrow L$	$\blacksquare_I : L \twoheadrightarrow S_I$	$\blacklozenge_C : L \twoheadrightarrow S_C$	$\circ_C : S_C \hookrightarrow L$
$\Box_\ell = \Box_C \cdot \bullet_C$	$\bullet_C \cdot \Box_C = id_{L_C}$	$\Diamond_\ell = \Diamond_I \cdot \bullet_I$	$\bullet_I \cdot \Diamond_I = id_{L_I}$
$\bullet_C : L \twoheadrightarrow L_C$	$\Box_C : L_C \hookrightarrow L$	$\Diamond_I : L_I \hookrightarrow L$	$\bullet_I : L \twoheadrightarrow L_I$

where $S_I := \Box_s[L]$, $S_C := \Diamond_s[L]$, $L_C := \Box_\ell[L]$, and $L_I := \Diamond_s[L]$, and such that for all $\alpha \in S_I, \delta \in S_C, a \in L, \pi \in L_I, \sigma \in L_C$,

$$\circ_I \alpha \le a \text{ iff } \alpha \le \blacksquare_I a \quad \blacklozenge_C a \le \delta \text{ iff } a \le \circ_C \delta \quad \bullet_C a \le \pi \text{ iff } a \le \Box_C \pi \quad \Diamond_I \sigma \le a \text{ iff } \sigma \le \bullet_I a. \tag{21}$$

Again similarly to what observed in [10], the lattice structure of L can be exported to each of the sets S_I, S_C, L_C and L_I via the corresponding pair of modal operators as follows.

Definition 8. *For any aKa A, the* strict interior kernel $S_I = (S_I, \cup_I, \cap_I, t_I, f_I)$ *and the* strict closure kernel $S_C = (S_C, \cup_C, \cap_C, t_C, f_C)$ *are such that, for all $\alpha, \beta \in S_I$, and all $\delta, \gamma \in S_C$,*

$$\alpha \cup_I \beta := \blacksquare_I(\circ_I \alpha \vee \circ_I \beta) \qquad \delta \cup_C \gamma := \blacklozenge_C(\circ_C \delta \vee \circ_C \gamma)$$
$$\alpha \cap_I \beta := \blacksquare_I(\circ_I \alpha \wedge \circ_I \beta) \qquad \delta \cap_C \gamma := \blacklozenge_C(\circ_C \delta \wedge \circ_C \gamma)$$
$$t_I := \blacksquare_I \top, \ f_I := \blacksquare_I \bot \qquad t_C := \blacklozenge_C \top, \ f_C = \blacklozenge_C \bot$$

The lax interior kernel $L_I = (L_I, \sqcup_I, \sqcap_I, 1_I, 0_I)$ *and the* lax closure kernel $L_C = (L_C, \sqcup_C, \sqcap_C, 1_C, 0_C)$ *are such that, for all $\pi, \xi \in L_I$, and all $\sigma, \tau \in L_C$,*

$$\pi \sqcup_I \xi := \bullet_I(\Diamond_I \pi \vee \Diamond_I \xi) \qquad \sigma \sqcup_C \tau := \bullet_C(\Box_C \sigma \vee \Box_C \tau)$$
$$\pi \sqcap_I \xi := \bullet_I(\Diamond_I \pi \wedge \Diamond_I \xi) \qquad \sigma \sqcap_C \tau := \bullet_C(\Box_C \sigma \wedge \Box_C \tau)$$
$$1_I := \bullet_I \top, \ 0_I := \bullet_I \bot \qquad 1_C := \bullet_C \top, \ 0_C = \bullet_C \bot$$

Similarly to what observed in [10], it is easy to verify that the algebras defined above are lattices, and the operations indicated with a circle (either black or white) are lattice homomorphisms (i.e. are both normal box-type and normal diamond-type operators). The construction above justifies the following definition of class of heterogeneous algebras equivalent to aKas:

Definition 9. *A* heterogeneous aKa *(haKa) is a tuple*

$$\mathbb{H} = (L, S_I, S_C, L_I, L_C, \circ_I, \blacksquare_I, \circ_C, \blacklozenge_C, \bullet_I, \Diamond_I, \bullet_C, \Box_C)$$

such that:

H1 L, S_I, S_C, L_I, L_C are bounded lattices;
H2 $\circ_I : S_I \hookrightarrow L, \circ_C : S_C \hookrightarrow L, \bullet_I : L \twoheadrightarrow L_I, \bullet_C : L \twoheadrightarrow L_C$ are lattice homomorphisms;
H3 $\circ_I \dashv \blacksquare_I \quad \blacklozenge_C \dashv \circ_C \quad \bullet_C \dashv \Box_C \quad \Diamond_I \dashv \bullet_I$;
H4 $\blacksquare_I \circ_I = id_{S_I} \quad \blacklozenge_C \circ_C = id_{S_C} \quad \bullet_C \Box_C = id_{L_C} \quad \bullet_I \Diamond_I = id_{L_I}$[4]

The haKas corresponding to the varieties of Definition 7 are defined as follows:

---

[4] Condition H3 implies that $\blacksquare_I : L \twoheadrightarrow S_I$ and $\Box_I : L_I \hookrightarrow L$ are $\wedge$-hemimorphisms and $\blacklozenge_C : L \twoheadrightarrow S_C$ and $\Diamond_C : L_C \hookrightarrow L$ are $\vee$-hemimorphisms; condition H4 implies that the black connectives are surjective and the white ones are injective.

Algebra	Acronym	Conditions
heterogeneous aKa5'	haKa5'	$\Diamond_I \pi \leq \circ_I \blacksquare_I \Diamond_I \pi$ $\circ_C \blacklozenge_C \Box_C \sigma \leq \Box_C \sigma$
		$\circ_I \alpha \leq \Diamond_I \bullet_I \circ_I \alpha$ $\Box_C \bullet_C \circ_C \delta \leq \circ_C \delta$
heterogeneous K-IA3$_s$	hK-IA3$_s$	$\blacksquare_I a \leq \blacksquare_I b$ and $\blacklozenge_C a \leq \blacklozenge_C b$ imply $a \leq b$
heterogeneous K-IA3$_\ell$	hK-IA3$_\ell$	$\Box_C \bullet_C a \leq \Box_C \bullet_C b$ and $\Diamond_I \bullet_I a \leq \Diamond_I \bullet_I b$ imply $a \leq b$

Notice that the inequalities defining haKa5' are all analytic inductive. A heterogeneous algebra $\mathbb{H}$ is perfect *if every lattice in the signature of $\mathbb{H}$ is perfect (cf. [4, Definition 1.8]), and every homomorphism (resp. hemimorphism) in the signature of $\mathbb{H}$ is a complete homomorphism (resp. hemimorphism).*

Similarly to what discussed in [10, Sect. 3], one can readily show that the classes of haKas defined above correspond to the varieties defined in Sect. 4. That is, for any aKa $\mathbb{A}$ one can define its corresponding haKa $\mathbb{A}^+$ using the factorizations described at the beginning of the present section and Definition 8, and conversely, given a haKa $\mathbb{H}$, one can define its corresponding aKa $\mathbb{H}_+$ by endowing its first domain $\mathbb{L}$ with modal operations defined by taking the appropriate compositions of pairs of heterogeneous maps of $\mathbb{H}$. Then, for every $\mathbb{K} \in \{aKa, aKa5', K\text{-}IA3_s, K\text{-}IA3_\ell\}$, letting $\mathbb{HK}$ denote its corresponding class of heterogeneous algebras, the following holds:

Proposition 2. *1. If $\mathbb{A} \in \mathbb{K}$, then $\mathbb{A}^+ \in \mathbb{HK}$.*
2. If $\mathbb{H} \in \mathbb{HK}$, then $\mathbb{H}_+ \in \mathbb{K}$.
3. $\mathbb{A} \cong (\mathbb{A}^+)_+$ and $\mathbb{H} \cong (\mathbb{H}_+)^+$.
4. The isomorphisms of the previous item restrict to perfect members of $\mathbb{K}$ and $\mathbb{HK}$.
5. If $\mathbb{A} \in \mathbb{K}$, then $\mathbb{A}^\delta \cong ((\mathbb{A}^+)^\delta)_+$ and if $\mathbb{H} \in \mathbb{HK}$, then $\mathbb{H}^\delta \cong ((\mathbb{H}_+)^\delta)^+$.

6 Multi-type Calculi for the Logics of Kent Algebras

In the present section, we introduce the multi-type calculi associated with each class of algebras $K \in \{aKa, aKa5', K\text{-}IA3_\ell\}$. The language of these logics matches the language of haKas, and is built up from structural and operational (i.e. logical) connectives. Each structural connective is denoted by decorating its corresponding logical connective with $\hat{}$ (resp. $\check{}$ or $\tilde{}$). In what follows, we will adopt the convention that unary connectives bind more strongly than binary ones.

general lattice L
$$A ::= p \mid \top \mid \bot \mid \circ_I \alpha \mid \circ_C \delta \mid \Diamond_I \pi \mid \Box_C \sigma \mid A \wedge A \mid A \vee A$$
$$X ::= A \mid \check{\bot} \mid \hat{\top} \mid \tilde{o}_I \Gamma \mid \tilde{o}_C \Delta \mid \hat{\Diamond}_I \Pi \mid \check{\Box}_I \Pi \mid \hat{\Diamond}_C \Sigma \mid \check{\Box}_C \Sigma \mid X \hat{\wedge} X \mid X \check{\vee} X$$

strict-interior kernel S_I
$$\alpha ::= \blacklozenge_I A \mid \blacksquare_I A$$
$$\Gamma ::= \alpha \mid \hat{\blacklozenge}_I X \mid \check{\blacksquare}_I X \mid \check{f}_I \mid \hat{f}_I \mid \Gamma \hat{\wedge}_I \Gamma \mid \Gamma \check{\vee}_I \Gamma$$

lax-interior kernel L_I
$$\pi ::= \bullet_I A$$
$$\Pi ::= \pi \mid \tilde{\bullet}_I X \mid \check{0}_I \mid \hat{1}_I \mid \Pi \hat{\wedge}_I \Pi \mid \Pi \check{\vee}_I \Pi$$

strict-closure kernel S_C
$$\delta ::= \blacklozenge_C A \mid \blacksquare_C A$$
$$\Delta ::= \delta \mid \hat{\blacklozenge}_C X \mid \check{\blacksquare}_C X \mid \check{f}_C \mid \hat{f}_C \mid \Delta \hat{\wedge}_C \Delta \mid \Delta \check{\vee}_C \Delta$$

lax-closure kernel L_C
$$\sigma ::= \bullet_C A$$
$$\Sigma ::= \sigma \mid \tilde{\bullet}_C X \mid \check{0}_C \mid \hat{1}_C \mid \Sigma \hat{\wedge}_C \Sigma \mid \Sigma \check{\vee}_C \Sigma$$

– Interpretation of structural connectives as their logical counterparts[5]

1. structural and operational pure L-type connectives:

structural operations	$\hat{\top}$	$\hat{\bot}$	$\hat{\wedge}$	$\check{\vee}$
logical operations	$\top$	$\bot$	$\wedge$	$\vee$

2. structural and operational pure S_I-type and S_C-type connectives:

structural operations	$\hat{t}_I$	$\hat{f}_I$	$\hat{\cap}_I$	$\check{\cup}_I$	$\hat{t}_C$	$\hat{f}_C$	$\hat{\cap}_C$	$\check{\cup}_C$
logical operations	t_I	f_I	$\cap_I$	$\cup_I$	t_C	f_C	$\cup_C$	$\cap_C$

3. structural and operational pure L_I-type and L_C-type connectives:

structural operations	$\hat{1}_I$	$\check{0}_I$	$\hat{\cap}_I$	$\check{\cup}_I$	$\hat{1}_C$	$\check{0}_C$	$\hat{\cap}_C$	$\check{\cup}_C$
logical operations	1_I	0_I	$\cap_I$	$\cup_I$	1_C	0_C	$\sqcup_C$	$\sqcap_C$

4. structural and operational multi-type strict connectives:

types	$L \rightarrow S_I$	$L \rightarrow S_C$	$S_I \rightarrow L$	$S_C \rightarrow L$
structural operations	$\hat{\blacklozenge}_I$ $\check{\blacksquare}_I$	$\hat{\blacklozenge}_C$ $\check{\blacksquare}_C$	$\tilde{\circ}_I$	$\tilde{\circ}_C$
logical operations	$\blacklozenge_I$ $\blacksquare_I$	$\blacklozenge_C$ $\blacksquare_C$	$\circ_I$	$\circ_C$

5. structural and operational multi-type lax connectives:

types	$L_I \rightarrow L$	$L_C \rightarrow L$	$L \rightarrow L_I$	$L \rightarrow L_C$
structural operations	$\hat{\lozenge}_I$ $\check{\square}_I$	$\hat{\lozenge}_C$ $\check{\square}_C$	$\tilde{\bullet}_I$	$\tilde{\bullet}_C$
logical operations	$\lozenge_I$ $\square_I$	$\lozenge_C$ $\square_C$	$\bullet_I$	$\bullet_C$

In what follows, we will use x, y, z as structural variables of arbitrary types, a, b, c as term variables of arbitrary types.

The calculus D.AKA consists of the following axiom and rules.

– Identity and Cut:

$$Id_L \; \frac{}{p \vdash p} \qquad \frac{x \vdash a \qquad a \vdash y}{x \vdash y} \; Cut$$

– Multi-type display rules (we omit the display rules capturing the adjunctions $\lozenge_I \dashv \bullet_I \dashv \square_I$ and $\lozenge_I \dashv \bullet_I \dashv \square_I$):

$$ad_{LS_I} \; \frac{\tilde{\circ}_I \Gamma \vdash X}{\Gamma \vdash \check{\blacksquare}_I X} \qquad \frac{X \vdash \tilde{\circ}_I \Gamma}{\hat{\blacklozenge}_I X \vdash \Gamma} \; ad_{LS_I} \qquad ad_{LS_C} \; \frac{X \vdash \tilde{\circ}_C \Delta}{\hat{\blacklozenge}_C X \vdash \Delta} \qquad \frac{\tilde{\circ}_C X \vdash \Delta}{X \vdash \check{\blacksquare}_C \Delta} \; ad_{LS_C}$$

– Multi-type structural rules for strict-kernel operators:

---

[5] The connectives which appear in a grey cell in the synoptic tables will only be included in the present language at the structural level.

$$\tilde{\partial}\,\hat{\iota}_I\ \frac{\tilde{\partial}_I\,\hat{\iota}_I\vdash X \qquad X\vdash \tilde{\partial}_I\,\check{\iota}_I}{\hat{\top}\vdash X \qquad X\vdash \check{\iota}}\ \tilde{\partial}_I\,\check{\iota}_I \qquad\qquad \tilde{\partial}_C\,\hat{\iota}_C\ \frac{\tilde{\partial}_C\,\hat{\iota}_C\vdash X \qquad X\vdash \tilde{\partial}_C\,\check{\iota}_C}{\hat{\top}\vdash X \qquad X\vdash \check{\iota}}\ \tilde{\partial}_C\,\check{\iota}_C$$

$$\hat{\blacklozenge}_I\,\tilde{\partial}_I\ \frac{\hat{\blacklozenge}_I\,\tilde{\partial}_I\,\Gamma\vdash \Gamma' \qquad \Gamma'\vdash \blacksquare_I\,\tilde{\partial}_I\,\Gamma}{\Gamma\vdash \Gamma' \qquad \Gamma'\vdash \Gamma}\ \blacksquare_I\,\tilde{\partial}_I \qquad \hat{\blacklozenge}_C\,\tilde{\partial}_C\ \frac{\hat{\blacklozenge}_C\,\tilde{\partial}_C\,\Delta\vdash \Delta' \qquad \Delta'\vdash \blacksquare_C\,\tilde{\partial}_C\,\Delta}{\Delta\vdash \Delta' \qquad \Delta'\vdash \Delta}\ \blacksquare_C\,\tilde{\partial}_C$$

$$\tilde{\partial}_I\,\hat{\blacklozenge}_I\ \frac{\tilde{\partial}_I\,\hat{\blacklozenge}_I\,X\vdash Y \qquad Y\vdash \tilde{\partial}_I\,\blacksquare_I\,X}{X\vdash Y \qquad Y\vdash X}\ \tilde{\partial}_I\,\blacksquare_I \qquad \tilde{\partial}_C\,\hat{\blacklozenge}_C\ \frac{\tilde{\partial}_C\,\hat{\blacklozenge}_C\,X\vdash Y \qquad Y\vdash \tilde{\partial}_C\,\blacksquare_C\,X}{X\vdash Y \qquad Y\vdash X}\ \tilde{\partial}_C\,\blacksquare_C$$

- Multi-type structural rules for lax-kernel operators:

$$\tilde{\bullet}_I\,\hat{\iota}_I\ \frac{\tilde{\bullet}_I\,\hat{\top}\vdash \Pi \qquad \Pi\vdash \tilde{\bullet}_I\,\check{\bot}}{\hat{\iota}_I\vdash \Pi \qquad \Pi\vdash \check{\partial}_I}\ \tilde{\bullet}_I\,\check{\partial}_I \qquad \tilde{\bullet}\,\hat{\iota}_C\ \frac{\tilde{\bullet}_C\,\hat{\top}\vdash \Sigma \qquad \Sigma\vdash \tilde{\bullet}_C\,\check{\bot}}{\hat{\iota}_C\vdash \Sigma \qquad \Sigma\vdash \check{\partial}_C}\ \tilde{\bullet}_C\,\check{\partial}_C$$

$$\hat{\diamond}_I\,\tilde{\bullet}_I\ \frac{\Pi\vdash \Pi' \qquad \Pi'\vdash \Pi}{\hat{\diamond}_I\,\tilde{\bullet}_I\,\Pi\vdash \Pi' \qquad \Pi'\vdash \check{\partial}_I\,\tilde{\bullet}_I\,\Pi}\ \check{\partial}_I\,\tilde{\bullet}_I \qquad \hat{\diamond}_C\,\tilde{\bullet}_C\ \frac{\Sigma\vdash \Sigma' \qquad \Sigma'\vdash \Sigma}{\hat{\diamond}_C\,\tilde{\bullet}_C\,\Sigma\vdash \Sigma' \qquad \Sigma'\vdash \check{\partial}_C\,\tilde{\bullet}_C\,\Sigma}\ \check{\partial}_C\,\tilde{\bullet}_C$$

$$\tilde{\bullet}_I\,\hat{\diamond}_I\ \frac{\tilde{\bullet}_I\,\hat{\diamond}_I\,\Pi\vdash \Pi' \qquad \Pi'\vdash \tilde{\bullet}_I\,\check{\partial}_I\,\Pi}{\Pi\vdash \Pi' \qquad \Pi'\vdash \Pi}\ \tilde{\bullet}_I\,\check{\partial}_I \qquad \tilde{\bullet}_C\,\hat{\diamond}_C\ \frac{\tilde{\bullet}_C\,\hat{\diamond}_C\,\Sigma\vdash \Sigma' \qquad \Sigma'\vdash \tilde{\bullet}_C\,\check{\partial}_C\,\Sigma}{\Sigma\vdash \Sigma' \qquad \Sigma'\vdash \Sigma}\ \tilde{\bullet}_C\,\check{\partial}_C$$

- Multi-type structural rules for the correspondence between kernels:

$$\tilde{\partial}\,\hat{\blacklozenge}\ \frac{\tilde{\partial}_I\,\hat{\blacklozenge}_I\,X\vdash Y \qquad Y\vdash \check{\partial}_I\,\tilde{\bullet}_I\,X}{\tilde{\partial}_C\,\hat{\blacklozenge}_C\,X\vdash Y \qquad Y\vdash \check{\partial}_C\,\tilde{\bullet}_C\,X}\ \check{\partial}\,\tilde{\bullet}$$

- Logical rules for multi-type connectives related to strict kernels:

$$\blacklozenge_I\ \frac{\hat{\blacklozenge}_I\,A\vdash \Gamma}{\blacklozenge_I\,A\vdash \Gamma} \qquad \frac{X\vdash A}{\hat{\blacklozenge}_I\,X\vdash \blacklozenge_I\,A}\ \blacklozenge_I \qquad\qquad \blacksquare_C\ \frac{A\vdash X}{\blacksquare_C\,A\vdash \blacksquare_C\,X} \qquad \frac{\Delta\vdash \blacksquare_C\,A}{\Delta\vdash \blacksquare_C\,A}\ \blacksquare_C$$

$$\circ_I\ \frac{\tilde{\partial}_I\,\alpha\vdash X}{\circ_I\,\alpha\vdash X} \qquad \frac{X\vdash \tilde{\partial}_I\,\alpha}{X\vdash \circ_I\,\alpha}\ \circ_I \qquad\qquad \circ_C\ \frac{\tilde{\partial}_C\,\delta\vdash X}{\circ_C\,\delta\vdash X} \qquad \frac{X\vdash \tilde{\partial}_C\,\delta}{X\vdash \circ_C\,\delta}\ \circ_C$$

- Logical rules for lattice connectives:

$$\diamond_I\ \frac{\hat{\diamond}_I\,\pi\vdash X}{\diamond_I\,\pi\vdash X} \qquad \frac{\Pi\vdash \pi}{\hat{\diamond}_I\,\Pi\vdash \diamond_I\,\pi}\ \diamond_I \qquad\qquad \Box_C\ \frac{\sigma\vdash \Sigma}{\Box_C\,\sigma\vdash \check{\partial}_C\,\Sigma} \qquad \frac{X\vdash \check{\partial}_C\,\sigma}{X\vdash \Box_C\,\sigma}\ \Box_C$$

$$\circ_I\ \frac{\tilde{\bullet}_I\,A\vdash \Pi}{\bullet_I\,A\vdash \Pi} \qquad \frac{\Pi\vdash \tilde{\bullet}_I\,A}{\Pi\vdash \bullet_I\,A}\ \bullet_I \qquad\qquad \bullet_C\ \frac{\tilde{\bullet}_C\,A\vdash \Sigma}{\bullet_C\,A\vdash \Sigma} \qquad \frac{\Sigma\vdash \tilde{\bullet}_C\,A}{\Sigma\vdash \bullet_C\,A}\ \bullet_C$$

- Logical rules for lattice connectives:

$$\top\ \frac{\hat{\top}\vdash X}{\top\vdash X} \qquad \frac{}{\hat{\top}\vdash \top}\ \top \qquad\qquad \bot\ \frac{}{\bot\vdash \check{\bot}} \qquad \frac{X\vdash \check{\bot}}{X\vdash \bot}\ \bot$$

$$\wedge_i\ \frac{A_{i\in\{1,2\}}\vdash X}{A_1\wedge A_2\vdash X} \qquad \frac{X\vdash A \qquad X\vdash B}{X\vdash A\wedge B}\ \wedge \qquad \vee\ \frac{A\vdash X \qquad B\vdash X}{A\vee B\vdash X} \qquad \frac{X\vdash A_{i\in\{1,2\}}}{X\vdash A_1\vee A_2}\ \vee_i$$

The proper display calculi for the subvarieties of aKa discussed in Sect. 4 are obtained by adding the following rules:

Logic	Calculus	Rules
H.aKa5'	D.aKa5'	$\check{\delta}_I\,\hat{\blacklozenge}_I\,\hat{\Diamond}_I\ \dfrac{\hat{\Diamond}_I\,\Pi \vdash X}{\check{\delta}_I\,\hat{\blacklozenge}_I\,\hat{\Diamond}_I\,\Pi \vdash X}\qquad \dfrac{X \vdash \check{\Box}_C\,\Sigma}{X \vdash \check{\delta}_C\,\check{\blacksquare}_C\,\check{\Box}_C\,\Sigma}\ \check{\delta}_C\,\check{\blacksquare}_C\,\check{\Box}_C$ $\hat{\Diamond}_I\,\check{\bullet}_I\,\check{\delta}_I\ \dfrac{\hat{\Diamond}_I\,\check{\bullet}_I\,\check{\delta}_I\,\Gamma \vdash X}{\check{\delta}_I\,\Gamma \vdash X}\qquad \dfrac{X \vdash \check{\Box}_C\,\check{\bullet}_C\,\check{\delta}_C\,\Delta}{X \vdash \check{\delta}_C\,\Delta}\ \check{\Box}_C\,\check{\bullet}_C\,\check{\delta}_C$
K-IA3$_\ell$	D.K-IA3$_\ell$	$\dfrac{X \vdash \check{\Box}_I\,\check{\bullet}_I\,Y \qquad \hat{\Diamond}_C\,\check{\bullet}_C\,X \vdash Y}{X \vdash Y}\ k\text{-}ia3_\ell$

These calculi enjoy the properties of soundness, completeness, conservativity, cut elimination and subformula property the verification of which is standard and follows from the general theory of proper display calculi (cf. [6, 11–14, 16, 23]). These verifications are discussed in the appendix.

A Properties

Throughout this section, we let $\mathsf{K} \in \{aKa, aKa5', K\text{-}IA3_\ell\}$, and HK the class of heterogeneous algebras corresponding to K. Further, we let $\mathsf{D.K}$ denote the multi-type calculus for the logic $\mathsf{H.K}$ canonically associated with K.

A.1 Soundness for Perfect HK Algebras

The verification of the soundness of the rules of $\mathsf{D.K}$ w.r.t. the semantics of *perfect* elements of HK (see Definition 9) is analogous to that of many other multi-type calculi (cf. [6, 11–14, 16, 23]). Here we only discuss the soundness of the rule $k\text{-}ia3_\ell$. By definition, the following quasi-inequality is valid on every K-IA3$_\ell$:

$$\Box_\ell a \leq \Box_\ell b \text{ and } \Diamond_\ell a \leq \Diamond_\ell b \text{ imply } a \leq b.$$

This quasi-inequality equivalently translates into the multi-type language as follows:

$$\Box_C \bullet_C a \leq \Box_C \bullet_C b \text{ and } \Diamond_I \bullet_I a \leq \Diamond_I \bullet_I b \text{ imply } a \leq b.$$

By adjunction, the quasi-inequality above can be equivalently rewritten as follows:

$$\Diamond_C \bullet_C \Box_C \bullet_C a \leq b \text{ and } a \leq \Box_I \bullet_I \Diamond_I \bullet_I b \text{ imply } a \leq b,$$

which, thanks to a well known property of adjoint maps, simplifies as:

$$\Diamond_C \bullet_C a \leq b \text{ and } a \leq \Box_I \bullet_I b \text{ imply } a \leq b.$$

Hence, the quasi-inequality above is equivalent to the following inequality:

$$a \wedge \Box_I \bullet_I b \leq \Diamond_C \bullet_C a \vee b.$$

The inequality above is analytic inductive (cf. [7, Definition 55]), and hence running ALBA on this inequality produces:

$$\forall a \forall b [a \wedge \Box_I \bullet_I b \leq \Diamond_C \bullet_C a \vee b]$$
$$\text{iff } \forall p \forall q \forall a \forall b [(p \leq a \wedge \Box_I \bullet_I b \ \& \ \Diamond_C \bullet_C a \vee b \leq q) \Rightarrow p \leq q]$$
$$\text{iff } \forall p \forall q \forall a \forall b [(p \leq a \ \& \ p \leq \Box_I \bullet_I b \ \& \ b \leq q \ \& \ \Diamond_C \bullet_C a \leq q) \Rightarrow p \leq q]$$
$$\text{iff } \forall p \forall q [(p \leq \Box_I \bullet_I q \ \& \ \Diamond_C \bullet_C p \leq q) \Rightarrow p \leq q].$$

The last quasi-inequality above is the semantic translation of the rule $k\text{-}ia3_\ell$:

$$\frac{X \vdash \check{\Box}_I \tilde{\bullet}_I Y \qquad \hat{\Diamond}_C \tilde{\bullet}_C X \vdash Y}{X \vdash Y} \ k\text{-}ia3_\ell$$

which we then proved to be sound on every perfect heterogeneous K-IA3$_\ell$, by the soundness of the ALBA steps. Likewise, the defining condition of K-IA3$_\ell$ translates into the inequality

$$a \wedge \circ_C \blacklozenge_C b \leq \circ_I \blacksquare_I a \vee b,$$

which, however, is not analytic inductive, and hence it cannot be transformed into an analytic rule via ALBA.

A.2   Completeness

Let $A^\tau \vdash B^\tau$ be the translation of any sequent $A \vdash B$ in the language of H.K into the language of D.K induced by the correspondence between K and HK described in Sect. 5.

Proposition 3. *For every H.K-derivable sequent $A \vdash B$, the sequent $A^\tau \vdash B^\tau$ is derivable in D.K.*

Below we provide the multi-type translations of the single-type sequents corresponding to inequalities (11). All of them are derivable in D.AKA by logical introduction rules, display rules, and the rules $\check{\Box} \ \tilde{\bullet}$ and $\tilde{\circ} \ \hat{\blacklozenge}$.

$$\Diamond_s A \vdash B \text{ iff } A \vdash \Box_s B \quad \leadsto \quad \circ_C \blacklozenge_C A \vdash B \text{ iff } A \vdash \circ_I \blacksquare_I B$$
$$\Diamond_\ell A \vdash B \text{ iff } A \vdash \Box_\ell B \quad \leadsto \quad \Diamond_I \bullet_I A \vdash B \text{ iff } A \vdash \Box_C \bullet_C B$$

Below we provide the multi-type translations of the single-type sequents corresponding to inequalities (12) and (13), respectively. All of them are derivable in D.AKA by logical introduction rules and display rules.

$\Box_s A \vdash A$	$\leadsto \quad \circ_I \blacksquare_I A \vdash A$	$\Box_s A \vdash \Box_s \Box_s A$	$\leadsto \quad \circ_I \blacksquare_I A \vdash \circ_I \blacksquare_I \circ_I \blacksquare_I A$
$A \vdash \Diamond_s A$	$\leadsto \quad A \vdash \circ_C \blacklozenge_C A$	$\Diamond_s \Diamond_s A \vdash \Diamond_s A$	$\leadsto \quad \circ_C \blacklozenge_C \circ_C \blacklozenge_C A \vdash \circ_C \blacklozenge_C A$
$A \vdash \Box_\ell A$	$\leadsto \quad A \vdash \Box_C \bullet_C A$	$\Box_\ell \Box_\ell A \vdash \Box_\ell A$	$\leadsto \quad \Box_C \bullet_C \Box_C \bullet_C A \vdash \Box_C \bullet_C A$
$\Diamond_\ell A \vdash A$	$\leadsto \quad \Diamond_I \bullet_I A \vdash A$	$\Diamond_\ell A \vdash \Diamond_\ell \Diamond_\ell A$	$\leadsto \quad \Diamond_I \bullet_I A \vdash \Diamond_I \bullet_I \Diamond_I \bullet_I A$

Below we provide the multi-type translation of the single-type sequents corresponding to inequalities (18). All of them are derivable in D.AKA5'.

$\Diamond_\ell A \vdash \Box_s \Diamond_\ell A$	$\leadsto \quad \Diamond_I \bullet_I A \vdash \circ_I \blacksquare_I \Diamond_I \bullet_I A$
$\Diamond_s \Box_\ell A \vdash \Box_\ell A$	$\leadsto \quad \circ_C \blacklozenge_C \Box_C \bullet_C A \vdash \Box_C \bullet_C A$
$\Box_s A \vdash \Diamond_\ell \Box_s A$	$\leadsto \quad \circ_I \blacksquare_I A \vdash \Diamond_I \bullet_I \circ_I \blacksquare_I A$
$\Box_\ell \Diamond_s A \vdash \Diamond_s A$	$\leadsto \quad \Box_C \bullet_C \circ_C \blacklozenge_C A \vdash \circ_C \blacklozenge_C A$

Below we provide the multi-type translations of the single-type rules corresponding to quasi-inequality (20), respectively.

$$\Diamond_\ell A \vdash \Diamond_\ell B \text{ and } \Box_\ell A \vdash \Box_\ell B \text{ imply } A \vdash B \quad \rightsquigarrow$$
$$\Diamond_I \bullet_I A \vdash \Diamond_I \bullet_I B \text{ and } \Box_C \bullet_C A \vdash \Box_C \bullet_C B \text{ imply } A \vdash B$$

Below, we derive (20). Firstly, $A \wedge \Box_C \bullet_C B \vdash \Diamond_I \bullet_I A \vee B$ is derivable via $k\text{-}ia3_\ell$ by means of the following derivation $\mathcal{D}$:

$$
\begin{array}{c}
\cfrac{
\cfrac{
\cfrac{
\cfrac{
\cfrac{
\cfrac{
\cfrac{
\cfrac{B \vdash B}{B \vdash \Diamond_I \bullet_I A \check{\vee} B}
}{B \vdash \Diamond_I \bullet_I A \vee B}
}{\tilde{\bullet}_C B \vdash \tilde{\bullet}_C(\Diamond_I \bullet_I A \vee B)}
}{\bullet_C B \vdash \tilde{\bullet}_C(\Diamond_I \bullet_I A \vee B)}
}{\Box_C \bullet_C B \vdash \check{\Box}_C \tilde{\bullet}_C(\Diamond_I \bullet_I A \vee B)}
}{A \wedge \Box_C \bullet_C B \vdash \check{\Box}_C \tilde{\bullet}_C(\Diamond_I \bullet_I A \vee B)}
}{A \wedge \Box_C \bullet_C B \vdash \check{\Box}_C \tilde{\bullet}_C(\Diamond_I \bullet_I A \vee B)}
\qquad
\cfrac{
\cfrac{
\cfrac{
\cfrac{
\cfrac{
\cfrac{
\cfrac{A \vdash A}{A \hat{\wedge} \Box_C \bullet_C B \vdash A}
}{A \wedge \Box_C \bullet_C B \vdash A}
}{\tilde{\bullet}_I(A \wedge \Box_C \bullet_C B) \vdash \tilde{\bullet}_I A}
}{\tilde{\bullet}_I(A \wedge \Box_C \bullet_C B) \vdash \bullet_I A}
}{\hat{\Diamond}_I \tilde{\bullet}_I(A \wedge \Box_C \bullet_C B) \vdash \Diamond_I \bullet_I A}
}{\hat{\Diamond}_I \tilde{\bullet}_I(A \wedge \Box_C \bullet_C B) \vdash \Diamond_I \bullet_I A \check{\vee} B}
}{\hat{\Diamond}_I \tilde{\bullet}_I(A \wedge \Box_C \bullet_C B) \vdash \Diamond_I \bullet_I A \vee B}
}{A \wedge \Box_C \bullet_C B \vdash \Diamond_I \bullet_I A \vee B} \, k\text{-}ia3_\ell
\end{array}
$$

Assuming $\Diamond_I \bullet_I A \vdash \Diamond_I \bullet_I B$ and $\Box_C \bullet_C A \vdash \Box_C \bullet_C B$, we derive $A \vdash B$ via cut as follows:

$$
\cfrac{
\cfrac{A \vdash A \qquad
\cfrac{
\cfrac{\Box_C \bullet_C A \vdash \Box_C \bullet_C B}{A \vdash \Box_C \bullet_C B}\, Id_L + \bullet_C + adj_{LS_C} + Cut_L}{A \vdash \Box_C \bullet_C B}\wedge
}{
\cfrac{A \hat{\wedge} A \vdash A \wedge \Box_C \bullet_C B}{A \vdash A \wedge \Box_C \bullet_C B}\, C_L}
\qquad
\cfrac{\mathcal{D}}{A \wedge \Box_C \bullet_C B \vdash \Diamond_I \bullet_I A \vee B}
\qquad
\cfrac{
\cfrac{
\cfrac{\Diamond_I \bullet_I A \vdash \Diamond_I \bullet_I B}{\Diamond_I \bullet_I A \vdash B}\, Id_L + \bullet_I + adj_{LS_I} + Cut_L \qquad B \vdash B}{
\cfrac{\Diamond_I \bullet_I A \vee B \vdash B \check{\vee} B}{\Diamond_I \bullet_I A \vee B \vdash B}\, C_L}\vee
}{A \wedge \Box_C \bullet_C B \vdash B}\, Cut_L
}{A \vdash B}\, Cut_L
$$

A.3   Conservativity

To argue that D.K is conservative w.r.t. H.K, we follow the standard proof strategy discussed in [7,9]. We need to show that, for all formulas A and B in the language of H.K, if $A^\tau \vdash B^\tau$ is a D.K-derivable sequent, then $A \vdash B$ is derivable in H.K. This claim can be proved using the following facts: (a) The rules of D.K are sound w.r.t. perfect members of HK (cf. Sect. A.1); (b) H.K is complete w.r.t. the class of perfect algebras in K; (c) A perfect element of K is equivalently presented as a perfect member of HK so that the semantic consequence relations arising from each type of structures preserve and reflect the translation. Let A, B be as above. If $A^\tau \vdash B^\tau$ is D.K-derivable, then by (a), $\models_{HK} A^\tau \vdash B^\tau$. By (c), this implies that $\models_K A \vdash B$, where $\models_K$ denotes the semantic consequence relation arising from the perfect members of class K. By (b), this implies that $A \vdash B$ is derivable in H.K, as required.

A.4   Cut Elimination and Subformula Property

Cut elimination and subformula property for each D.K are obtained by verifying the assumptions of [5, Theorem 4.1]. All of them except C'_8 are readily satisfied by inspecting the rules. Condition C'_8 requires to check that reduction steps can be performed for every application of cut in which both cut-formulas are principal, which either remove the original cut altogether or replace it by one or more cuts on formulas of strictly lower complexity. In what follows, we only show C'_8 for some heterogeneous connectives.

$$
\cfrac{
  \cfrac{\vdots \pi_1}{\Gamma \vdash \blacksquare_I A} \qquad \cfrac{\vdots \pi_2}{\Gamma \vdash \blacksquare_I A \qquad \blacksquare_I A \vdash \check{\blacksquare}_I Y}
}{\Gamma \vdash \check{\blacksquare}_I Y}
\quad \leadsto \quad
\cfrac{
  \cfrac{\cfrac{\vdots \pi_1}{\Gamma \vdash \check{\blacksquare}_I A}}{\tilde{o}_I \Gamma \vdash A} \qquad \cfrac{\vdots \pi_2}{A \vdash Y}
}{\cfrac{\tilde{o}_I \Gamma \vdash Y}{\Gamma \vdash \check{\blacksquare}_I Y}}
$$

$$
\cfrac{
  \cfrac{\vdots \pi_1}{X \vdash \tilde{o}_I \alpha} \qquad \cfrac{\vdots \pi_2}{\tilde{o}_I \alpha \vdash Y}
}{X \vdash Y}
\quad \leadsto \quad
\cfrac{
  \cfrac{\cfrac{\vdots \pi_1}{X \vdash \tilde{o}_I \alpha}}{\blacklozenge_I X \vdash \alpha} \qquad \cfrac{\cfrac{\vdots \pi_2}{\tilde{o}_I \alpha \vdash Y}}{\alpha \vdash \check{\blacksquare}_I Y}
}{\cfrac{\blacklozenge_I X \vdash \check{\blacksquare}_I Y}{X \vdash Y}}
$$

The remaining cases are analogous.

References

1. Banerjee, M., Chakraborty, M.K.: Rough sets through algebraic logic. Fundamenta Informaticae **28**(3, 4), 211–221 (1996)
2. Conradie, W., Frittella, S., Palmigiano, A., Piazzai, M., Tzimoulis, A., Wijnberg, N.M.: Categories: how I learned to stop worrying and love two sorts. In: Väänänen, J., Hirvonen, Å., de Queiroz, R. (eds.) WoLLIC 2016. LNCS, vol. 9803, pp. 145–164. Springer, Heidelberg (2016). https://doi.org/10.1007/978-3-662-52921-8_10
3. Conradie, W., Frittella, S., Palmigiano, A., Piazzai, M., Tzimoulis, A., Wijnberg, N.M.: Toward an epistemic-logical theory of categorization. In: Proceedings of the TARK 2017. EPTCS, vol. 251, pp. 167–186 (2017)
4. Conradie, W., Palmigiano, A.: Algorithmic correspondence and canonicity for non-distributive logics. Annals of Pure and Applied Logic ArXiv:1603.08515
5. Frittella, S., Greco, G., Kurz, A., Palmigiano, A., Sikimić, V.: Multi-type sequent calculi. In: Indrzejczak, A., et al. (eds.) Proceedings of the Trends in Logic XIII, pp. 81–93 (2014)
6. Frittella, S., Greco, G., Palmigiano, A., Yang, F.: A multi-type calculus for inquisitive logic. In: Väänänen, J., Hirvonen, Å., de Queiroz, R. (eds.) WoLLIC 2016. LNCS, vol. 9803, pp. 215–233. Springer, Heidelberg (2016). https://doi.org/10.1007/978-3-662-52921-8_14
7. Greco, G., Ma, M., Palmigiano, A., Tzimoulis, A., Zhao, Z.: Unified correspondence as a proof-theoretic tool. J. Logic Comput. **28**(7), 1367–1442 (2018)
8. Ganter, B., Wille, R.: Formal Concept Analysis: Mathematical Foundations. Springer, Heidelberg (1999). https://doi.org/10.1007/978-3-642-59830-2
9. Greco, G., Kurz, A., Palmigiano, A.: Dynamic epistemic logic displayed. In: Grossi, D., Roy, O., Huang, H. (eds.) LORI 2013. LNCS, vol. 8196, pp. 135–148. Springer, Heidelberg (2013). https://doi.org/10.1007/978-3-642-40948-6_11
10. Greco, G., Liang, F., Manoorkar, K., Palmigiano, A.: Proper multi-type display calculi for rough algebras. arXiv preprint 1808.07278

11. Greco, G., Liang, F., Moshier, M.A., Palmigiano, A.: Multi-type display calculus for semi De Morgan logic. In: Kennedy, J., de Queiroz, R.J.G.B. (eds.) WoLLIC 2017. LNCS, vol. 10388, pp. 199–215. Springer, Heidelberg (2017). https://doi.org/10.1007/978-3-662-55386-2_14

12. Greco, G., Liang, F., Palmigiano, A., Rivieccio, U.: Bilattice logic properly displayed. Fuzzy Sets Syst. (2018). https://doi.org/10.1016/j.fss.2018.05.007

13. Greco, G., Palmigiano, A.: Linear logic properly displayed. arXiv preprint: 1611.04184

14. Greco, G., Palmigiano, A.: Lattice logic properly displayed. In: Kennedy, J., de Queiroz, R.J.G.B. (eds.) WoLLIC 2017. LNCS, vol. 10388, pp. 153–169. Springer, Heidelberg (2017). https://doi.org/10.1007/978-3-662-55386-2_11

15. Kent, R.E.: Rough concept analysis: a synthesis of rough sets and formal concept analysis. Fundamenta Informaticae **27**(2, 3), 169–181 (1996)

16. Liang, F.: Multi-type algebraic proof theory. Ph.D. thesis, TU Delft (2018)

17. Ma, M., Chakraborty, M.K., Lin, Z.: Sequent calculi for varieties of topological quasi-Boolean algebras. In: Nguyen, H.S., Ha, Q.-T., Li, T., Przybyła-Kasperek, M. (eds.) IJCRS 2018. LNCS (LNAI), vol. 11103, pp. 309–322. Springer, Cham (2018). https://doi.org/10.1007/978-3-319-99368-3_24

18. Manoorkar, K., Nazari, S., Palmigiano, A., Tzimoulis, A., Wijnberg, N.M.: Rough concepts (2018, Submitted)

19. Pawlak, Z.: Rough set theory and its applications to data analysis. Cybern. Syst. **29**(7), 661–688 (1998)

20. Saha, A., Sen, J., Chakraborty, M.K.: Algebraic structures in the vicinity of pre-rough algebra and their logics. Inf. Sci. **282**, 296–320 (2014)

21. Saha, A., Sen, J., Chakraborty, M.K.: Algebraic structures in the vicinity of pre-rough algebra and their logics II. Inf. Sci. **333**, 44–60 (2016)

22. Sen, J., Chakraborty, M.K.: A study of interconnections between rough and 3-valued Łukasiewicz logics. Fundamenta Informaticae **51**(3), 311–324 (2002)

23. Tzimoulis, A.: Algebraic and proof-theoretic foundations of the logics for social behaviour. Ph.D. thesis, TUDelft (2018)

24. Wansing, H.: Sequent systems for modal logics. In: Gabbay, D.M., Guenthner, F. (eds.) Handbook of Philosophical Logic, vol. 8, pp. 61–145. Springer, Dordrecht (2002). https://doi.org/10.1007/978-94-010-0387-2_2

A Fix-Point Characterization of Herbrand Equivalence of Expressions in Data Flow Frameworks

Jasine Babu[1]($\boxtimes$), Karunakaran Murali Krishnan[2], and Vineeth Paleri[2]

[1] Indian Institute of Technology Palakkad, Palakkad, India
jasine@iitpkd.ac.in
[2] National Institute of Technology Calicut, Kozhikode, India
{kmurali,vpaleri}@nitc.ac.in

Abstract. Computing Herbrand equivalences of terms in data flow frameworks is well studied in program analysis. While algorithms use iterative fix-point computation on some abstract lattice of expressions relevant to the flow graph, the definition of Herbrand equivalences is based on an equivalence over all program paths formulation, on the (infinite) set of all expressions. The aim of this paper is to develop a lattice theoretic fix-point formulation of Herbrand equivalence on a concrete lattice defined over the set of all terms constructible from variables, constants and operators of a program. This new formulation makes explicit the underlying lattice framework as well as the monotone function involved in computing Herbrand equivalences. We introduce the notion of Herbrand congruence and define an (infinite) concrete lattice of Herbrand congruences. Herbrand equivalence is defined as the maximum fix-point of a composite transfer function defined over an appropriate product lattice of the above concrete lattice. We then reformulate the traditional meet over all paths definition in our lattice theoretic framework. and prove that the maximum fix-point (MFP) and the meet-over-all-paths (MOP) formulations coincide as expected.

Keywords: Herbrand equivalence · Data flow framework · Fix-point

1 Introduction

Data flow frameworks are abstract representations of programs, used in program analysis and compiler optimizations [1,2]. As detection of semantic equivalence of expressions at program points is unsolvable [3], algorithms try to detect a weaker, syntactic notion of equivalence called *Herbrand equivalence* [4–8]. Herbrand equivalence treats operators as uninterpreted functions, and expressions obtained by applying the same operator on equivalent operands are treated equivalent.

Kildall [9] used abstract interpretation [10] to compute Herbrand equivalences at program points using an iterative fix point algorithm over a meet semi-lattice

© Springer-Verlag GmbH Germany, part of Springer Nature 2019
Md. A. Khan and A. Manuel (Eds.): ICLA 2019, LNCS 11600, pp. 160–172, 2019.
https://doi.org/10.1007/978-3-662-58771-3_15

[2, 3, 11]. Several algorithms for computation of Herbrand equivalence of program expressions are known [4, 5, 7, 8, 12, 13]. Most of these algorithms use iterative fix-point computation on an abstract lattice defined over a *working set* of expressions relevant to the program. It is known that the working set needs to include certain non-program expressions as well [9].

The theoretical question of describing the concrete lattice of all expression equivalences and the monotone function whose fix point defines the Herbrand equivalence of expressions at program points seems to be unaddressed in the literature. This is possibly due the fact that the traditional definition of Herbrand equivalence of expressions (see [5, p. 393]) is based on an equivalence over all program paths formulation, rather than a lattice based formulation. In this work, we axiomatize the notion of congruence of all expressions over the variables, constants and operators in a program. We then define a concrete lattice of all congruences. We show that this lattice is complete, which ensures that every monotone function f over the lattice indeed has a maximum fix-point [14], though the lattice has infinite height.

Given a data flow framework, we define transfer functions and indefinite assignments as certain monotone functions on the lattice described above. By a standard product lattice construction, we then define a composite transfer function [15, 16], which is monotone over the product lattice. We define the Herbrand equivalence at each program point as component maps (projections) of the maximum fix-point of this composite transfer function. Finally, to validate the new formulation against the existing definition of Herbrand equivalence, we re-formulate the standard definition of Herbrand equivalence in [5] in our lattice framework and show it to be equivalent to the fix-point formulation developed in this paper.

Many of the standard proofs for the equivalence of fix-point and meet over all paths formulations assume that the lattice is either of finite height or that every chain has finite height [1, 3, 16], where equivalence holds whenever the transfer function under consideration is a meet-morphism. Even though chains are of infinite height, the transfer function being a complete-meet-morphism guarantees that the equivalence still holds true for the formulation presented in this paper. A proof of this equivalence is presented in Sect. 8.

2 Terminology

Let C be a set of constants and X be a set of variables. For simplicity, we assume that the set of operators $Op = \{+\}$. (More operators can be added without any difficulty). The set of all terms over $C \cup X$, $\mathcal{T} = \mathcal{T}(X, C)$ is defined by $t ::= c \mid x \mid (t + t)$, with $c \in C$ and $x \in X$. (Parenthesis is avoided when there is no confusion.) Let $\mathcal{P}$ be a partition of $\mathcal{T}$. Let $[t]_{\mathcal{P}}$ (or simply $[t]$ when there is no confusion) denote the equivalence class from $\mathcal{P}$ containing the term $t \in \mathcal{T}$. If $t' \in [t]_{\mathcal{P}}$, we write $t \cong_{\mathcal{P}} t'$ (or simply $t \cong t'$). For any $A \subseteq \mathcal{T}$, $A(x)$ denotes the set of all terms in A in which the variable x appears and $\overline{A}(x)$ denotes the set of all terms in A in which x does not appear. In particular, for any $x \in X$, $\mathcal{T}(x)$ is the set of all terms containing the variable x and $\overline{\mathcal{T}}(x) = \mathcal{T} \setminus \mathcal{T}(x)$.

Definition 1 (Substitution). *For* $t, \alpha \in \mathcal{T}$, $x \in X$, *substitution of* x *with* α *in* t, *denoted by* $t[x \leftarrow \alpha]$ *is defined by: (1) If* $t = x$, *then* $t[x \leftarrow \alpha] = \alpha$. *(2) If* $t \notin \mathcal{T}(x)$, $t[x \leftarrow \alpha] = t$. *(3) If* $t = t_1 + t_2$ *then* $t[x \leftarrow \alpha] = t_1[x \leftarrow \alpha] + t_2[x \leftarrow \alpha]$.

For proofs of statements left unproven in the main text and proofs of some elementary results of lattice theory which are used in the main text, the reader may refer to the preprint [17].

3   Congruences of Terms

Definition 2 (Congruence of Terms). *Let* $\mathcal{P}$ *be a partition of* $\mathcal{T}$. $\mathcal{P}$ *is a Congruence (of terms) if the following conditions hold:*

1. *For* $t, t', s, s' \in \mathcal{T}$, $t' \cong t$ *and* $s' \cong s$ *if and only if* $t' + s' \cong t + s$. *(Congruences respect operators).*
2. *For any* $c \in C$, $t \in \mathcal{T}$, *if* $t \cong c$ *then either* $t = c$ *or* $t \in X$. *(The only non-constant terms that are allowed to be congruent to a constant are variables).*

Given a data flow framework, we will associate a congruence with each program point. Each iteration refines the present congruence at each program point till a fix-point is reached. The Herbrand equivalence at each program point will be defined as this fix-point congruence.

Definition 3. *The set of all congruences over* $\mathcal{T}$ *is denoted by* $\mathcal{G}(\mathcal{T})$.

We define a binary *confluence operation* on the set of congruences, $\mathcal{G}(\mathcal{T})$.

Definition 4 (Confluence). *Let* $\mathcal{P}_1 = \{A_i\}_{i \in I}$ *and* $\mathcal{P}_2 = \{B_j\}_{j \in J}$ *be two congruences. For all* $i \in I$ *and* $j \in J$, *define* $C_{i,j} = A_i \cap B_j$. *The confluence of* $\mathcal{P}_1$ *and* $\mathcal{P}_2$ *is defined by:* $\mathcal{P}_1 \wedge \mathcal{P}_2 = \{C_{i,j} : i \in I, j \in J, C_{i,j} \neq \emptyset\}$.

Theorem 5. *If* $\mathcal{P}_1$ *and* $\mathcal{P}_2$ *are congruences, then* $\mathcal{P}_1 \wedge \mathcal{P}_2$ *is a congruence.*

We next define an ordering on the set $\mathcal{G}(\mathcal{T})$ and extend it to a complete lattice.

Definition 6 (Refinement of a Congruence). *Let* $\mathcal{P}, \mathcal{P}'$ *be congruences over* $\mathcal{T}$. *We say* $P \preceq P'$ *(read* P *is a refinement of* P'*) if for each equivalence class* $A \in \mathcal{P}$, *there exists an equivalence class* $A' \in \mathcal{P}'$ *such that* $A \subseteq A'$.

Definition 7. *The partition in which each term in* $\mathcal{T}$ *belongs to a different class is defined as:* $\bot = \{\{t\} : t \in \mathcal{T}\}$.

The following observation is a direct consequence of the definition of $\bot$.

Observation 8. $\bot$ *is a congruence. Moreover for any* $\mathcal{P} \in \mathcal{G}(\mathcal{T})$, $\bot \wedge \mathcal{P} = \bot$.

Lemma 9. *Every non-empty subset of* $(\mathcal{G}(\mathcal{T}), \preceq)$ *has a greatest lower bound.*

Next, we extend $(\mathcal{G}(\mathcal{T}), \preceq)$ by artificially adding a top element $\top$, so that the greatest lower bound of the empty set is also well defined.

Definition 10. *The lattice* $(\overline{\mathcal{G}(\mathcal{T})}, \preceq, \bot, \top)$ *is defined over the set* $\overline{\mathcal{G}(\mathcal{T})} = \mathcal{G}(\mathcal{T}) \cup \{\top\}$ *with* $\mathcal{P} \preceq \top$ *for each* $\mathcal{P} \in \mathcal{G}(\mathcal{T})$. *In particular,* $\top$ *is the greatest lower bound of* $\emptyset$ *and* $\top \wedge \mathcal{P} = \mathcal{P}$ *for every* $\mathcal{P} \in \overline{\mathcal{G}(\mathcal{T})}$.

Hereafter, we will be referring to the element $\top$ as a congruence. Combining Lemma 9, Definition 10 and standard results in lattice theory, we get:

Theorem 11. $(\overline{\mathcal{G}(\mathcal{T})}, \preceq, \bot, \top)$ *is a complete lattice.*

Definition 12 (Infimum). *Let* $S = \{\mathcal{P}_i\}_{i \in I}$ *be an arbitrary collection of congruences in* $\overline{\mathcal{G}(\mathcal{T})}$ *(S may be empty or may contain* $\top$*). The infimum of* S, *denoted by* $\bigwedge S$ *or* $\bigwedge_{i \in I} \mathcal{P}_i$, *is defined as the greatest lower bound of the set* $\{\mathcal{P}_i\}_{i \in I}$.

Remark 13. The results in this paper only assumes the meet-completeness of $(\mathcal{G}(\mathcal{T}), \preceq, \bot, \top)$ and the existence of a top element. Though the lattice is also join-complete, our proofs do not rely on this property.

4   Transfer Functions

A *transfer function* describes the effect of an assignment on a congruence.

Definition 14 (Transfer Functions). *Let* $y \in X$ *and* $\beta \in \overline{\mathcal{T}}(y)$. *(Note that* y *does not appear in* β*). The transfer function* $f_{y=\beta}$ *transforms a congruence* $\mathcal{P} = \{A_i\}_{i \in I}$ *to* $f_{y=\beta}(\mathcal{P})$, *a collection of subsets of* $\mathcal{T}$, *given by the following:*

- *If* $\mathcal{P} = \{A_i\}_{i \in I}$, *then let* $B_i = \{t \in \mathcal{T} : t[y \leftarrow \beta] \in A_i\}$, *for each* $i \in I$.
- $f_{y=\beta}(\mathcal{P}) = \{B_i : i \in I, B_i \neq \emptyset\}$.

See Fig. 1 for an example.

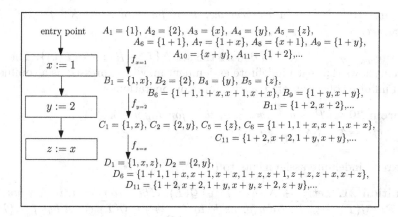

Fig. 1. Application of transfer functions

Note that statements of the form $y := y + c$ can be transformed to the form permitted by Definition 14. We write $f(\mathcal{P})$ instead of $f_{y=\beta}(\mathcal{P})$ to avoid cumbersome notation.

Theorem 15. *If $\mathcal{P}$ is a congruence, then for any $y \in X$, $\beta \in \overline{\mathcal{T}}(y)$, $f_{y=\beta}(\mathcal{P})$ is a congruence.*

Next, we extend the definition of transfer functions to $(\overline{\mathcal{G}(\mathcal{T})}, \preceq, \perp, \top)$.

Definition 16. *Let $y \in X$ and $\beta \in \overline{\mathcal{T}}(y)$. Let $\mathcal{P} \in \overline{\mathcal{G}(\mathcal{T})}$. The extended transfer function $\overline{f}_{y=\beta}(\mathcal{P}) : \overline{\mathcal{G}(\mathcal{T})} \longrightarrow \overline{\mathcal{G}(\mathcal{T})}$ transforms $\mathcal{P}$ to $\overline{f}_{y=\beta}(\mathcal{P}) \in \overline{\mathcal{G}(\mathcal{T})}$ given by $\overline{f}_{y=\beta}(\mathcal{P}) = f_{y=\beta}(\mathcal{P})$ for all $\mathcal{P} \in \mathcal{G}(\mathcal{T})$ and $\overline{f}_{y=\beta}(\top) = \top$.*

We will write $f_{y=\beta}(\mathcal{P})$ (or $f(\mathcal{P})$) instead of $\overline{f}_{y=\beta}(\mathcal{P})$ and call it a transfer function.

We next show that transfer functions are complete-meet-morphisms over the complete lattice $(\overline{\mathcal{G}(\mathcal{T})}, \preceq, \perp, \top)$. Let $f = f_{y=\beta}$, where $y \in X$, $\beta \in \overline{\mathcal{T}}(y)$. For arbitrary collections of congruences $S \subseteq \overline{\mathcal{G}(\mathcal{T})}$, The notation $f(S)$ denotes the set $\{f(s) : s \in S\}$.

Theorem 17. *f is a complete-meet-morphism. That is, for any $\emptyset \neq S \subseteq \overline{\mathcal{G}(\mathcal{T})}$, $f(\bigwedge S) = \bigwedge f(S)$.*

5   Indefinite Assignment

Indefinite (or non-deterministic) assignments model input statements in programs.

Definition 18. *Let $y \in X$. The transfer function $f_{y=*}$ transforms a congruence $\mathcal{P} \in \mathcal{G}(\mathcal{T})$ to $f(\mathcal{P}) = f_{y=*}(\mathcal{P})$, a collection of subsets of $\mathcal{T}$ given by: for every $t, t' \in \mathcal{T}$, $t \cong_{f(\mathcal{P})} t'$ if and only if both the following conditions are satisfied: (1) $t \cong_{\mathcal{P}} t'$. (2) For every $\beta \in \mathcal{T} \setminus \mathcal{T}(y), t[y \leftarrow \beta] \cong_{\mathcal{P}} t'[y \leftarrow \beta]$.*

Theorem 19. *If $\mathcal{P}$ is a congruence, then for any $y \in X$, $f_{y=*}(\mathcal{P})$ is a congruence.*

We write $\bigwedge_{\beta \in \overline{\mathcal{T}}(y)} f_{y=\beta}(\mathcal{P})$ to denote the set $\bigwedge\{f_{y=\beta}(\mathcal{P}) : \beta \in \overline{\mathcal{T}}(y)\}$. The next theorem shows that each indefinite assignment may be expressed as a confluence of (an infinite collection of) transfer function operations.

Theorem 20. *If $\mathcal{P}$ is a congruence, then for any $y \in X$,*

$$f_{y=*}(\mathcal{P}) = \mathcal{P} \wedge \left(\bigwedge_{\beta \in \overline{\mathcal{T}}(y)} f_{y=\beta}(\mathcal{P}) \right).$$

We extend indefinite assignments to $(\overline{\mathcal{G}(\mathcal{T})}, \preceq, \perp, \top)$.

Definition 21. *Let $y \in X$ and $\mathcal{P} \in \overline{\mathcal{G}(\mathcal{T})}$. The extended transfer function $\overline{f}_{y=*}(\mathcal{P}) : \overline{\mathcal{G}(\mathcal{T})} \mapsto \overline{\mathcal{G}(\mathcal{T})}$ transforms $\mathcal{P}$ to $\overline{f}_{y=*}(\mathcal{P}) \in \overline{\mathcal{G}(\mathcal{T})}$ given by: $\overline{f}_{y=*}(\mathcal{P}) = f_{y=*}(\mathcal{P})$ for all $\mathcal{P} \in \mathcal{G}(\mathcal{T})$, and $\overline{f}_{y=*}(\top) = \top$.*

We will write $f_{y=*}(\mathcal{P})$ instead of $\overline{f}_{y=*}(\mathcal{P})$ to simplify notation.

Theorem 22. *$f_{y=*}$ is a complete-meet-morphism. That is, for any $\emptyset \neq S \subseteq \overline{\mathcal{G}(\mathcal{T})}$, $f_{y=*}(\bigwedge S) = \bigwedge f_{y=*}(S)$, where $f_{y=*}(S) = \{f_{y=*}(s) : s \in S\}$.*

We show that indefinite assignments can be characterized in terms of just three congruences (instead of dealing with infinitely many as in Theorem 20).

Theorem 23. *Let $\mathcal{P} \in \mathcal{G}(\mathcal{T})$ and let $c_1, c_2 \in C$, $c_1 \neq c_2$. Then, for any $y \in X$,*
$f_{y=*}(\mathcal{P}) = \mathcal{P} \wedge f_{y \leftarrow c_1}(\mathcal{P}) \wedge f_{y \leftarrow c_2}(\mathcal{P})$.

6   Data Flow Analysis Frameworks

We next formalize the notion of a data flow framework and apply the formalism developed above to characterize Herbrand equivalence at each point in a program.

Definition 24. *A control flow graph $G(V, E)$ is a directed graph over the vertex set $V = \{1, 2, \ldots, n\}$ for some $n \geq 1$ satisfying the following properties:*

- *$1 \in V$ is called the entry point and has no predecessors.*
- *Every vertex $i \in V$, $i \neq 1$ is reachable from 1 and has at least one predecessor and at most two predecessors.*
- *Vertices with two predecessors are called confluence points.*
- *Vertices with a single predecessor are called (transfer) function points.*

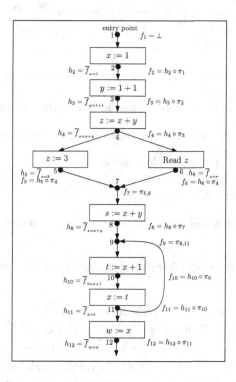

Fig. 2. Component maps of the composite transfer function

Definition 25. *A data flow framework over $\mathcal{T}$ is a pair $D = (G, F)$, where $G(V, E)$ is a control flow graph on the vertex set $V = \{1, 2, \ldots, n\}$ and F is a mapping from the set of function points in V to the set of transfer functions over $\mathcal{G}(\mathcal{T})$. The transfer function associated with function point i will be denoted as h_i.*

Data flow frameworks can be used to represent programs. An example is given in Fig. 2. In the sections that follow, we will use h_i to denote the extended transfer function $\overline{h}_i$ (see Definitions 16 and 21) without further explanation.

7   Herbrand Equivalence

Let $D = (G, F)$ be a data flow framework over $\mathcal{T}$. In the following, we will define the Herbrand Congruence function $H_D : V(G) \mapsto \overline{\mathcal{G}(\mathcal{T})}$. For each vertex $i \in V(G)$, the congruence $H_D(i)$ will be called the *Herbrand Congruence* associated with the vertex i of the data flow framework D. The function H_D will be defined as the maximum fix-point of a complete-meet-morphism $f_D : \overline{\mathcal{G}(\mathcal{T})}^n \mapsto \overline{\mathcal{G}(\mathcal{T})}^n$. The function f_D will be called the *composite transfer function* associated with the data flow framework D.

Definition 26 (Product Lattice). *Let n be a positive integer. The product lattice, $(\overline{\mathcal{G}(\mathcal{T})}^n, \preceq_n, \perp_n, \top_n)$ is defined as follows: for $\overline{\mathcal{P}} = (\mathcal{P}_1, \mathcal{P}_2, \ldots, \mathcal{P}_n)$, $\overline{\mathcal{Q}} = (\mathcal{Q}_1, \mathcal{Q}_2, \ldots, \mathcal{Q}_n) \in \overline{\mathcal{G}(\mathcal{T})}^n$, $\overline{\mathcal{P}} \preceq_n \overline{\mathcal{Q}}$ if $\mathcal{P}_i \preceq \mathcal{Q}_i$ for each $1 \leq i \leq n$, $\perp_n = (\perp, \perp, \ldots, \perp)$ and $\top_n = (\top, \top, \ldots, \top)$.*

For $S \subset \overline{\mathcal{G}(\mathcal{T})}^n$, the notation $\bigwedge S$ will be used to denote the greatest lower bound of S in the product lattice.

By Theorem 11, and standard results in lattice theory, we have:

Theorem 27. *The product lattice satisfies the following properties:*

1. *$(\overline{\mathcal{G}(\mathcal{T})}^n, \preceq_n, \perp_n, \top_n)$ is a complete lattice.*
2. *If $\tilde{S} \subseteq \overline{\mathcal{G}(\mathcal{T})}^n$ is non-empty, with $\tilde{S} = S_1 \times S_2 \times \cdots \times S_n$, where $S_i \subseteq \overline{\mathcal{G}(\mathcal{T})}$ for $1 \leq i \leq n$. Then $\bigwedge \tilde{S} = (\bigwedge S_1, \bigwedge S_2, \ldots, \bigwedge S_n)$.*

As preparation for defining the composite transfer function, we introduce the following functions:

Definition 28 (Projection Maps). *Let n be a positive integer. For each $i \in \{1, 2, \ldots, n\}$,*

- *The projection map to the ith co-ordinate, $\pi_i : \overline{\mathcal{G}(\mathcal{T})}^n \mapsto \overline{\mathcal{G}(\mathcal{T})}$ is defined by $\pi_i(\mathcal{P}_1, \mathcal{P}_2, \ldots, \mathcal{P}_n) = \mathcal{P}_i$ for any $(\mathcal{P}_1, \mathcal{P}_2, \ldots, \mathcal{P}_n) \in \overline{\mathcal{G}(\mathcal{T})}^n$.*
- *The confluence map $\pi_{i,j} : \overline{\mathcal{G}(\mathcal{T})}^n \mapsto \overline{\mathcal{G}(\mathcal{T})}$ is defined by $\pi_{i,j}(\mathcal{P}_1, \mathcal{P}_2, \ldots, \mathcal{P}_n) = \mathcal{P}_i \wedge \mathcal{P}_j$ for any $(\mathcal{P}_1, \mathcal{P}_2, \ldots, \mathcal{P}_n) \in \overline{\mathcal{G}(\mathcal{T})}^n$.*

In addition to the above functions, we will also use the constant map which maps each element in $\overline{\mathcal{G}(\mathcal{T})}^n$ to $\bot$. The following observation is a consequence of standard results in lattice theory.

Observation 29. *Constant maps, projection maps and confluence maps are complete-meet-morphisms.*

For each $k \in V(G)$, $\mathrm{Pred}(k)$ denotes the set of predecessors of the vertex k in the control flow graph G.

Definition 30 (Composite Transfer Function). *Let $D = (G, F)$ be a data flow framework over $\mathcal{T}$. For each $k \in V(G)$, define the component map $f_k :$ $\overline{\mathcal{G}(\mathcal{T})}^n \mapsto \overline{\mathcal{G}(\mathcal{T})}$ as follows:*

1. *If $k = 1$, the entry point, then $f_k = \bot$. (f_1 is the constant function that always returns the value $\bot$).*
2. *If k is a function point with $\mathrm{Pred}(k) = \{j\}$, then $f_k = h_k \circ \pi_j$, where h_k is the (extended) transfer function corresponding to the function point k, and π_j the projection map to the jth coordinate as defined in Definition 28.*
3. *If k is a confluence point with $\mathrm{Pred}(k) = \{i, j\}$, then $f_k = \pi_{i,j}$, where $\pi_{i,j}$ is the confluence map as defined in Definition 28.*

The composite transfer function of D is defined to be the unique function f_D satisfying $\pi_k \circ f_D = f_k$ for each $k \in V(G)$.

The purpose of defining f_D is the following. Suppose we have associated a congruence with each program point in a data flow framework. Then f_D specifies how a simultaneous and synchronous application of all the transfer functions/confluence operations at the respective program points modifies the congruences at each program point. The definition of f_D conservatively sets the confluence at the entry point to $\bot$, treating terms in $\mathcal{G}(\mathcal{T})$ to be inequivalent to each other at the entry point. See Fig. 2 for an example. The following observation is a direct consequence of the above definition.

Observation 31. *The composite transfer function f_D (Definition 30) satisfies the following properties:*

1. *If $k = 1$, the entry point, then $\pi_k \circ f_D = \bot$.*
2. *If k is a function point with $\mathrm{Pred}(k) = \{j\}$, then $f_k = \pi_k \circ f_D = h_k \circ \pi_j$, where h_k is the (extended) transfer function corresponding to the function point k.*
3. *If k is a confluence point with $\mathrm{Pred}(k) = \{i, j\}$, then $f_k = \pi_k \circ f_D = \pi_{i,j}$.*

The following lemma is a consequence of Observation 31.

Lemma 32. *Let $D = (G, F)$ be a data flow framework over $\mathcal{T}$ and $k \in V(G)$. Let $S = \{f_D(\top_n), f_D^2(\top_n), \ldots\}$, where f_D is the composite transfer function of D.*

1. *If $k = 1$, the entry point, then $\pi_k \circ f_D^l(\top_n) = \bot$ for all $l \geq 1$, hence $\pi_k(\bigwedge S) = \bot$.*

2. *If k is a function point with $\mathrm{Pred}(k) = \{j\}$, then for all $l \geq 1$,*

$$(\pi_k \circ f_D^l)(\top_n) = (\pi_k \circ f_D)(f_D^{l-1}(\top_n))$$
$$= (h_k \circ \pi_j \circ f_D^{l-1})(\top_n)$$

3. *If k is a confluence point with $\mathrm{Pred}(k) = \{i, j\}$, then for all $l \geq 1$,*

$$(\pi_k \circ f_D^l)(\top_n) = (\pi_k \circ f_D)(f_D^{l-1}(\top_n))$$
$$= (\pi_{i,j})(f_D^{l-1}(\top_n))$$
$$= ((\pi_i \circ f_D^{l-1})(\top_n)) \wedge ((\pi_j \circ f_D^{l-1})(\top_n))$$

By standard facts in lattice theory, we have:

Theorem 33. *The following properties hold for the composite transfer function f_D (Definition 30):*

1. *f_D is monotone, and is a complete-meet-morphism.*
2. *The component maps $f_k = \pi_k \circ f_D$ are complete-meet-morphisms for all $k \in \{1, 2, \ldots, n\}$.*
3. *f_D has a maximum fix-point.*
4. *If $S = \{\top, f_D(\top_n), f_D^2(\top_n), \ldots\}$, then $\bigwedge S$ is the maximum fix-point of f_D.*

The objective of defining Herbrand Congruence as the maximum fix-point of the composite transfer function is possible now.

Definition 34 (Herbrand Congruence). *Given a data flow framework $D = (G, F)$ over $\mathcal{T}$, the Herbrand Congruence function $H_D : V(G) \mapsto \overline{\mathcal{G}(\mathcal{T})}$ is defined as the maximum fix-point of the composite transfer function f_D. For each $k \in V(G)$, the value $H_D(k) \in \overline{\mathcal{G}(\mathcal{T})}$ is referred to as the Herbrand Congruence at program point k.*

The following is a consequence of Theorem 33 and the definition of Herbrand Congruence.

Observation 35. *For each $k \in V(G)$, $H_D(k) = \bigwedge\{(\pi_k \circ f_D{}^l)(\top_n) : l \geq 0\}$.*

Proof

$$H_D(k) = \pi_k(\bigwedge{}_n\{f_D{}^l(\top_n) : l \geq 0\}) \text{ (by Theorem 33)}$$
$$= \bigwedge\{\pi_k(f_D{}^l(\top_n)) : l \geq 0\} \text{ (because } \pi_k \text{ is a complete-meet-morphism)}$$
$$= \bigwedge\{(\pi_k \circ f_D{}^l)(\top_n) : l \geq 0\}$$

$\square$

The definition of Herbrand congruence must be shown to be consistent with the traditional meet-over-all-paths description of Herbrand equivalence of terms in a data flow framework. The next section addresses this issue.

8   MOP Characterization

In this section, we give a meet over all paths characterization for the Herbrand Congruence at each program point. This is essentially a lattice theoretic reformulation of the characterization by Steffen et al. [5, p. 393]. Consider a data flow framework $D = (G, F)$ over $\mathcal{T}$, with $V(G) = \{1, 2, \ldots, n\}$.

Definition 36 (Path). *For any non-negative integer l, a program path (or simply a path) of length l to a vertex $k \in V(G)$ is a sequence $\alpha = (v_0, v_2, \ldots v_l)$ satisfying $v_0 = 1$, $v_l = k$ and $(v_{i-1}, v_i) \in E(G)$ for each $i \in \{1, 2, \ldots l\}$. For each $i \in \{0, 1, \ldots, l\}$, α_i denotes the initial segment of α up to the ith vertex, given by $(v_0, v_1, \ldots, v_i)$.*

Next, we associate a congruence in $\overline{\mathcal{G}(\mathcal{T})}$ with each path in D. The path function captures the effect of application of transfer functions along the path on the initial congruence $\perp$, in the order in which the transfer functions appear along the path.

Definition 37 (Path Congruence). *Let $\alpha = (v_0, v_1, \ldots, v_l)$ be a path of length l to vertex $k \in V(G)$ for some $l \geq 0$. We define:*

1. *When $i = 0$, $m_{\alpha_i} = \perp$.*
2. *If $i > 0$ and $v_i = j$, where $j \in V(G)$ is a function point, then $m_{\alpha_i} = h_j(m_{\alpha_{i-1}})$, where $h_j \in F$ is the extended transfer function associated with the function point j.*
3. *If $i > 0$ and v_i is a confluence point, then $m_{\alpha_i} = m_{\alpha_{i-1}}$.*
4. *The path congruence associated with α, $m_\alpha = m_{\alpha_l}$.*

For $k \in V(G)$ and $l \geq 0$, let $\Phi_l(k)$ denote the set of all paths of length *less than* l from the entry point 1 to the vertex k. In particular, $\Phi_0(k) = \emptyset$, for all $k \in V(G)$. The following observation is a consequence of the definition of $\Phi_l(k)$.

Observation 38. *If $k \in V(G)$ and $l \geq 1$,*

1. *If k is the entry point, then $\Phi_l(k) = \{(1)\}$, the set containing only the path of length zero, starting and ending at vertex 1.*
2. *If k is a function point with $\mathrm{Pred}(k) = \{j\}$, then*
   $\{\alpha_{l-1} : \alpha \in \Phi_l(k)\} = \{\alpha' : \alpha' \in \Phi_{l-1}(j)\} = \Phi_{l-1}(j)$.
3. *If k is a confluence point with $\mathrm{Pred}(k) = \{i, j\}$, then*
   $\{\alpha_{l-1} : \alpha \in \Phi_l(k)\} = \{\alpha' : \alpha' \in \Phi_{l-1}(i)\} \cup \{\alpha' : \alpha' \in \Phi_{l-1}(j)\} = \Phi_{l-1}(i) \cup \Phi_{l-1}(j)$.

For $l \geq 0$, we define the congruence $M_l(k)$ to be the meet of all path congruences associated with paths of length less than l from the entry point to vertex k in G. Stated formally,

$$M_l(k) = \bigwedge \{m_\alpha : \alpha \in \Phi_l(k)\}, \text{ for } l \geq 0$$

Observation 39. *If $l = 0$, $\Phi_l(k) = \emptyset$ and hence $M_0(k) = \top$, for all $k \in V(G)$. Further, $M_1(1) = \bot$ and $M_1(k) = \top$, for $k \neq 1$. In general, $M_l(k) = \top$ if there are no paths of length less than l from 1 to k in G.*

We define Φ_k to be the set of all paths from vertex 1 to vertex k in G, i.e., $\Phi_k = \bigcup_{l \geq 1} \Phi_l(k)$ and $MOP(k) = \bigwedge\{m_\alpha : \alpha \in \Phi(k)\} = \bigwedge\{M_l(k) : l \geq 0\}$. (The second equality follows from standard results in lattice theory and Observation 39.) The congruence $MOP(k)$ is the meet of all path congruences associated with paths in Φ_k.

Our objective is to prove $MOP(k) = H_D(k)$ for each $k \in \{1, 2, \ldots, n\}$ so that H_D captures the meet over all paths information about equivalence of expressions in $\mathcal{T}$. As noted in the introduction, the proof does not immediately follow from the transfer function being a meet-morphism, as in [1,3,16] since the lattice is neither of finite height nor the chains in the lattice stabilize at finite height.

We begin with the following observations.

Lemma 40. *For each $k \in V(G)$ and $l \geq 1$*

1. *If $k = 1$, the entry point, then $M_l(k) = \bot$.*
2. *If k is a function point with $\mathrm{Pred}(k) = \{j\}$, then $M_l(k) = h_k(M_{l-1}(j))$, where h_k is the (extended) transfer function corresponding to the function point k.*
3. *If k is a confluence point with $\mathrm{Pred}(k) = \{i, j\}$, then $M_l(k) = M_{l-1}(i) \wedge M_{l-1}(j)$.*

Lemma 41. *For each $k \in V(G)$ and $l \geq 0$, $M_l(k) = (\pi_k \circ f_D{}^l)(\top_n)$.*

Finally, we show that the iterative fix-point characterization of Herbrand equivalence and the meet over all paths characterization coincide.

Theorem 42. *Let $D = (G, F)$ be a data flow framework. Then, for each $k \in V(G)$, $MOP(k) = H_D(k)$.*

Proof

$$MOP(k) = \bigwedge\{m_\alpha : \alpha \in \Phi(k)\}$$
$$= \bigwedge\{M_l(k) : l \geq 0\} \text{ (by Observation 39)}$$
$$= \bigwedge\{(\pi_k \circ f_D{}^l)(\top_n) : l \geq 0\} \text{ (by Lemma 41)}$$
$$= H_D(k) \text{ (by Observation 35)}$$

$\square$

Note that the proof of Observation 35 requires the composite transfer function to be a complete-meet-morphism.

9   Conclusion

We have shown that Herbrand equivalences of terms at program points in a data flow framework can be formulated in terms of the maximum fix-point of a composite transfer function defined over an infinite complete lattice of congruences. The standard definition of Herbrand equivalence [5] is reformulated as a meet over all paths definition in this new lattice framework and is shown to be equivalent to the fix-point formulation. The equivalence of the MFP characterization with the standard formulation provides a theoretical justification for the use of fix-point algorithms used in practice for computing Herbrand equivalences. The new fix-point formulation permits us to view the existing working set based fix-point algorithms as instances of abstract interpretation from the ideal concrete lattice into appropriately defined abstract lattices. We hope that this view can help to make correctness proofs of working set algorithms more transparent.

References

1. Nielson, F., Nielson, H.R., Hankin, C.: Principles of Program Analysis, vol. 1. Springer, Heidelberg (1999). https://doi.org/10.1007/978-3-662-03811-6
2. Aho, A.V., Lam, M.S., Sethi, R., Ullman, J.D.: Compilers: Principles, Techniques, and Tools, vol. 2. Addison-Wesley Reading, Boston (2006)
3. Kam, J.B., Ullman, J.D.: Monotone data flow analysis frameworks. Acta Informatica **7**(3), 305–317 (1977)
4. Rüthing, O., Knoop, J., Steffen, B.: Detecting equalities of variables: combining efficiency with precision. In: Cortesi, A., Filé, G. (eds.) SAS 1999. LNCS, vol. 1694, pp. 232–247. Springer, Heidelberg (1999). https://doi.org/10.1007/3-540-48294-6_15
5. Steffen, B., Knoop, J., Rüthing, O.: The value flow graph: a program representation for optimal program transformations. In: Jones, N. (ed.) ESOP 1990. LNCS, vol. 432, pp. 389–405. Springer, Heidelberg (1990). https://doi.org/10.1007/3-540-52592-0_76
6. Müller-Olm, M., Rüthing, O., Seidl, H.: Checking herbrand equalities and beyond. In: Cousot, R. (ed.) VMCAI 2005. LNCS, vol. 3385, pp. 79–96. Springer, Heidelberg (2005). https://doi.org/10.1007/978-3-540-30579-8_6
7. Gulwani, S., Necula, G.C.: A polynomial-time algorithm for global value numbering. Sci. Comput. Program. **64**(1), 97–114 (2007)
8. Pai, R.R.: Detection of redundant expressions: a precise, efficient, and pragmatic algorithm in SSA. Comput. Lang. Syst. Struct. **46**, 167–181 (2016)
9. Kildall, G.A.: A unified approach to global program optimization. In: Proceedings of the 1st Annual ACM SIGACT-SIGPLAN Symposium on Principles of Programming Languages, POPL 1973, pp. 194–206. ACM (1973)
10. Cousot, P., Cousot, R.: Abstract interpretation: a unified lattice model for static analysis of programs by construction or approximation of fixpoints. In: Proceedings of the 4th ACM SIGACT-SIGPLAN Symposium on Principles of Programming Languages, POPL 1977, pp. 238–252. ACM (1977)
11. Kam, J.B., Ullman, J.D.: Global data flow analysis and iterative algorithms. J. ACM **23**(1), 158–171 (1976)

12. Alpern, B., Wegman, M.N., Zadeck, F.K.: Detecting equality of variables in programs. In: Proceedings of the 15th ACM SIGPLAN-SIGACT Symposium on Principles of Programming Languages, POPL 1988, pp. 1–11. ACM (1988)

13. Rosen, B.K., Wegman, M.N., Zadeck, F.K.: Global value numbers and redundant computations. In: Proceedings of the 15th ACM SIGPLAN-SIGACT Symposium on Principles of Programming Languages, POPL 1988, pp. 12–27. ACM (1988)

14. Tarski, A.: A lattice-theoretical fixpoint theorem and its applications. Pacific J. Math. **5**(2), 285–309 (1955)

15. Cousot, P.: Asynchronous iterative methods for solving a fixed point system of monotone equations in a complete lattice. In: Research Report R.R. 88, Laboratoire IMAG, Université scientifique et médicale de Grenoble, Grenoble, France, 15 p., September 1977

16. Geser, A., Knoop, J., Lüttgen, G., Rüthing, O., Steffen, B.: Non-monotone fixpoint iterations to resolve second order effects. In: Gyimóthy, T. (ed.) CC 1996. LNCS, vol. 1060, pp. 106–118. Springer, Heidelberg (1996). https://doi.org/10.1007/3-540-61053-7_56

17. Babu, J., Krishnan, K.M., Paleri, V.: A fix-point characterization of herbrand equivalence of expressions in data flow frameworks. arXiv:1708.04976 (2017)

Logic Without Language

Rohit Parikh[✉]

City University of New York, Brooklyn College and CUNY Graduate Center,
New York, USA
rparikh@gc.cuny.edu

Abstract. Does a dog think? Does a prelingual child think? Creatures
without language seem to be making some logical inferences which allow
them to make decisions. We offer a utility and perception based account
which allows us to deal with this phenomenon formally. We offer the
suggestion that non-lingual creatures have a certain perception of the
world and that they make the best decisions relative to that perception.
Logic may be "used" to infer non-perceived facts from perceived facts.

1 Davidson on Animals

*Neither an infant one week old nor a snail is a rational creature. If the infant
survives long enough he will probably become rational while this is not true of
the snail....*

*The difference consists, it is argued, in the having of propositional attitudes
such as belief, desire, intention, and shame. This raises the question of how to
tell when a creature has propositional attitudes. Snails, we may agree, do not but
how about dogs or chimpanzees?...*

*It is next contended that language is a necessary concommitant of any of the
propositional attitudes. This idea is not new, but there seem to be few arguments
in its favor in the literature. One is attempted here.*

Donald Davidson, Rational Animals, *Dialectica,* 1982

2 But Is Davidson Correct?

Here is Rescorla, citing Sextus Empiricus: Sextus Empiricus, who credits the
argument to Chrysippus, presents it as follows: *[Chrysippus] declares that the
dog makes use of the fifth complex indemonstrable[1] syllogism when, on arriving
at a spot where three ways meet..., after smelling at the two roads by which the
quarry did not pass, he rushes off at once by the third without stopping to smell.
For, says the old writer, the dog implicitly reasons thus: "The animal went either
by this road, or by that, or by the other: but it did not go by this or that, therefore .
he went the other way."*

---

[1] (modus tollendo ponens, or disjunctive syllogism).

© Springer-Verlag GmbH Germany, part of Springer Nature 2019
Md. A. Khan and A. Manuel (Eds.): ICLA 2019, LNCS 11600, pp. 173–182, 2019.
https://doi.org/10.1007/978-3-662-58771-3_16

And here is Jakob von Uexküll: explaining his notion of the *Umwelt* or 'personal world'. or[2]

This little monograph does not claim to point the way to a new science. Perhaps it should be called strolls into unfamiliar worlds, *worlds strange to us but known to other creatures manifold and varied as the animals themselves. The best time to set out on such an adventure is on a sunny day. The place, a flower strewn meadow humming with insects, fluttering with butterflies. Here we may glimpse the worlds of the lowly dwellers of the meadow. To do so we must first blow in fancy a soap bubble around each creature to represent its own world filled with the perceptions of which it alone knows.*

When we ourselves step into one of these bubbles, The familiar meadow is transformed. Many of its colorful features disappear, others no longer belong together but appear in new relationships. A new world comes into being. Through the bubble we see the world of the burrowing worm, of the butterfly, or of the field mouse; the world as it appears to the animals themselves, not as it appears to us. This we may call the phenomenal world or the self-world of the animal.

Jakob von Uexküll, *Forays into the worlds of animals and men,* [8]

The purpose of this paper is to make the contrast between Davidson and Uexküll more explicit by bringing in utility based reasoning, which is appropriate for both humans and animals, and relating this reasoning to a logical language.[3]

For those of us technically minded, the Umwelt is the semantics (or semiotics) of the agent. If we see this agent as having beliefs and desires (in the BDI sense, see [4].) then we need to understand what world these beliefs and desires are about. Logics for action and belief need to use the real semantics of such agents. We will offer a path towards formalizing such logics.

And then we can understand what actions will come about from these beliefs and desires.

3   Introduction

Suppose that Aruna has a sofa in her living room. If you ask her if she knows that she has a sofa in her living room she will say, "Are you crazy? Of course I know." but if you say to her "How many pounds of air are in your apartment?" She would have no idea. (It could be about 750 pounds in a typical apartment.)

The sofa is in her apartment and so is the air so why does she know about the one but not the other? Aren't they both part of her world? But the sofa is part of her Umwelt and the weight of the air is not.

· Why does the fly get caught in the spider's web? Because a thread in the web is too fine for the fly's vision. So it does not know that the web is there. Once caught, it knows quite well because it is no longer using its eyes but its sense of touch.

---

[2] Uexküll seems here to anticipate the discussion in Nagel [3].

[3] See Vigo and Allen [7] for a related discussion.

There are certain things that we are all supposed to know like whether there is a sofa in our living room but we do not usually know about the air, even though it too is in our living room.

Now for us humans, our individual Umwelt is supplemented by the community Umwelts which include information from the Umwelts of others, and also from science. The sun and the moon *look* to us as if they are at the same distance but science *tells* us that the sun is much further.

And we certainly did not send a man to the moon using just the phenomenal world (Umwelt). But animals and young humans tend to act primarily or entirely in terms of their phenomenal worlds.

4   Agent? Or Machine?

The mechanists have pieced together the sensory and motor organs of animals, like so many parts of a machine, ignoring their real functions of perceiving and acting, and have gone on to mechanize man himself. According to the behaviorists, man's own sensations and will are mere appearance, to be considered, if at all, only as disturbing static. But we who still hold that our sense organs serve our perceptions, and our motor organs our actions, see in animals as well not only the mechanical structure, but also the operator, who is built into their organs as we are into our bodies.

(Uexküll [8])

5   Two Computer Scientists Respond

On this basis we shall say that an entity is intelligent if it has an adequate model of the world (including the intellectual world of mathematics, understanding of its own goals and other mental processes), if it is clever enough to answer a wide variety of questions on the basis of this model, if it can get additional information from the external world when required, and can perform such tasks in the external world as its goals demand and its physical abilities permit.

(McCarthy and Hayes, Some philosophical problems from the point of view of AI, [2] 1969)

6   And a Psychologist Offers His View

If the organism carries a 'small-scale model' of external reality and of its own possible actions within its head, it is able to try out various alternatives, conclude which is the best of them, react to future situations before they arise, utilize the knowledge of past events in dealing with the present and future, and in every way to react in a much fuller, safer, and more competent manner to the emergencies which face it.

(Kenneth Craik, *The Nature of Explanation*, 1943: 61)

7   Umwelts

We can think of the Umwelt as a homomorphic image of the real world. And that means that some information is *missing*. A child does not know that the moon is much farther than a tree. In view of this missing information the best action for an agent is not always the same as the *apparent* best action[4]. When an agent is in a world w, then for all it knows, it could be in some other world w', indistinguishable from w. The best action is that action which has the highest utility over all these other possible worlds. But it might not be the best action for the actual world w.

Now the expected value of the apparent best action increases when more information is received. But in order to receive more information the animal needs to develop tools for getting that information, and these tools incur a cost, so unless the cost is less than the gain the improvement will not be sought. The fly could have had better eyesight and be caught less often, but that more sophisticated eye would be expensive to maintain.

In this context, reconsider Uexküll's account of the life story of a tick. *A tick has three perceptions. And three effectors (or actions).*

The typical tick climbs on a grass blade or something similar and waits.[5]

When a mammal passes under the grass blade, its skin releases butyric acid which the tick detects and it drops onto the mammal. It knows it is a mammal because of the warmth.

Then it moves around in the mammal's skin until it finds a bald spot. It sucks blood and then drops to the ground where it lays its eggs and dies.

So the tick needs three perceptions,

(1) the sunlight to know which direction is up and so to rise,
(2) the smell of butyric acid which tells it when to drop, and
(3) the feeling of warmth which enables it to know that precious blood is available.

It also has three actions, (1) rising, (2) dropping, (3) sucking blood and then (again) (2) dropping.

The tick can be easily represented by a transducer finite automaton. The perception and abilities of the tick are strikingly similar to that of the agent in the Wumpus world, see [6], Chap. 7. Dennett's *The intentional Stance* [1] has asimilar considerations.

It also uses *default reasoning, see* [5], because it does not (bother to) distinguish between blood and some other warm liquid supplied to it by an experimenter. Under normal circumstances it *is* blood and the tick does not need expensive equipment to distinguish blood from fake blood.

---

[4] A child who thinks the moon is close by may ask her father to bring the moon to her. The request which we may find amusingly irrational is in fact rational in the child's Umwelt.

[5] Apparently a tick can wait for several years without starving to death.

Uexküll has lots of examples of creatures being fooled in this way when the best action in their Umwelt is not the best action in the real world.

Default reasoning is a rational strategy when we would incur too high a cost to deviate from it. It's cheaper to assume that what you see is what you expect to see.

Uexküll is skeptical of the idea that there is the "real world" which we see imperfectly. We shall not follow him or ask the reader to. Rather our representation will assume that *there is* a real world which is perceived imperfectly by every creature, whether a bat or a dog or a child.

Thus each creature sees a homomorphism from real world to its personal world.

8   A Little Bit of Mathematics

Definition 1. *An Umwelt U consists of two parts. A homomorphism H (many one mapping) from the actual world to the perceived world. And a set A of possible actions. Thus $U = (H, A)$. In addition each creature has a utility function u, so that $u(a, w) = x$ is the utility of action a performed when the world is w. We will assume that x is a real number. (In actuality it could be some level of satisfaction for us humans, or the expected number of progeny for animals).*

Given a world w, the best action $b(w)$ for the creature is that a which maximizes the expected utility $u(a, w')$ over the set $\{w'|H(w') = H(w)\}$. (There is an implicit probability distribution here which we will not specify). The expected value $E(U)$ of the Umwelt U is the expected value of the random variable a.

Definition 2. *Umwelt $U' = (H', A')$ refines Umwelt $U = (H, A)$ if*

(a) $H'(w) = H'(w') \rightarrow H(w) = H(w')$ and
(b) $A \subseteq A'$.

Thus H' has more information and more abilities than H.

Theorem 1. *If U' refines U then $E(U) \le E(U')$.*

The more you know and the more you can do, the better off you are.

Proof. U' can refine U in two ways. Either by having more actions or by having more information. Clearly if there are more actions, then the best action from a larger set will be at least as good as the best action from a smaller set. What about more information? (let us keep the set of actions constant for the moment.)

Suppose that w is world, let the cell of w according to U be $C = \{w'|H(w') = H(w)\}$. Let a be the best U-action over C. Now, since U' has more information than U, C according to U' will break up into a number of cells $C, ..., C_k$. Now, the best action a_i over C_i will be at least as good as a (over C_i). U' will give rise to action a_i over C_i and their average over C will be at least as good as the value of a over C. Hence U' will give a better average overall, and $E(U')$ will be at least as large as $E(U)$.

Now suppose that U' has both more actions and more information. Then let U'' have the same actions as U and the same information as U'. Then the preceding two arguments show that $E(U) \leq E(U'')$ as well as $E(U'') \leq E(U')$ This gives $E(U) \leq E(U'')$.   □

Here is the intuitive idea. Suppose I am driving to New Jersey and can take either the tunnel or the bridge. Normally the tunnel is better as it is closer. But it might be closed.

The procedure *if the tunnel is open then take the tunnel else take the bridge* has a higher average utility than either *just take the tunnel* or *just take the bridge*. But that *if then else* procedure can only be carried out in the refined Umwelt where the question about the tunnel has been answered.

Thus it pays to know more and it also pays to have more options for action.

9   Uninformed Agent

Here A and B are incompatible conditions which might obtain. X and Y are possible actions of the agent.

	A	B
Action X	−100, 25	10
Action Y	6	−50, 15

This is a decision theoretic matrix. In condition A, the agent does not know whether the payoff will be 25 or −100 if action X is performed. Thus if A is true then the best action is Y and if B is true then the best action is X. The utility of the Umwelt is

$$(6 + 10)/2 = 8.$$

In this more detailed table P is an additional condition about which the agent could find out.

10   Better Informed Agent

	A and P	A and -P	B and P	B and -P
Action X	−100	25	10	10
Action Y	6	6	15	−50

So the best action is (if A&P then Y or else if A&-P then X or if B&P then Y or else if B&-P then X). The utility of the Umwelt is now $56/4 = 14$. Knowing about P has pushed the utility up by 6 and so one could say that the knowledge of P is worth 6 units.

11   Learning More

Why then does a creature not have a maximal U where H is the identity function and A is enormous?

Because acquiring more information and acquiring more possible actions has a cost and the benefit may not justify the cost.

And for Darwinian creatures which rely on evolution to 'learn,' the entire species has to have the extra sensory ability so that *one* creature may benefit. The cost summed up over the entire species may not be justified by the benefit to one member or a few members of the species.

If I have an Umwelt U and I ask a question Q then the H becomes refined to a finer H'. The utility of the new Umwelt will be greater but the question will have a cost. To ask the question requires me to make sure that the cost is less than the utility gain.

If I am at a train station and ask the agent what time my train is leaving, I will benefit from the answer.

But if I ask how many dishes are available in the Dining Car, the agent's rudeness will be too high a price to pay for any answer.

Similarly for an increase in actions. If I am going mountain climbing then it makes sense for me to undergo training so that I have more actions available while on the mountain. But if I am not going mountain climbing then the effort gains me nothing.

12   Symbiosis

Suppose that two different creatures have two different Umwelts.[6] For example a man with eyesight but no legs, riding another man without vision but with legs.[7] Or it could be a dog leading a man who is blind. in that case the combined Umwelt would be to the benefit of both. What is essential in that case that the Umwelts supplement each other and that their utilities align.

Consider two creatures with Umwelts U and U' and a common utility[8].

Then the two together have joint Umwelt U'' whose H'' is the least upper bound of H and H' and whose action set A'' is $A \cup A'$. I.e. $H''(w) = H''(w')$ iff $H(w) = H(w')$ and $H'(w) = H'(w')$.

Then the joint Umwelt refines both the individual Umwelts and (with a reasonable bargain) yields a higher utility for both creatures. This explains why we have cases of symbiosis among animals and massive cooperation among humans[9].

---

[6] While the Umwelt is personal, the homomorphism is objective and two different homomorphisms can be combined.

[7] Something like this happens in one of the Sinbad stories.

[8] The utility need not be common but the two utilities must be compatible. See e.g. John Nash's work on *The Bargaining Problem, Econometrica* 1950.

[9] There is also the issue of *compatible* utilities. A leopard and a deer do not have compatible utilities unless we think of the leopard as having the *job* of keeping the deer herd under control.

Here is an example. In the ocean, certain species, like shrimps and gobies, will clean fish. They remove parasites, dead tissue, and mucus.

13   Animal Logic

A tiger watches a deer going towards a bush from the left. Then the deer is not seen any more. And it has not emerged on the other side. So the tiger knows and believes that the deer is behind the bush.

The tiger is inferring the presence of the deer behind the bush, which it does not see, from the previous appearance of the deer to the left of the bush, and from the non-appearance of the deer to the right of the bush.

Thus it is inferring some variable free sentences which it does not experience, from other variable free sentences which it *has* experienced.

14   A General Framework: Theories Enter

Suppose we are given a first order theory T with plenty of constants and variable free terms. T defines a relation R between finite sets X of variable free sentences and other sets Y of variable free sentences as follows:

$$R(X,Y) \text{ iff } T \cup X \models \phi \text{ for all } \phi \in Y.$$

Clearly R is monotonic in X, in T, and anti-monotonic in Y.

Suppose the tiger's behavior shows awareness of Y on the basis of X.

Does the tiger then *believe* T?

Not necessarily. There are many such theories which will work. And the tiger may be using some other means to infer Y.

But it can be harmless if we attribute to the tiger such a theory T as long as we are aware that this is merely a *facon de parler*. We try to enter the tiger's world and reason as we think it does. This is fine as long as our predictions of the tiger's behavior match its actual behavior.

Thus it is fine for **us** to say, "the tiger acts as if it believes T".

Question: For which relations R does there exist a **finite** first order theory which 'explains' R?

One could also ask which R are computable in polytime or even in linear time.

15   The Usefulness of a Theory

Suppose an agent knows some atomic formulas U. It also knows a theory T. Then using T it can infer other atomic formulas V (where we assume, using reflexivity that $U \subseteq V$). Let X be the set of worlds satisfying U and Y the set of worlds satisfying V. Then $Y \subseteq X$ and the best action b over Y may be different from the best action a over X. The utility of b over Y minus the utility of a over Y is the gain from knowing theory T. It is easy to see that if $T \subseteq T'$ then the utility of T' will be greater than the utility of T.

Presumably Chrysippus' dog knew $A \vee B \vee C$ where A, B, C are the three roads. On finding $\neg A, \neg B$ the dog infers C.

Here the dog merely makes an inference which is logically necessary. But the tiger inferring that the deer is behind the bush is not making an inference which is *logically* necessary. Rather the inference depends on properties of the physical world. A driverless car may even make inferences which are not necessitated by the laws of physics but by traffic laws or even by plausible inferences about other cars.

16   Can Logic Exist Without Language?

Two kinds of agents considered by AI are as follows.

1. Stimulus response creatures. These are creatures whose reactions are fixed given what they perceive. In AI they are represented by means of a table. And indeed the head of a Turing machine is just that. It sees something on a square and it acts.
2. Creatures with a 'knowledge base'. These are creatures who have some cache of 'facts' which they revise and which they use to infer other facts.

It is easy to see that the second kind of creature could well do with some logic.

But does this logic need to be expressed in language? Perhaps not, but if we see our logic as reflected in the behavior of a creature without language then we have added to our understanding.

17   Conclusion

We have made a start towards formalizing some ideas implicit in Uexküll, Dennett and Nagel as well as others. Such a preliminary effort must leave many loose ends untied. Here is an example.

The logic of a creature need not use an *average* utility over some set of possible worlds. It may well have some *default* worlds which are assumed to be the only possible ones and the creature's actions will then be in terms of these default worlds.

But these are also issues for a sequel.

Acknowledgments. We thank Priya Chakraborty, John Greenwood, Peter Godfrey-Smith, Steve Pinker, Vaughan Pratt, Jesse Prinz, Yunqi Xue and two referees for useful suggestions to an earlier version of this research.

References

1. Dennett, D.C.: The Intentional Stance. MIT Press, Cambridge (1989)
2. McCarthy, J., Hayes, P.J.: Some philosophical problems from the standpoint of artificial intelligence. Stanford University, Computer Science Department (1969)
3. Nagel, T.: What is it like to be a bat? Philos. Rev. **83**(4), 435–450 (1974)
4. Rao, A.S., Georgeff, M.P.: BDI agents: from theory to practice. In: ICMAS 1995 (1995)
5. Reiter, R.: A logic for default reasoning. Artif. Intell. **13**(1–2), 81–132 (1980)
6. Russell, S.J., Norvig, P.: Artificial Intelligence: A Modern Approach. Pearson Education Limited, New Delhi (2016)
7. Vigo, R., Allen, C.: How to reason without words: inference as categorization. Cogn. Process. **10**(1), 77–88 (2009)
8. Von Uexküll, J.: A stroll through the worlds of animals and men: a picture book of invisible worlds. Semiotica **89**(4), 319–391 (1992). There are other versions, 1934, and 1957, of the same paper

Towards a Constructive Formalization
of Perfect Graph Theorems

Abhishek Kr Singh[✉] and Raja Natarajan[✉]

Tata Institute of Fundamental Research, Mumbai, India
{abhishek.singh,raja}@tifr.res.in

Abstract. Interaction between clique number $\omega(G)$ and chromatic number $\chi(G)$ of a graph is a well studied topic in graph theory. Perfect Graph Theorems are probably the most important results in this direction. Graph G is called *perfect* if $\chi(H) = \omega(H)$ for every induced subgraph H of G. The Strong Perfect Graph Theorem (SPGT) states that a graph is perfect if and only if it does not contain an odd hole (or an odd anti-hole) as its induced subgraph. The Weak Perfect Graph Theorem (WPGT) states that a graph is perfect if and only if its complement is perfect. In this paper, we present a formal framework for verifying these results. We model finite simple graphs in the constructive type theory of Coq Proof Assistant without adding any axiom to it. Finally, we use this framework to present a constructive proof of the Lovász Replication Lemma, which is the central idea in the proof of Weak Perfect Graph Theorem.

1 Introduction

The chromatic number $\chi(G)$ of a graph G is the minimum number of colours needed to colour the vertices so that every two adjacent vertices get distinct colours. Finding out the chromatic number of a graph is NP-Hard [5]. However, one obvious lower bound is clique number $\omega(G)$, the size of the biggest clique in G. Consider the graphs shown in Fig. 1.

In each of these cases $\chi(G) = \omega(G)$, i.e. the number of colours needed is the minimum we can hope. Can we always hope $\chi(G) = \omega(G)$ for every graph G? The answer is no. Consider the cycle of odd length 5 and its complement shown in Fig. 2. In this case one can see that $\chi(G) = 3$ and $\omega(G) = 2$ (i.e. $\chi(G) > \omega(G)$). A cycle of odd length greater than or equal to 5 is called an *odd hole*. Complement of an odd hole is called an *odd anti-hole*. Indeed, the gap between $\chi(G)$ and $\omega(G)$ can be made arbitrarily large. Consider the other graph shown in Fig. 2 which consist of two disjoint 5-cycles with all possible edges between the two cycles.

This graph is a special case of the general construction where we have k disjoint 5-cycles with all possible edges between any two copies. In this case one can show [7] that $\chi(G) = 3k$ but $\omega(G) = 2k$. In fact, there is an even stronger result [9] which constructs triangle-free Micielski graph M_k that satisfies $\chi(M_k) = k$.

© Springer-Verlag GmbH Germany, part of Springer Nature 2019
Md. A. Khan and A. Manuel (Eds.): ICLA 2019, LNCS 11600, pp. 183–194, 2019.
https://doi.org/10.1007/978-3-662-58771-3_17

Fig. 1. Some graphs where $\chi(G) = \omega(G)$

Fig. 2. Some graphs where, $\chi(G) > \omega(G)$. For k disjoint 5-cycles $\chi(G) = 3k$ and $\omega(G) = 2k$.

In 1961, Claude Berge noticed the presence of odd holes (or odd anti-holes) as induced subgraph in all the graphs presented to him that does not have a nice colouring, i.e. $\chi(G) > \omega(G)$. However, he also observed some graphs containing odd holes, where $\chi(G) = \omega(G)$. Consider the graphs shown in Fig. 3.

Fig. 3. Graphs with $\chi(G) = \omega(G)$, and having odd hole as induced subgraph.

A good way to avoid such artificial construction is to make the notion of nice colouring hereditary. We say that a graph H is an induced subgraph of G, if H is a subgraph of G and $E(H) = \{uv \in E(G) \mid u, v \in V(H)\}$. A graph G is then called a *perfect graph* if $\chi(H) = \omega(H)$ for all of its induced subgraphs H.

Berge observed that the presence of odd holes (or odd anti-holes) as induced subgraphs is the only possible obstruction for a graph to be perfect. These observations led Berge to the conjecture that a graph is perfect if and only if it does not contain an odd hole (or an odd anti-hole) as its induced subgraph. This was soon known as the Strong Perfect Graph Conjecture (SPGC). Berge thought this conjecture to be a hard goal to prove and gave a weaker statement referred to as the Weak Perfect Graph Conjecture (WPGC): a graph is perfect if and only if its complement is perfect. Both the conjectures are theorems now. In 1972, Lovász proved a result [8] (known as Lovász Replication Lemma) which finally helped him to prove the WPGC. It took however three more decades to come up with a proof for SPGC. The proof of Strong Perfect Graph Conjecture was announced in 2002 by Chudnovsky et al. and finally published [3] in a 178-page paper in 2006.

In this paper, we present a formal framework for verifying these results. In Sect. 2 we provide an overview of the Lovász Replication Lemma which is the key result used in the proof of WPGT. In Sect. 3 we present a constructive formalization of finite simple graphs. In this constructive setting, we present a formal proof of the Lovász Replication Lemma (Sect. 4). We summarise the work in Sect. 5 with an overview of possible future works. The Coq formalization for this paper is available at [1].

2 Overview of Lovász Replication Lemma

Let G be a graph and $v \in V(G)$. We say that G' is obtained from G by repeating vertex v if G' is obtained from G by adding a new vertex v' such that v' is connected to v and to all the neighbours of v in G. For example, consider the graphs shown in Fig. 4 obtained by repeating different vertices of a cycle of length 5.

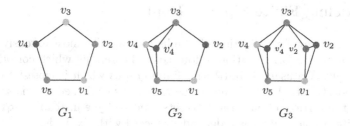

Fig. 4. Graph G_2 is obtained from G_1 by repeating vertex v_4 whereas the graph G_3 is obtained from G_2 by repeating vertex v_2. Note that $\chi(G_2) = \omega(G_2) = 3$ but $\chi(G_3) > \omega(G_3)$.

Note that the graph G_2 has a nice coloring (i.e. $\chi(G_2) = \omega(G_2)$), however the graph G_3 which is obtained by repeating vertex v_2 in graph G_2, does not have a nice coloring (i.e. $\chi(G_3) > \omega(G_3)$). Thus, property $\chi(G) = \omega(G)$ is not preserved by replication. Although, the property $\chi(G) = \omega(G)$ is not preserved, Lovász in 1971 came up with a surprising result which says that perfectness is preserved by replication. It states that *if G' is obtained from a perfect graph G by replicating a vertex, then G' is perfect*. Note that this result does not apply to any graph shown in Fig. 4, since none of them is a perfect graph. All of these graphs has an induced subgraph (odd hole of length 5) which does not admit a nice coloring.

The process of replication can be continued to obtain a graph where each vertex is replaced with a complete graph of arbitrary size (Fig. 5). This gives us a generalised version of the Lovász Replication lemma.

Let G be a perfect graph and $f : V(G) \rightarrow N$. Let G' be the graph obtained by replacing each vertex v_i of the graph G with a complete graph of order $f(v_i)$. Then G' is a perfect graph. For example, consider the graphs shown in Fig. 5. Vertex a in G_1 is replaced by a complete graph V_a of order 2 to obtain

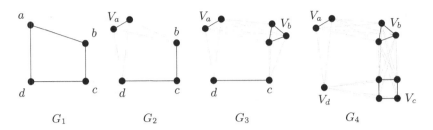

Fig. 5. The graphs resulting from repeated replication of vertices a, b and c of G_1 to form cliques V_a, V_b and V_c of sizes 2, 3 and 4 respectively.

G_2. Similarly, vertex b in G_2 is replaced by a clique V_b of size 3 to obtain G_3. Since G_1 is perfect, all the other graphs (G_2, G_3 and G_4) obtained by repeated replications are also perfect.

3   Modeling Finite Simple Graphs

There are very few formalization of graphs in Coq. The most extensive among these is due to the formalization of four color theorem [6] which considers only planar graphs. We use a definition for finite graphs which is closest to the one used by Doczkal et. al. [4]. However due to reasons explained in this section we represent the vertices of our graphs as sets over ordType instead of eqType. We define finite simple graphs as a dependent record with six fields.

```
Record UG (A:ordType) : Type:= Build_UG {
nodes :> list A;    nodes_IsOrd : IsOrd nodes;    edg: A -> A -> bool;
edg_irefl: irefl edg;    no_edg: edg only_at nodes;    edg_sym: sym edg }.
Variable G: UG A.
```

The last line in the above code declares a finite graph G whose vertices come from an infinite domain A. The first field of G can be accessed using the term (nodes G). It is a list that represents the set of vertices of G. The second field of G ensures that the list of vertices can be considered as a set (details in Sect. 3.1). The third field, which is accessed using the term (edg G), is a binary relation representing the edges of the graph G. The terms (edg_irefl G) and (edg_sym G) are proof terms whose type ensures that the edge relation of G is irreflexive and symmetric. These restrictions on edge makes the graph G simple and undirected. For simplicity in reasoning, the edge relation is considered false everywhere outside the vertices of G. This fact is represented by the proof term (no_edg G).

3.1   Vertices as Constructive Sets

In our work we only consider finite graphs. Vertices of a finite graph can be represented using a finite set. The Mathematical Components library [6] describes an efficient way of working with finite sets. Finite sets are implemented using finite functions (**ffun**) built over a finite type (**finType**). Since all the elements of a set now come from a finite domain (i.e. **finType**) almost every property on the set can be represented using computable (boolean) functions. These boolean functions can be used to do case analysis on different properties of a finite set in a constructive way.

The proof of WPGT involves expansion of graph in which the vertices of the initial graph are replaced with cliques of different sizes. Therefore, in our formalization we can't assume that the vertices of our initial graph are sets over some **finType**. Instead, we represent the set of vertices of a graph G as a list whose elements come from an infinite domain (defined as **ordType**).

Reflection, eqType and ordType. The **finType** in ssreflect is defined on a base type called **eqType**. The **eqType** is defined as a dependent record which packs together a type (**T:Type**) and a boolean comparison operator (**eqb: T → T → bool**) that can be used to check the equality of any two elements of type **T**. Therefore, it tries to capture the notion of a domain with decidable equality. For example, consider the following canonical instance which connects natural numbers to the theory of **eqType**.

```
Canonical nat_eqType: eqType:=
{|Decidable.E:= nat; Decidable.eqb:= Nat.eqb;  Decidable.eqP:= nat_eqP|}.
```

Here, **Nat.eqb** is a boolean function that checks the equality of two natural number and the term **nat_eqP** is name of a lemma which ensures that the function eqb evaluates equality in correct way.

Lemma 1. nat_eqP (m n:nat): reflect (m = n)(Nat.eqb m n).

Note, the use of **reflect** predicate to specify a boolean function. It is a common practice in the ssreflect library. Once we connect a proposition P with a boolean B using the **reflect** predicate we can easily navigate between them. This makes case analysis on P possible even though the Excluded Middle principle is not provable for an arbitrary proposition Q in the constructive type theory of Coq. Consider the following lemma (from **GenReflect.v**), which makes case analysis possible on a predicate P.

Lemma 2. reflect_EM (P: Prop)(b:bool): reflect P b -> P $\lor \neg$ P.

To keep the library constructive we follow this style of proof development. All the basic predicates on sets are connected with their corresponding boolean functions using reflection lemmas. For example, consider the predicates mentioned in Table 1.

These lemmas can be used to do case analysis on any statement about sets containing elements of **eqType**. However, in this framework we can't be constructive while reasoning about properties of power sets. Hence, we base our set theory on **ordType** which is defined as a subtype of **eqType**.

The **ordType** inherits all the fields of **eqType** and has an extra boolean operator which we call the less than boolean operator (i.e. **ltb**). This new operator represents the notion of ordering among elements of **ordType**.

Table 1. Some decidable predicates on sets from the file `SetReflect.v`

Propositions (P:Prop)	Boolean functions (b:bool)	Reflection lemmas
In a l	memb a l	membP
Equal l s	equal l s	equalP
∃ x, (In x l ∧ f x)	existsb f l	existsbP
∀ x, (In x l -> f x)	forallb f l	forallbP

Sets as Ordered Lists. Let `T` be an `ordType`. Sets on domain `T` is then defined as a dependent record with two fields. The first field is a list of elements of type `T` and the second field ensures that the list is ordered using the `ltb` relation of `T`.

```
Record set_on  (T:ordType): Type := { S_of :> list T;  IsOrd_S : IsOrd S_of }.
```

All the basic operations on sets (e.g. union, intersection and set difference) are implemented using functions on ordered lists which outputs an ordered list. An important consequence of representing sets using ordered list is the following lemma (from `OrdList.v`) which states that the element wise equal sets can be substituted for each other in any context.

Lemma 3. `set_equal (A: ordType)(l s: set_on A): Equal l s -> l = s.`

Another advantage of representing sets using ordered list is that we can now enumerate all the subsets of a set `S` in a list using the function `pw(S)`. Moreover, we have following lemma which states that the list generated by `pw(S)` is a set. The details of function `pw(S)` can be found in the file `Powerset.v`.

Lemma 4. `pw_is_ord (S: list A): IsOrd (pw S).`

Since all the subsets of `S` are present in the list `pw(S)` we can express any predicate on power set using a boolean function on list. This gives us a constructive framework for reasoning about properties of sets as well as their power sets. For example, consider the following definition of a boolean function `forall_xyb` and its corresponding reflection lemma `forall_xyP` from the file `SetReflect.v`.

```
Definition forall_xyb (g:A->A->bool)(l:list A):=
              (forallb (fun x=> (forallb (fun y => g x y) l )) l).
  Lemma forall_xyP (g:A->A->bool) (l:list A):
    reflect (forall x y, In x l-> In y l-> g x y)  (forall_xyb g l).
```

3.2   Decidable Edge Relation

The edges of graph `G` are represented using a decidable binary relation on the vertices of `G`. Hence, one can check the presence of an edge between vertices u and v by evaluating the expression (`edg G u v`). The decidability of edge relation is useful for defining many other important properties of graphs as decidable predicates.

Cliques, Stable Set and Graph Coloring. Consider the following definition of a complete graph K present in the graph G. Note that the proposition `Cliq G K` is decidable because it is connected to a term of type bool (i.e. `cliq G K`) by the reflection lemma `cliqP`.

```
Definition cliq(G:UG)(K:list A):=forall_xyb (fun x y=> (x==y) || edg G x y) K.
Definition Cliq(G:UG)(K:list A):=(forall x y,In x K->In y K-> x=y \/ edg G x y).
Lemma cliqP (G: UG)(K: list A): reflect (Cliq G K) (cliq G K).
```

In a similar way we also define independence set (or stable set) and graph coloring using decidable predicates. The details can be found in the file `UG.v` and `MoreUG.v`. Most of these these definitions together with their reflection lemmas are listed in Table 2.

Table 2. Decidable predicates on finite graphs (from `UG.v` and `MoreUG.v`).

Propositions (P:Prop)	Boolean functions (b:bool)	Reflection lemmas
Subgraph G1 G2	subgraph G1 G2	subgraphP
Ind_Subgraph G1 G2	ind_subgraph G1 G2	ind_subgraphP
Stable G I	stable G I	stableP
Max_I_in G I	max_I_in G I	max_I_inP
Cliq G K	cliq G K	cliqP
Max_K_in G K	max_K_in G K	max_K_inP
Coloring_of G f	coloring_of G f	coloring_ofP

We call a graph G to be a nice graph if $\chi(G) = \omega(G)$. A graph G is then called a perfect graph if every induced subgraph of it is a nice graph.

```
Definition Nice (G: UG): Prop:= forall n, cliq_num G n -> chrom_num G n.
Definition Perfect (G: UG): Prop:= forall H, Ind_subgraph H G -> Nice H.
```

In this setting we have the following lemma establishing the obvious relationship between $\chi(G)$ and $\omega(G)$. Here the expression (`clrs_of f G`) represents the set containing all colors used by `f` to color the vertices of G.

Lemma 5. more_clrs_than_cliq_size (G: UG)(K: list A)(f: A -> nat): `Cliq_in G K-> Coloring_of G f -> |K| <= |clrs_of f G|`.

Lemma 6. more_clrs_than_cliq_num (G: UG) (n:nat)(f: A->nat): `cliq_num G n-> Coloring_of G f -> n <= |clrs_of f G|`.

3.3 Graph Isomorphism

It is typically assumed in any proof involving graphs that isomorphic graphs have exactly the same properties. However, in a formal setting we need a proper representation for graph isomorphism to claim the exact behaviour of isomorphic graphs.

```
Definition iso_using (f: A->A)(G G': @UG A) := (forall x, f (f x) = x) /\
   (nodes G') = (img f G)  /\   (forall x y, edg G x y = edg G' (f x) (f y)).
Definition iso (G G': @UG A) := exists (f: A->A), iso_using f G G'.
```

Consider the following lemmas which shows the symmetric nature of graph isomorphism.

Lemma 7. iso_sym (G G': UG): iso G G' -> iso G' G.

Note the self invertible nature of f which makes it injective on both G and G'. The second condition (i.e. (nodes G') = (img f G)) expresses the fact that f maps all the vertices of G to the vertices of G'.

Lemma 8. iso_one_one (G G': UG)(f: A-> A): iso_using f G G'-> one_one_on G f.

Lemma 9. iso_subgraphs (G G' H :UG) (f: A-> A) : iso_using f G G'-> Ind_subgraph H G -> (∃ H', Ind_subgraph H' G' ∧ iso_using f H H').

For the graphs G and G' Lemma 9 states that every induced subgraph H of G has an isomorphic counterpart H' in G'. In a similar way we can prove that the stable sets and cliques in G has isomorphic counterparts in G'. For example, consider the following lemmas from IsoUG.v summarising these results.

Lemma 10. iso_cliq (G G': UG)(f:A-> A)(K:list A):iso_using f G G'-> Cliq G K -> Cliq G' (img f K).

Lemma 11. iso_stable (G G': UG)(f: A-> A)(I: list A):iso_using f G G'-> Stable G I-> Stable G' (img f I).

Lemma 12. iso_coloring(G G':UG)(f:A->A)(C: A->nat):iso_using f G G' -> Coloring_of G C -> Coloring_of G' (fun (x:A) => C (f x)).

Lemma 13. perfect_G' (G G':UG): iso G G'-> Perfect G -> Perfect G'.

3.4   Graph Constructions

Adding (or removing) edges in an existing graph to obtain a new graph is a common procedure in proofs involving graphs. In such circumstances an explicit specification of all the fields of the new graph becomes a tedious job.

For example, consider the definition of following function (nw_edg G a a').

```
Definition nw_edg(G:UG)(a a':A):= fun(x y:A)=> match (x==a), (y==a') with
                                    | _ , false => (edg G) x y
                                    | true, true => true
                                    |false, true => (edg G) x a
                                  end.
```

The term (nw_edg G a a') can be used to describe the edge relation of graph G' shown in Fig. 6, which is obtained from G by repeating the vertex a to a'.

This function has a simple definition and hence it is easy to prove various properties about it. For example, we can prove following results establishing connections between the edges of G and G'.

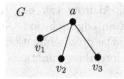 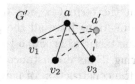

Fig. 6. Graph G' is obtained from G by repeating vertex a to a'.

Lemma 14. `nw_edg_xa_xa`' `(G: UG)(a a' x:A):` `(edg G) x a ->` `(nw_edg G a a')` `x a'`.

Lemma 15. `nw_edg_xy_xy (G: UG)(a a' x y:A)(P':` ¬ `In a' G):` `(edg G) x y -> (nw_edg G a a') x y`

Lemma 16.
`nw_edg_xy_xy4 (G: UG)(a a' x y:A)(P: In a G)(P': ¬In a' G):`
`y` ≠ `a' -> (edg G) x y = (nw_edg G a a') x y`.

Although, the term (`nw_edg G a a'`) contains all the essential properties of the construction it doesn't have the irreflexive and symmetric properties necessary for an edge relation. Hence, we can't use this term for edge relation while declaring G' as an instance of UG. To ensure these properties one can add more branches to the match statement and provide an extra term P of type a ≠ a' as argument to the function. However, this will result in a more complex function and proving even the essential properties of the new function becomes hard.

Instead of writing complex edge relations every time we define functions namely `mk_irefl`, `mk_sym` and `E_res_to` which make minimum changes and convert any binary relation on vertices into an edge relation. For example consider the following specification lemmas for the functions `mk_irefl` and `mk_sym`.

Lemma 17. `mk_ireflP (E: A -> A-> bool): irefl (mk_irefl E)`.

Lemma 18. `mk_symP (E: A-> A-> bool): sym (mk_sym E)`.

Lemma 19. `irefl_inv_for_mk_sym (E: A-> A-> bool): irefl E -> irefl (mk_sym E)`.

Lemma 20. `sym_inv_for_mk_irefl (E: A->A-> bool): sym E -> sym (mk_irefl E)`.

Note that these functions do not change the properties ensured by each other. The file UG.v contains other invariance results about these functions proving that these functions work well when used together. For example, consider the following declaration of G' as an instance of UG.

```
Definition ex_edg(G:UG)(a a':A):=
                        mk_sym(mk_irefl((nw_edg G a a') at_ (add a' G))).
Variable G: UG.
Definition G':= refine({| nodes:= add a' G; edg:= (ex_edg G a a');
                        |}); unfold ex_edg. all: auto. Defined.
```

Note that the term (`ex_edg G a a'`) obtained from (`nw_edg G a a'`) by using these functions have all the properties of an edge relation. Now, we can simply use the tactic `all: auto` to discharge all the proof obligations generated while declaring `G'` as an instance of `UG`. Therefore these functions can significantly ease the construction of new graphs.

All the important properties of the final edge relation (i.e. `ex_edg G a a'`) can now be derived from the properties of `nw_edg G a a'` by using the specification lemmas for the functions `mk_irefl`, `mk_sym` and `E_res_to`. For example consider following lemmas (from `Lovasz.v`) which describes the final edge relation (i.e. `ex_edg`) of graph `G'`.

Lemma 21. `Exy_E'xy` (x y:A)(P: In a G)(P': ¬ In a' G): edg G x y -> edg G' x y.

Lemma 22. `In_Exy_eq_E'xy` (x y:A)(P: In a G)(P': ¬ In a' G): In x G-> In y G-> edg G x y=edg G' x y.

Lemma 23. `Exy_eq_E'xy` (x y:A)(P: In a G)(P': ¬ In a' G): x ≠ a'-> y ≠ a'-> edg G x y = edg G' x y.

Lemma 24. `Exa_eq_E'xa'` (x:A)(P: In a G)(P': ¬ In a' G): x ≠ a-> x ≠ a'-> edg G x a = edg G' x a'.

Lemma 25. `Eay_eq_E'a'y` (y:A)(P: In a G)(P': ¬ In a' G): y ≠ a -> y ≠ a' -> edg G a y = edg G' a' y.

4   Constructive Proof of Lovász Replication Lemma

Let `G` and `G'` be the graphs discussed in Sect. 3.4, where `G'` is obtained from `G` by repeating the vertex `a` to `a'`. Then we have the following lemma.

Lemma 26. `ReplicationLemma` **Perfect G -> Perfect G'.**

Proof: We prove this result using induction on the size of graph `G`.

- Induction hypothesis (IH): ∀ X, |X|<|G|-> Perfect X -> Perfect X'

Let `H'` be an arbitrary induced subgraph of `G'`, then we need to prove that $\chi(H') = \omega(H')$. We prove this equality in both of the following cases.

- **Case-1** (H' ≠ G'): In this case H' is strictly included in G'. We further do case analysis on the proposition (a ∈ H').
  - **Case-1A** (a ∉ H'): In this case if a' ∉ H' then H' is an induced subgraph of G and hence $\chi(H') = \omega(H')$. Now consider the case when a' ∈ H'. Let H be the induced subgraph of H' restricted to the vertex-set (H'\a') ∪ {a}. Note that H' is isomorphic to H and H is an induced subgraph of G. Hence H' is a perfect graph and we have $\chi(H') = \omega(H')$.

- **Case-1B** (a $\in$ H'): Again in this case if a' $\notin$ H' then H' is an induced subgraph of G and hence χ(H') = ω(H'). Now we are in the case where a $\in$ H', a' $\in$ H' and H' is strictly included in G'. Therefore, the set H'\a' is strictly included in G. Let H be the induced subgraph of G with vertex set H'\a'. Note that H' can be obtained by repeating a to a' in H. But, we know that |H| < |G|, hence H' is a perfect graph by induction hypothesis (IH) and we have χ(H') = ω(H').

- **Case-2** (H' = G'): In this case we need to prove χ(G') = ω(G'). We further split this case into two sub cases.
  - **Case-2A**: In this case we assume that there exists a clique K of size ω(G) such that a $\in$ K. Hence K gets extended to a clique of size ω(G)+1 in G' and ω(G')= ω(G)+1. Now we can assign a new color to the vertex a' which is different from all the colors assigned to G. Hence χ(G')= χ(G)+1=ω(G)+1=ω(G').
  - **Case-2B**: In this case we assume that a does not belong to any clique K of size ω(G). Since G is a perfect graph let f be a coloring of graph G which uses exactly ω(G) colors. Let G* = { v$\in$ G: f(v) $\neq$ f(a) $\lor$ v=a }. For the subgraph G* we can then show that ω(G*) < ω(G). Hence there must exist a coloring f* which uses ω(G*) colors for coloring G*. Since ω(G*) < χ(G) we can safely assume that f* does not use the color f (a) for coloring the vertices of G*. Now consider a coloring f' which assigns a vertex x color f*(x) if x belongs to G* otherwise f' (x) = f (a). Note that the number of colors used by f is at most ω(G). Hence χ(G') = ω(G').

Note that all the cases in the above proof correspond to predicates on sets and finite graphs. Since we have decidable representations for all of these predicates, we could do case analysis on them without assuming any axiom.   □

5   Conclusions and Future Work

Formal verification of a mathematical theory can often lead to a deeper understanding of the verified results and hence increases our confidence in the theory. However, the task of formalization soon becomes overwhelming because the length of formal proofs blows up significantly. In such circumstances having a library of facts on commonly occurring mathematical structures can be really helpful. The main contribution of this paper is a constructive formalisation of finite simple graphs in the Coq proof assistant [2]. This formalization can be used as a framework to verify other important results on finite graphs. To keep the formalisation constructive we follow a proof style similar to the small scale reflections technique of the ssreflect. We use small boolean functions in a systematic way to represent various predicates over sets and graphs. These functions together with their specification lemmas can help in avoiding the use of Excluded-Middle in the proof development. We also describe functions to ease the process of new graph construction. These functions can help in discharging most of the proof obligation generated while creating a new instance of finite graph. Finally, we use this framework to present a fully constructive proof of the Lovász Replication Lemma [8], which is the central idea in the proof of Weak Perfect Graph Theorem. One can immediately extend this work by formally verifying Weak Perfect Graph Theorem in the same framework. Another direction of work could be to add in the present framework all the basic classes of graphs and decompositions involved in the proof of Strong Perfect Graph Theorem. This can finally result in a constructive formalisation of strong Perfect Graph Theorem in the Coq proof assistant.

References

1. Coq formalization. https://github.com/Abhishek-TIFR/List-Set
2. The Coq Standard Library. https://coq.inria.fr/library/
3. Chudnovsky, M., Robertson, N., Seymour, P., Thomas, R.: The strong perfect graph theorem. Annal. Math. **164**, 51–229 (2006)
4. Doczkal, C., Combette, G., Pous, D.: A formal proof of the minor-exclusion property for treewidth-two graphs. In: Avigad, J., Mahboubi, A. (eds.) ITP 2018. LNCS, vol. 10895, pp. 178–195. Springer, Cham (2018). https://doi.org/10.1007/978-3-319-94821-8_11
5. Garey, M.R., Johnson, D.S.: Computers and Intractability: A Guide to the Theory of NP-Completeness. W. H. Freeman & Co., New York (1990)
6. Gonthier, G., Mahboubi, A.: An introduction to small scale reflection in Coq. J. Formalized Reason. **3**(2), 95–152 (2010)
7. Gyarfas, A., Sebo, A., Trotignon, N.: The chromatic gap and its extremes. J. Comb. Theory, Series B **102**(5), 1155–1178 (2012)
8. Lovász, L.: Normal hypergraphs and the perfect graph conjecture. Discrete Math. **2**(3), 253–267 (1972)
9. Mycielski, J.: Sur le coloriage des graphs. Colloquium Mathematicae **3**(2), 161–162 (1955)

Author Index

Printed in the United States
By Bookmasters